TX
571
P4
A52
2008
c. 1

Analysis of Pesticides in Food and Environmental Samples

Analysis of Pesticides in Food and Environmental Samples

Edited by José L. Tadeo

CRC Press
Taylor & Francis Group
Boca Raton   London   New York

CRC Press is an imprint of the
Taylor & Francis Group, an **informa** business

CRC Press
Taylor & Francis Group
6000 Broken Sound Parkway NW, Suite 300
Boca Raton, FL 33487-2742

© 2008 by Taylor & Francis Group, LLC
CRC Press is an imprint of Taylor & Francis Group, an Informa business

No claim to original U.S. Government works
Printed in the United States of America on acid-free paper
10 9 8 7 6 5 4 3 2 1

International Standard Book Number-13: 978-0-8493-7552-1 (Hardcover)

This book contains information obtained from authentic and highly regarded sources. Reprinted material is quoted with permission, and sources are indicated. A wide variety of references are listed. Reasonable efforts have been made to publish reliable data and information, but the author and the publisher cannot assume responsibility for the validity of all materials or for the consequences of their use.

No part of this book may be reprinted, reproduced, transmitted, or utilized in any form by any electronic, mechanical, or other means, now known or hereafter invented, including photocopying, microfilming, and recording, or in any information storage or retrieval system, without written permission from the publishers.

For permission to photocopy or use material electronically from this work, please access www.copyright.com (http://www.copyright.com/) or contact the Copyright Clearance Center, Inc. (CCC) 222 Rosewood Drive, Danvers, MA 01923, 978-750-8400. CCC is a not-for-profit organization that provides licenses and registration for a variety of users. For organizations that have been granted a photocopy license by the CCC, a separate system of payment has been arranged.

Trademark Notice: Product or corporate names may be trademarks or registered trademarks, and are used only for identification and explanation without intent to infringe.

Library of Congress Cataloging-in-Publication Data

Analysis of pesticides in food and environmental samples / editor, Jose L. Tadeo.
    p. cm.
Includes bibliographical references and index.
ISBN 978-0-8493-7552-1 (alk. paper)
    1. Pesticide residues in food. 2. Food--Analysis. 3. Pesticides I. Tadeo, Jose L.

TX571.P4A52 2008
664'.07--dc22
                                           2007030736

Visit the Taylor & Francis Web site at
http://www.taylorandfrancis.com

and the CRC Press Web site at
http://www.crcpress.com

Contents

Preface

You should go on learning for as long as your ignorance lasts;
and, if the proverb is to be believed, *for the whole of your life.*

Lucius Annaeus Seneca

Consumer concerns on food safety and society awareness of chemical contaminants in the environment have increased in the past few years. As a consequence, more restrictions in the use of chemical products have been imposed at national and international levels.

Pesticides are widely used for the control of weeds, diseases, and pests of cultivated plants all over the world, mainly since after Second World War, with the discovery of some organic compounds with good insecticide or herbicide activity. At present, around 2.5 million tons of pesticides are used annually and the number of registered active substances is higher than 500.

However, as pesticides are toxic substances that may have undesirable effects, their use has to be regulated. Risk assessment of pesticides requires information on the toxicological and ecotoxicological properties of these compounds as well as on their levels in food and environmental compartments. Therefore, reliable analytical methods are needed to carry out the monitoring of pesticide residues in those matrices.

Analysis of Pesticides in Food and Environmental Samples focuses on the analytical methodologies developed for the determination of these compounds and on their levels in food and in the environment. It includes information on the different pesticides used, sample preparation methods, quality assurance, chromatographic techniques, immunoassays, pesticide determination in food, soil, water, and air, and the results of their monitoring in food and environmental compartments. I think that this timely and up-to-date work can significantly improve the information in this research area and contribute to a better understanding of the behavior of pesticides that will lead to an improvement of their use.

My sincere thanks to everyone who has contributed and particularly to all the contributors of the different chapters of *Analysis of Pesticides in Food and Environmental Samples*.

This work is dedicated to Teresa, my wife.

José L. Tadeo

Editor

José L. Tadeo, PhD in chemistry, is a senior researcher at the National Institute for Agricultural and Food Research and Technology, Instituto Nacional de Investigación y Tecnología Agraria y Alimentaria in Madrid, Spain. He graduated with a degree in chemistry in June 1972 from the University of Valencia and began his research career at the Institute of Agrochemistry and Food Technology, Spanish Council for Scientific Research, in Valencia, investigating natural components of plants with insecticide activity. In 1976, he was engaged in research of analytical methodologies for the determination of pesticide residues in food, water, and soil at the Jealott's Hill Research Station in the United Kingdom.

In 1977, Dr. Tadeo was a research scientist at the Institute for Agricultural Research in Valencia where his work focused on the study of the chemical composition of citrus fruits and the behavior of fungicides used during postharvest of fruits.

In 1988, he became a senior researcher at the Instituto Nacional de Investigación y Tecnología Agraria y Alimentaria. During his stay at the Plant Protection Department, the main research lines were the analysis of herbicide residues and the study of their persistence and mobility in soil.

His current research at the Environment Department of the Instituto Nacional de Investigación y Tecnología Agraria y Alimentaria is the analysis of pesticides and other contaminants in food and environmental matrices and the evaluation of exposure to biocides and existing chemicals. He has published numerous scientific papers, monographs, and book chapters on these topics. He has been a member of national and international working groups for the evaluation of chemicals, and he is currently involved in the assessment of biocides at the international level.

Contributors

Triantafyllos Albanis
Department of Chemistry
University of Ioannina
Ioannina, Greece

Beatriz Albero
Department of Environment
Instituto Nacional de Investigación y
Tecnología Agraria y Alimentaria
Madrid, Spain

Árpád Ambrus
Hungarian Food Safety Office
Budapest, Hungary

Svetlana Bondarenko
Department of Environmental Sciences
University of California
Riverside, California

Jane C. Chuang
Battelle
Columbus, Ohio

Adrian Covaci
Toxicological Centre
University of Antwerp
Wilrijk, Belgium

Kilian Dill
Antara Biosciences
Mountain View, California

Jay Gan
Department of Environmental Sciences
University of California
Riverside, California

Antonia Garrido Frenich
Department of Analytical Chemistry
University of Almeria
Almeria, Spain

Lorena González
Department of Environment
Instituto Nacional de Investigación y
Tecnología Agraria y Alimentaria
Madrid, Spain

Kit Granby
The National Food Institute
Technical University of Denmark
Søborg, Denmark

Dimitra Hela
Department of Business Administration
of Agricultural Products and Food
University of Ioannina
Agrinio, Greece

Susan S. Herrmann
The National Food Institute
Technical University of Denmark
Søborg, Denmark

Simon Hird
Central Science Laboratory
Sand Hutton, York, United Kingdom

Ioannis Konstantinou
Department of Environmental and
Natural Resources Management
University of Ioannina
Agrinio, Greece

Dimitra Lambropoulou
Department of Chemistry
University of Ioannina
Ioannina, Greece

Antonio Martín-Esteban
Department of Environment
Instituto Nacional de Investigación y
    Tecnología Agraria y Alimentaria
Madrid, Spain

Jose Luis Martinez
Department of Analytical Chemistry
University of Almeria
Almeria, Spain

Maurice Millet
Laboratoire de Physico-Chimie de
    l'Atmosphère
Centre de Géochimie de la Surface
Université Louis Pasteur
Strasbourg, France

Annette Petersen
The National Food Institute
Technical University of Denmark
Søborg, Denmark

Mette Erecius Poulsen
The National Food Institute
Technical University of Denmark
Søborg, Denmark

Consuelo Sánchez-Brunete
Department of Environment
Instituto Nacional de Investigación y
    Tecnología Agraria y Alimentaria
Madrid, Spain

Frank J. Schenck
Southeast Regional Laboratory
U.S. Food and Drug Administration
    Office of Regulatory Affairs
Atlanta, Georgia

José L. Tadeo
Department of Environment
Instituto Nacional de Investigación y
    Tecnología Agraria y Alimentaria
Madrid, Spain

Esther Turiel
Department of Environment
Instituto Nacional de Investigación y
    Tecnología Agraria y Alimentaria
Madrid, Spain

Jeanette M. Van Emon
National Exposure Research Laboratory
U.S. Environmental Protection Agency
Las Vegas, Nevada

Jon W. Wong
Center for Food Safety and Applied
    Nutrition
U.S. Food and Drug Administration
College Park, Maryland

Guohua Xiong
National Exposure Research Laboratory
U.S. Environmental Protection Agency
Las Vegas, Nevada

1 Pesticides: Classification and Properties

José L. Tadeo, Consuelo Sánchez-Brunete, and Lorena González

CONTENTS

1.1   INTRODUCTION

A pesticide is any substance or mixture of substances, natural or synthetic, formulated to control or repel any pest that competes with humans for food, destroys property, and spreads disease. The term pest includes insects, weeds, mammals, and microbes, among others [1].

Pesticides are usually chemical substances, although they can be sometimes biological agents such as virus or bacteria. The active portion of a pesticide, known as the active ingredient, is generally formulated by the manufacturer as emulsifiable concentrates or in solid particles (dust, granules, soluble powder, or wettable powder). Many commercial formulations have to be diluted with water before use and contain adjuvants to improve pesticide retention and absorption by leaves or shoots.

There are different classes of pesticides according to their type of use. The main pesticide groups are herbicides, used to kill weeds and other plants growing in places where they are unwanted; insecticides, employed to kill insects and other arthropods; and fungicides, used to kill fungi. Other types of pesticides are acaricides, molluscicides, nematicides, pheromones, plant growth regulators, repellents, and rodenticides.

Chemical substances have been used by human to control pests from the beginning of agriculture. Initially, inorganic compounds such as sulfur, arsenic, mercury, and lead were used. The discovery of dichlorodiphenyltrichloroethane (DDT) as an insecticide by Paul Müller in 1939 caused a great impact in the control of pests and soon became widely used in the world. At that time, pesticides had a good reputation mainly due to the control of diseases like malaria transmitted by mosquitoes and the bubonic plague transmitted by fleas, both killing millions of people over time. Nevertheless, this opinion changed after knowing the toxic effects of DDT on birds, particularly after the publication of the book *Silent Spring* by Rachel Carson in 1962 [2]. At present, due to the possible toxic effects of pesticides on human health and on the environment, there are strict regulations for their registration and use all over the world, especially in developed countries. However, although some progress is achieved in the biological control and in the development of resistance of plants to pests, pesticides are still indispensable for feeding and protecting the world population from diseases. It has been estimated that around one-third of the crop production would be lost if pesticides were not applied.

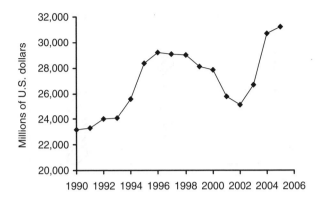

FIGURE 1.1 World market of pesticides since 1990. Values are expressed in millions of U.S. dollars. (From European Crop Protection Association (ECPA) Review 2005–2006, http://www.ecpa.be.)

Pesticide use has increased 50-fold since 1950 and around 2.5 million tons of industrial pesticides per year are used nowadays. Figure 1.1 shows the time course of pesticide sales during the last years.

According to the European Crop Protection Association (ECPA) Annual Report 2001–2002, the main agricultural areas of pesticide usage are North America, Europe, and Asia with 31.9%, 23.8%, and 22.6%, respectively, in 2001 (Figure 1.2). These percentages of pesticide sales are expressed in millions of euros and, although the mentioned regions are the most important agricultural areas in the global pesticide market, their relative position may vary due to changes in the currency exchange rates, climatic conditions, and national policies on agricultural support and regulations.

The amount of pesticides applied in a determined geographical area depends on the climatic conditions and on the outbreak of pests and diseases of a particular year. Nevertheless, herbicides are the main group of pesticides used worldwide, followed by insecticides and fungicides (Figure 1.3).

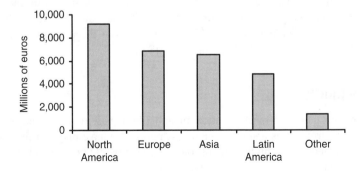

FIGURE 1.2 Regional pesticide sales expressed in millions of euros. (From ECPA Annual Report 2001–2002, http://www.ecpa.be.)

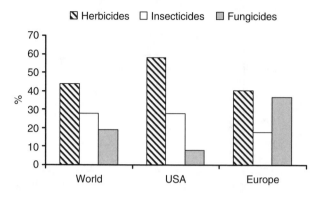

FIGURE 1.3 Distribution of the market (%) per pesticide type. (From Environmental Protection Agency (USEPA), pesticides industry sales and usage, 2001, http://www.epa.gov/oppbead1/pestsales/ and ECPA Annual Report 2001–2002, http://www.ecpa.be.)

The development of a new chemical as a pesticide takes at present nearly 15 years and around $20 million, and only one compound out of 10,000 compounds initially tested might reach, on average, final commercial production. The registration of a pesticide for its application on a particular crop requires a complete set of data to prove its efficacy and safe use. This normally includes data on physicochemical properties, analytical methods, efficacy, toxicology, ecotoxicology, and fate and behavior in the environment. Residues left on crops after pesticide application have been restricted in developed countries to guaranty a safe food consumption. The maximum residue levels (MRLs) in different foods have been established according to good agricultural practices, the observed toxic effects of the pesticide, and the amount of food consumed. MRLs are normally fixed in relation with the admissible daily intake (ADI) of pesticides, which is the amount of pesticide that can be ingested daily during the whole life without showing an appreciable adverse effect. MRLs are proposed by the Joint FAO/WHO Meeting on Pesticide Residues (JMPR) and recommended for adoption by the Codex Committee on Pesticide Residues [3,4].

In the following sections of this chapter, the main classes of pesticides (herbicides, insecticides, and fungicides) will be described together with their main physicochemical properties and principal uses. These data have been gathered mainly from *The Pesticide Manual* [5] as well as from the primary manufacture sources [6,7] and other available publications [8,9].

1.2 HERBICIDES

The implementation of mechanization in agriculture has increased the ability of human to control weeds and cultivate crops; herbicides have played a main part in this development; and a higher proportion of farmers would be needed if herbicides were not used.

Herbicides can be classified as soil- or foliage-applied compounds, which are normally absorbed by roots or leaf tissues, respectively. These compounds can be

total or selective herbicides. Total herbicides can kill all vegetation, whereas selective herbicides can control weeds without affecting the crop. These chemical substances may be applied at different crop stages, such as presowing and pre- or
postemergence, and these different treatments will be used depending on the weed
needed to be controlled in a particular crop. The selectivity of a herbicide may
depend on a differential plant uptake, translocation, or metabolism, as well as on
differences at the site of action. A knowledge of physicochemical properties, that is,
vapor pressure (V.p.), octanol/water partition coefficient (K_{ow}, expressed in the
logarithmic form log P), and solubility in water allows the fate and behavior of
such chemicals in the environment to be predicted.

In addition, herbicides can be classified according to their chemical composition.
The principal physicochemical properties, together with the field persistence and
major uses of representative herbicides, grouped in their main chemical classes, are
described later.

1.2.1 AMIDES

A large variety of compounds form this group of herbicides, which have the
following general formula: R_1–CO–N–(R_2,R_3).

The key components of this group are the N-substituted chloroacetamides and
the substituted anilides.

Propanil   Alachlor

The chloroacetamides are effective preemergence herbicides for annual grasses and
annual broad-leaved weeds but they also have foliar contact activity. In general,
these compounds are soil applied and used in various horticultural crops, such as
maize, soybean, and sugarcane. These herbicides are normally absorbed by shoots
and roots and they are, in general, nonpersistent compounds in soil (Table 1.1).

1.2.2 BENZOIC ACIDS

This group is mainly formed by chlorinated derivatives of substituted benzoic acids.

Dicamba

TABLE 1.1

Chemical Names and Properties of Amide Herbicides

Common Name	IUPAC Name	Vapor Pressure mPa (25°C)	K_{ow} log P	Water Solubility mg/L (25°C)	Half-Life in Soil (Days)
Acetochlor $C_{14}H_{20}ClNO_2$	2-Chloro-*N*-ethoxymethyl-6'-ethylacet-*o*-toluidide	0.005	4.14	223	8–18
Alachlor $C_{14}H_{20}ClNO_2$	2-Chloro-2',6'-diethyl-*N*-methoxymethylacetanilide	2.0	3.09	170[a]	1–30
Butachlor $C_{17}H_{26}ClNO_2$	*N*-Butoxymethyl-2-chloro-2',6'-diethylacetanilide	0.24	—	23[a]	12
Metolachlor $C_{15}H_{22}ClNO_2$	2-Chloro-6-ethyl-*N*-(2-methoxy-1-methylethyl)acet-*o*-toluidide	4.2	2.9	488	20
Propachlor $C_{11}H_{14}ClNO$	2-Chloro-*N*-isopropyl acetanilide	10	1.4–2.3	580	4
Propanil $C_9H_9Cl_2NO$	3',4'-Dichloro propionanilide	0.05	3.3	130[a]	2–3

Sources: Data from Tomlin, C. (Ed.) in *The Pesticide Manual*, British Crop Protection Council, 2000; http://ec.europa.eu/food/plant/protection/evaluation/exist_subs_rep_en.htm; http://www.epa.gov/opprd001/factsheets/; Hornsby, A.G., Wauchope, R.D., and Herner, A.E. in *Pesticide Properties in the Environment*, Springer-Verlag, New York, 1996; De Liñan, C. in *Farmacología Vegetal*, Ediciones Agrotecnicas S.L., 1997.

[a] 20°C.

The benzoic acid herbicides are known to have growth regulating and auxin activity properties. These compounds are especially used to control deep-rooted perennial weeds and applied as salts or esters (Table 1.2).

1.2.3 CARBAMATES

Carbamates are esters of the carbamic acid (R_1–O–CO–NR_2R_3) and together with thiocarbamates (R_1–S–CO–NR_2R_3) represent a broad group of herbicides, frequently applied to soil in preemergence.

—NHCO₂CH(CH₃)₂   [CH₃(CH₂)₂]₂ NC(O)SCH₂CH₃

Propham   EPTC

TABLE 1.2

Chemical Names and Properties of Benzoic Acid Herbicides

Common Name	IUPAC Name	Vapor Pressure mPa (25°C)	K_{ow} log P	Water Solubility g/L (25°C)	Half-Life in Soil (Days)
Chloramben $C_7H_5Cl_2NO_2$	3-Amino-2,5-dichlorobenzoic acid	—	—	0.7	14–21
Chlorthal-dimethyl $C_{10}H_6Cl_4O_4$	Dimethyl tetrachloroterephthalate	0.21	4.28	0.5×10^{-3}	33
Dicamba $C_8H_6Cl_2O_3$	3,6-Dichloro-*o*-methoxybenzoic acid	1.67	−1.88	6.1	<14

Sources: Data from Tomlin, C. (Ed.) in *The Pesticide Manual*, British Crop Protection Council, 2000; http://ec.europa.eu/food/plant/protection/evaluation/exist_subs_rep_en.htm; http://www.epa.gov/opprd001/factsheets/; Hornsby, A.G., Wauchope, R.D., and Herner, A.E. in *Pesticide Properties in the Environment*, Springer-Verlag, New York, 1996; De Liñan, C. in *Farmacología Vegetal*, Ediciones Agrotecnicas S.L., 1997.

These compounds are root or shoot absorbed and are frequently used to control annual grasses and broad-leaved weeds in peas, beet, and other horticultural crops. These herbicides are normally decomposed by soil microorganisms in 3–5 weeks. Their main physicochemical properties are summarized in Table 1.3.

1.2.4 NITRILES

Bromoxynil and ioxynil are the hydroxybenzonitriles used as herbicides.

Bromoxynil   Ioxynil

They are formulated as salts or octanoate esters and foliage applied to control broad-leaved weeds in cereals and horticultural crops. These compounds are used in postemergence and frequently applied in combination with other herbicides to extend the spectrum of weed species to be controlled. They have a low persistence in soil (Table 1.4).

TABLE 1.3

Chemical Names and Properties of Carbamate Herbicides

Common Name	IUPAC Name	Vapor Pressure mPa (25°C)	K_{ow} log P	Water Solubility mg/L (25°C)	Half-Life in Soil (Days)
Chlorpropham $C_{10}H_{12}ClNO_2$	Isopropyl-3-chlorocarbanilate	1.3	3.76	89	30–65
Desmedipham $C_{16}H_{16}N_2O_4$	Ethyl-3-phenylcarbamoyloxy phenylcarbamate	4×10^{-5}	3.39	7[a]	34
EPTC $C_9H_{19}NOS$	S-Ethyl dipropylthiocarbamate	0.01	3.2	375	6–30
Molinate $C_9H_{17}NOS$	S-Ethyl azepane-1-carbothioate	746	2.88	970	8–25
Phenmedipham $C_{16}H_{16}N_2O_4$	Methyl-3-(3-methylcarbaniloyloxy) carbanilate	1.3×10^{-6}	3.59	4.7	25
Propham $C_{10}H_{13}NO_2$	Isopropyl phenylcarbamate	Sublimes slowly	—	250[a]	5–15
Thiobencarb $C_{12}H_{16}ClNOS$	S-4-Chlorobenzyl diethylthiocarbamate	2.93	3.42	30[a]	14–21
Triallate $C_{10}H_{16}Cl_3NOS$	S-2,3,3-Trichloroallyl diisopropyl(thiocarbamate)	16	4.6	4	56–77

Sources: Data from Tomlin, C. (Ed.) in *The Pesticide Manual*, British Crop Protection Council, 2000; http://ec.europa.eu/food/plant/protection/evaluation/exist_subs_rep_en.htm; http://www.epa.gov/opprd001/factsheets/; Hornsby, A.G., Wauchope, R.D., and Herner, A.E. in *Pesticide Properties in the Environment*, Springer-Verlag, New York, 1996; De Liñan, C. in *Farmacología Vegetal*, Ediciones Agrotecnicas S.L., 1997.

[a]   20°C.

1.2.5   NITROANILINES

These compounds are derivatives of 2,6-dinitroaniline.

Pendimethalin

Nitroanilines are a group of herbicides with similar physicochemical properties, such as low water solubility and high octanol–water partition coefficient. These compounds are soil-applied herbicides used to control annual grasses and many broad-leaved

TABLE 1.4
Chemical Names and Properties of Nitrile Herbicides

Common Name	IUPAC Name	Vapor Pressure mPa (20°C)	K_{ow} log P	Water Solubility mg/L (20°C)	Half-Life in Soil (Days)
Bromoxynil $C_7H_3Br_2NO$	3,5-Dibromo-4-hydroxybenzonitrile	6.3×10^{-3}	2.8	130	10
Ioxynil $C_7H_3I_2NO$	4-Hydroxy-3,5-diiodobenzonitrile	<1	3.43	50	10

Sources: Data from Tomlin, C. (Ed.) in *The Pesticide Manual*, British Crop Protection Council, 2000; http://ec.europa.eu/food/plant/protection/evaluation/exist_subs_rep_en.htm; http://www.epa.gov/opprd001/factsheets/; Hornsby, A.G., Wauchope, R.D., and Herner, A.E. in *Pesticide Properties in the Environment*, Springer-Verlag, New York, 1996; De Liñan, C. in *Farmacología Vegetal*, Ediciones Agrotecnicas S.L., 1997.

weeds in a wide variety of crops. The 2,6-dinitroanilines possess a marked general herbicide activity. Substitution at the third and/or fourth position of the ring or on the amino group modifies the degree of herbicidal activity. In general, they have a certain persistence in soil and are normally soil incorporated due to their significant vapor pressure (Table 1.5).

TABLE 1.5
Chemical Names and Properties of Nitroaniline Herbicides

Common Name	IUPAC Name	Vapor Pressure mPa (25°C)	K_{ow} log P	Water Solubility mg/L (25°C)	Half-Life in Soil (Days)
Butralin $C_{14}H_{21}N_3O_4$	*N-sec*-Butyl-4-*tert*-butyl-2,6-dinitroaniline	0.77	4.93	1	14
Ethalfluralin $C_{13}H_{14}F_3N_3O_4$	*N*-Ethyl-α,α,α-trifluoro-*N*-(2-methylallyl)-2,6-dinitro-*p*-toluidine	11.7	5.11	0.3	25–46
Pendimethalin $C_{13}H_{19}N_3O_4$	*N*-(1-Ethylpropyl)-2,6-dinitro-3,4-xylidine	4	5.18	0.3[a]	90–120
Trifluralin $C_{13}H_{16}F_3N_3O_4$	α,α,α-Trifluoro-2,6-dinitro-*N,N*-dipropyl-*p*-toluidine	6.1	4.83[a]	0.22	57–126

Sources: Data from Tomlin, C. (Ed.) in *The Pesticide Manual*, British Crop Protection Council, 2000; http://ec.europa.eu/food/plant/protection/evaluation/exist_subs_rep_en.htm; http://www.epa.gov/opprd001/factsheets/; Hornsby, A.G., Wauchope, R.D., and Herner, A.E. in *Pesticide Properties in the Environment*, Springer-Verlag, New York, 1996; De Liñan, C. in *Farmacología Vegetal*, Ediciones Agrotecnicas S.L., 1997.

[a] 20°C.

1.2.6 ORGANOPHOSPHORUS

Glyphosate   Glufosinate

Glyphosate and glufosinate are broad spectrum, nonselective, postemergence contact herbicides active only for foliar application. They are extensively used in various applications for weed control in aquatic systems and vegetation control in noncrop areas. Aminomethylphosphonic acid (AMPA) is the major degradation product of glyphosate found in plants, water, and soil. The main properties of these compounds are shown in Table 1.6.

1.2.7 PHENOXY ACIDS

Phenoxy acids are a common name given to a group of compounds formed by a phenoxy radical linked to a low carbon number alkanoic acid, such as 2,4-dichlorophe-noxyacetic acid (2,4-D, acetic acid) or mecoprop (propionic acid). Some herbicides of this group are formed by stereoisomers, which are commercialized as single enanthiomers or racemic mixtures.

TABLE 1.6

Chemical Names and Properties of Organophosphorus Herbicides

Common Name	IUPAC Name	Vapor Pressure mPa (25°C)	K_{ow} log P	Water Solubility g/L (25°C)	Half-Life in Soil (Days)
Glyphosate $C_3H_8NO_5P$	N-(Phosphonomethyl) glycine	1.3×10^{-2}	<-3.2	11.6	3–174
Glufosinate-ammonium $C_5H_{15}N_2O_4P$	Ammonium 4-[hydroxy(methyl) phosphinoyl]-DL-homoalaninate	$<0.1^a$	<0.1	1370	7–20

Sources: Data from Tomlin, C. (Ed.) in *The Pesticide Manual*, British Crop Protection Council, 2000; http://ec.europa.eu/food/plant/protection/evaluation/exist_subs_rep_en.htm; http://www.epa.gov/opprd001/factsheets/; Hornsby, A.G., Wauchope, R.D., and Herner, A.E. in *Pesticide Properties in the Environment*, Springer-Verlag, New York, 1996; De Liñan, C. in *Farmacología Vegetal*, Ediciones Agrotecnicas S.L., 1997.

[a] 20°C.

Cl—⬡—OCH₂CO₂H

Cl

2,4-D

Cl—⬡—O—⬡—O—CHCO₂H (CH₃)

Cl

Diclofop

These hormone type herbicides were discovered during the Second World War and, some years later, the phenoxy–phenoxy acids like diclofop were introduced to overcome the problem of selective control of grass weeds in cereal crops. These compounds are active by contact and by translocation from leaves to roots of perennial weeds and they are also used in preemergence applications to the soil for the control of young seedlings. The chlorophenoxy compounds are selective against broad-leaved annual weeds in cereal and grass crops. In general, they have a short persistence in soil (Table 1.7).

TABLE 1.7
Chemical Names and Properties of Phenoxy Acid Herbicides

Common Name	IUPAC Name	Vapor Pressure mPa (25°C)	K_{ow} log P	Water Solubility mg/L (20°C)	Half-Life in Soil (Days)
2,4-D $C_8H_6Cl_2O_3$	2,4-Dichlorophenoxy acetic acid	1.86×10^{-2}	0.04	23,180	<7
Diclofop $C_{15}H_{12}Cl_2O_4$	(RS)-2-[4-(2,4-Dichlorophenoxy) phenoxy]propionic acid	9.7×10^{-6}	2.81	122,700	30
Fenoxaprop-P $C_{16}H_{12}ClNO_5$	(R)-2-[4-(6-Chloro-1,3-benzoxazol -2-yloxy)phenoxy]propionic acid	1.8×10^{-1a}	1.83	61,000	1–10
Fluazifop-P $C_{15}H_{12}F_3NO_4$	(R)-2-[4-(5-Trifluoromethyl-2- pyridyloxy)phenoxy]propionic acid	7.9×10^{-4a}	−0.8	780	<32
MCPA $C_9H_9ClO_3$	4-Chloro-(2-methylphenoxy)acetic acid	2.3×10^{-2a}	−0.71	274[b]	<7
Mecoprop-P $C_{10}H_{11}ClO_3$	(R)-2-(4-Chloro-o-tolyloxy) propionic acid	0.4[a]	0.02	860	3–13
Quizalofop- P-ethyl $C_{19}H_{17}ClN_2O_4$	Ethyl(R)-2-[4- (6-chloroquinoxalin-2-yloxy) phenoxy]propionate	1.1×10^{-4a}	4.66	0.61	≤1
Triclopyr $C_7H_4Cl_3NO_3$	3,5,6-Trichloro-2-pyridyloxyacetic acid	0.2	−0.45	8.10	46

Sources: Data from Tomlin, C. (Ed.) in *The Pesticide Manual*, British Crop Protection Council, 2000; http://ec.europa.eu/food/plant/protection/evaluation/exist_subs_rep_en.htm; http://www.epa .gov/opprd001/factsheets/; Hornsby, A.G., Wauchope, R.D., and Herner, A.E. in *Pesticide Properties in the Environment*, Springer-Verlag, New York, 1996; De Liñan, C. in *Farmacología Vegetal*, Ediciones Agrotecnicas S.L., 1997.

[a] 20°C.
[b] 25°C.

1.2.8 PYRIDINES AND QUATERNARY AMMONIUM COMPOUNDS

The herbicide group of pyridines, also named bipyridylium, is formed by paraquat and diquat. These compounds were developed as the result of observations that quaternary ammonium germicides, such as cetyl trimethylammonium bromide, desiccated young plants. Other quaternary ammonium compounds, like chlormequat and mepiquat, have been developed and used as plant growth regulators to increase yields in cereals, promote flowering in ornamental plants, and improve fruit setting in horticultural plants and trees.

Paraquat   Diquat

Paraquat and diquat are broad spectrum herbicides absorbed by leaves, but they are not translocated in sufficient quantities to kill the roots of perennial weeds. These compounds are very strong bases because of their quaternary ammonium structures and are rapidly adsorbed and inactivated in soil. Therefore, these compounds are not effective as preemergence herbicides. They have a high water solubility and low octanol–water partition coefficient (Table 1.8), and are available commercially as

TABLE 1.8
Chemical Names and Properties of Pyridine Herbicides and Quaternary Ammonium Compounds

Common Name	IUPAC Name	Vapor Pressure mPa (20°C)	K_{ow} log P	Water Solubility g/L (20°C)	Half-Life in Soil (Days)
Diquat dibromide $C_{12}H_{12}Br_2N_2$	1,1'-Ethylene-2,2'-bipyridyldiylium dibromide	<0.013	−4.6	700	<7
Paraquat dichloride $C_{12}H_{14}Cl_2N_2$	1,1'-Dimethyl-4,4'-bipyridinium dichloride	<0.01[a]	−4.5	620	<7
Chlormequat chloride $C_5H_{13}Cl_2N$	2-Chloroethyl trimethyl ammonium	<0.01	−1.59	1000	1–28
Mepiquat chloride $C_7H_{16}ClN$	1,1'-Dimethyl-piperidinium chloride	<0.01	−2.82	500	10–97

Sources: Data from Tomlin, C. (Ed.) in *The Pesticide Manual*, British Crop Protection Council, 2000; http://ec.europa.eu/food/plant/protection/evaluation/exist_subs_rep_en.htm; http://www.epa.gov/opprd001/factsheets/; Hornsby, A.G., Wauchope, R.D., and Herner, A.E. in *Pesticide Properties in the Environment*, Springer-Verlag, New York, 1996; De Liñan, C. in *Farmacología Vegetal*, Ediciones Agrotecnicas S.L., 1997.

[a] 25°C.

dibromide or dichloride salts. These herbicides are strongly adsorbed in soil, requiring acid digestion for several hours for their desorption.

1.2.9 PYRIDAZINES AND PYRIDAZINONES

Pyridate and pyridazinones, like norflurazon and chloridazon, are included in this group.

Norflurazon   Pyridate

They are contact-selective herbicides with foliar activity and are used in pre- or postemergence to control annual grasses, broad-leaved weeds, and grassy weeds on cereals, maize, rice, and some other crops. In general, the pyridazinone herbicides are long lasting in soil (Table 1.9).

TABLE 1.9
Chemical Names and Properties of Pyridazine and Pyridazinone Herbicides

Common Name	IUPAC Name	Vapor Pressure mPa (25°C)	K_{ow} log P (25°C)	Water Solubility mg/L (20°C)	Half-Life in Soil (Days)
Chloridazon $C_{10}H_8ClN_3O$	5-Amino-4-chloro-2-phenylpyridazin-3(2H)-one	<0.01[a]	1.19	340	21–76
Norflurazon $C_{12}H_9ClF_3N_3O$	4-Chloro-5-methylamino-2-(α,α,α-trifluoro-m-tolyl)pyridazin-3(2H)-one	3.8×10^{-3}	2.45	34[b]	45–180
Pyridate $C_{19}H_{23}ClN_2O_2S$	6-Chloro-3-phenylpyridazin-4-yl-S-octylthiocarbonate	4.8×10^{-4a}	4.01	ca. 1.5	<3

Sources: Data from Tomlin, C. (Ed.) in *The Pesticide Manual*, British Crop Protection Council, 2000; http://ec.europa.eu/food/plant/protection/evaluation/exist_subs_rep_en.htm; http://www.epa.gov/opprd001/factsheets/; Hornsby, A.G., Wauchope, R.D., and Herner, A.E. in *Pesticide Properties in the Environment*, Springer-Verlag, New York, 1996; De Liñan, C. in *Farmacología Vegetal*, Ediciones Agrotecnicas S.L., 1997.

[a] 20°C.
[b] 25°C.

1.2.10 TRIAZINES

Simazine

Metribuzin

A wide range of triazines have been synthesized over time to control annual and broad-leaved weeds in a variety of crops as well as in noncropped land. They are effective, at low dosages, in killing broad-leaved weeds in corn and other crops and they can be used in high dosages as soil sterilants. In general, these herbicides are applied in pre- or postemergence and they are absorbed by the roots or by the foliage, respectively. In some cases, they are used in combination with other herbicides to broaden the spectrum of activity. These compounds have an appreciable persistence in soil (Table 1.10).

TABLE 1.10
Chemical Names and Properties of Triazine Herbicides

Common Name	IUPAC Name	Vapor Pressure mPa (25°C)	K_{ow} log P (25°C)	Water Solubility mg/L (25°C)	Half-Life in Soil (Days)
Atrazine $C_8H_{14}ClN_5$	6-Chloro-N^2-ethyl-N^4-isopropyl-1,3,5-triazine-2,4-diamine	3.8×10^{-2}	2.5	33[a]	35–50
Cyanazine $C_9H_{13}ClN_6$	2-(4-Chloro-6-ethylamino-1,3,5-triazin-2-ylamino)-2-methylpropionitrile	2.0×10^{-4a}	2.1	171	ca. 14
Metribuzin $C_8H_{14}N_4OS$	4-Amino-6-$tert$-butyl-4,5-dihydro-3-methylthio-1,2,4-triazin-5-one	0.058[a]	1.6[a]	1050[a]	40
Prometryn $C_{10}H_{19}N_5S$	N^2,N^4-Diisopropyl-6-methylthio-1,3,5-triazine-2,4-diamine	0.165	3.1	33	50
Simazine $C_7H_{12}ClN_5$	6-Chloro-N^2,N^4-diethyl-1,3,5-triazine-2,4-diamine	2.9×10^{-3}	2.1	6.2[a]	27–102
Terbutryn $C_{10}H_{19}N_5S$	N^2-$tert$-Butyl-N^4-ethyl-6-methylthio-1,3,5-triazine-2,4-diamine	0.225	3.65	22	14–50

Sources: Data from Tomlin, C. (Ed.) in *The Pesticide Manual*, British Crop Protection Council, 2000; http://ec.europa.eu/food/plant/protection/evaluation/exist_subs_rep_en.htm; http://www.epa.gov/opprd001/factsheets/; Hornsby, A.G., Wauchope, R.D., and Herner, A.E. in *Pesticide Properties in the Environment*, Springer-Verlag, New York, 1996; De Liñan, C. in *Farmacología Vegetal*, Ediciones Agrotecnicas S.L., 1997.

[a] 20°C.

1.2.11 UREAS

1.2.11.1 Phenylureas

The urea herbicides may be considered as derivatives of urea, $H_2NC(=O)NH_2$.

Fenuron   Linuron

Phenylureas belong to a numerous group of substituted ureas directly applied to soil in preemergence to control annual grasses in various crops. These compounds have a range of specific selectivity as well as variable persistence in soil according to their chemical composition (Table 1.11).

TABLE 1.11
Chemical Names and Properties of Phenyl Urea Herbicides

Common Name	IUPAC Name	Vapor Pressure mPa (25°C)	K_{ow} log P (25°C)	Water Solubility mg/L (25°C)	Half-Life in Soil (Days)
Chlorotoluron $C_{10}H_{13}ClN_2O$	3-(3-Chloro-p-tolyl)-1,1-dimethylurea	0.005	2.5	74	30–40
Diuron $C_9H_{10}Cl_2N_2O$	3-(3,4-Dichlorophenyl)-1,1-dimethylurea	1.1×10^{-3}	2.85	36	90–180
Fenuron $C_9H_{12}N_2O$	1,1-Dimethyl-3-phenylurea	21[a]	—	3850	60
Isoproturon $C_{12}H_{18}N_2O$	3-(4-Isopropylphenyl)-1,1-dimethylurea	8.1×10^{-3}	2.5[b]	65	6–28
Linuron $C_9H_{10}Cl_2N_2O_2$	3-(3,4-Dichlorophenyl)-1-methoxy-1-methylurea	0.051[b]	3.0	63.8[b]	38–67

Sources: Data from Tomlin, C. (Ed.) in *The Pesticide Manual*, British Crop Protection Council, 2000; http://ec.europa.eu/food/plant/protection/evaluation/exist_subs_rep_en.htm; http://www.epa.gov/opprd001/factsheets/; Hornsby, A.G., Wauchope, R.D., and Herner, A.E. in *Pesticide Properties in the Environment*, Springer-Verlag, New York, 1996; De Liñan, C. in *Farmacología Vegetal*, Ediciones Agrotecnicas S.L., 1997.

[a] 60°C.
[b] 20°C.

1.2.11.2   Sulfonylureas

Chlorsulfuron   Triasulfuron

This group of substituted ureas has been developed more recently and they have, in general, a herbicidal activity higher than the phenylurea herbicides, with application rates in the range of gram/hectare instead of kilogram/hectare. They can be absorbed by foliage and roots. They are normally applied in postemergence and in some cases may have a noticeable field persistence (Table 1.12).

1.3   INSECTICIDES

Horticultural crops may be affected by various pests causing serious damages to plants and consequently important yield reductions. Therefore, insecticides are widely used to control pests in crops. These compounds may be applied to the soil to kill soilborne pests or to the aerial part of the plant.

A major part of the applied insecticides reaches the soil, either by direct applications to the soil or indirectly by runoff from leaves and stems.

1.3.1   BENZOYLUREAS

Teflubenzuron

A new insecticide activity acting on the moulting process of insects was discovered in the study of biological activity of some benzoylurea derivatives. Benzoylureas act as insect growth regulators, interfering with the chitin formation in the vital insect exoskeleton. Most benzoylureas used as insecticides contain fluorine atoms and have high molecular weights. Table 1.13 summarizes the physicochemical properties of these compounds.

1.3.2   CARBAMATES

The N-methyl and N,N-dimethyl carbamic esters of a variety of phenols possess useful insecticidal properties. Aromatic N-methylcarbamates are derivatives of

TABLE 1.12
Chemical Names and Properties of Sulfonylurea Herbicides

Common Name	IUPAC Name	Vapor Pressure mPa (25°C)	K_{ow} log P (25°C)	Water Solubility mg/L (25°C)	Half-Life in Soil (Days)
Azimsulfuron $C_{13}H_{16}N_{10}O_5S$	1-(4,6-Dimethoxypyrimidin-2-yl)-3-[1-methyl-4-(2-methyl-2H-tetrazol-5-yl)-pyrazol-5-ylsulfonyl]urea	4.0×10^{-6}	-1.37	1050[a]	—
Chlorsulfuron $C_{12}H_{12}ClN_5O_4S$	1-(2-Chlorophenylsulfonyl)-3-(4-methoxy-6-methyl-1,3,5-triazin-2-yl)urea	3×10^{-6}	-0.99	7000	28-42
Flazasulfuron $C_{13}H_{12}F_3N_5O_5S$	1-(4,6-Dimethoxypyrimidin-2-yl)-3-(3-trifluoromethyl-2-pyridylsulfonyl)urea	<0.013	-0.06	2100	<7
Imazosulfuron $C_{14}H_{13}ClN_6O_5S$	1-(2-Chloroimidazo[1,2-a]pyridin-3-ylsulfonyl)-3-(4,6-dimethoxypyrimidin-2-yl)urea	4.5×10^{-5}	0.049	308	—
Metsulfuron-methyl $C_{14}H_{15}N_5O_6S$	Methyl-2-(4-methoxy-6-methyl-1,3,5-triazin-2-ylcarbamoylsulfamoyl)benzoate	3.3×10^{-7}	-1.74	2790	7-35
Rimsulfuron $C_{14}H_{17}N_5O_7S_2$	1-(4,6-Dimethoxypyrimidin-2-yl)-3-(3-ethylsulfonyl-2-pyridylsulfonyl)urea	1.5×10^{-3}	-1.47	7300	10-20
Thifensulfuron-methyl $C_{12}H_{13}N_5O_6S_2$	Methyl 3-(4-methoxy-6-methyl-1,3,5-triazin-2-ylcarbamoylsulfamoyl)thiophen-2-carboxylate	1.7×10^{-5}	0.02	6270	6-12
Triasulfuron $C_{14}H_{16}ClN_5O_5S$	1-[2-(2-Chloroethoxy)phenylsulfonyl]-3-(4-methoxy-6-methyl-1,3,5-triazin-2-yl)urea	$<2 \times 10^{-3}$	-0.59	815	19
Tribenuron-methyl $C_{15}H_{17}N_5O_6S$	Methyl 2-[4-methoxy-6-methyl-1,3,5-triazin-2-yl(methyl)carbamoylsulfamoyl]benzoate	5.2×10^{-5}	-0.44	2040[a]	1-7

Sources: Data from Tomlin, C. (Ed.) in *The Pesticide Manual*, British Crop Protection Council, 2000; http://ec.europa.eu/food/plant/protection/evaluation/exist_subs_rep_en.htm; http://www.epa.gov/opprd001/factsheets/; Hornsby, A.G., Wauchope, R.D., and Herner, A.E. in *Pesticide Properties in the Environment*, Springer-Verlag, New York, 1996; De Liñan, C. in *Farmacología Vegetal*, Ediciones Agrotecnicas S.L., 1997.

[a] 20°C.

TABLE 1.13

Chemical Names and Properties of Benzoylurea Insecticides

Common Name	IUPAC Name	Vapor Pressure mPa (20°C)	K_{ow} log P	Water Solubility mg/L (25°C)	Half-Life in Soil (Days)
Diflubenzuron $C_{14}H_9ClF_2N_2O_2$	1-(4-Chlorophenyl)-3-(2,6-difluorobenzoyl)urea	1.2×10^{-4a}	3.89	0.08	<7
Hexaflumuron $C_{16}H_8Cl_2F_6N_2O_3$	1-[3,5-Dichloro-4-(1,1,2,2-tetrafluoroethoxy)phenyl]-3-(2,6-difluorobenzoyl)urea	5.9×10^{-2a}	5.68	0.027^b	50–64
Teflubenzuron $C_{14}H_6Cl_2F_4N_2O_2$	1-(3,5-Dichloro-2,4-difluorophenyl)-3-(2,6-difluorobenzoyl)urea	0.8×10^{-6}	4.3	0.019^b	14–84
Triflumuron $C_{15}H_{10}ClF_3N_2O_3$	1-(2-Chlorobenzoyl)-3-(4-trifluoromethoxyphenyl)urea	4×10^{-5}	4.91	0.025^b	112

Sources: Data from Tomlin, C. (Ed.) in *The Pesticide Manual*, British Crop Protection Council, 2000; http://ec.europa.eu/food/plant/protection/evaluation/exist_subs_rep_en.htm; http://www.epa.gov/opprd001/factsheets/; Hornsby, A.G., Wauchope, R.D., and Herner, A.E. in *Pesticide Properties in the Environment*, Springer-Verlag, New York, 1996; De Liñan, C. in *Farmacología Vegetal*, Ediciones Agrotecnicas S.L., 1997.

[a] 25°C.

[b] 18°C–23°C.

phenyl *N*-methylcarbamate with a great variety of chloride, alkyl, alkylthio, alkoxy, and dialkylamino side chains. Some carbamate insecticides contain a sulfur atom in their molecule.

Carbaryl   Methomyl

These compounds have a very broad spectrum of action, and they are particularly effective on lepidopterous larvae and on ornamental pests including snails, slugs, and household pests. Some of them exhibit systemic characteristics (Table 1.14).

TABLE 1.14
Chemical Names and Properties of Carbamate Insecticides

Common Name	IUPAC Name	Vapor Pressure mPa (20°C)	K_{ow} log P	Water Solubility mg/L (20°C)	Half-Life in Soil (Days)
Aldicarb $C_7H_{14}N_2O_2S$	2-Methyl-2-(methylthio) propionaldehyde O-methylcarbamoyloxime	13	—	4930	30
Carbaryl $C_{12}H_{11}NO_2$	1-Naphthyl methylcarbamate	4.1×10^{-2a}	1.59	120	7–28
Carbofuran $C_{12}H_{15}NO_3$	2,3-Dihydro-2,2-dimethyl benzofuran-7-yl methylcarbamate	0.031	1.52	320	30–60
Carbosulfan $C_{20}H_{32}N_2O_3S$	2,3-Dihydro-2,2-dimethyl benzofuran-7-yl(dibutylaminothio) methylcarbamate	0.041^a		0.35^a	2–5
Fenoxycarb $C_{17}H_{19}NO_4$	Ethyl-2-(4-phenoxyphenoxy) ethylcarbamate	8.67×10^{-4a}	4.07	7.9^a	31
Methomyl $C_5H_{10}N_2O_2S$	S-Methyl N-(methylcarba-moyloxy) thioacetamidate	0.72^a	0.093	$57,900^a$	5–45
Oxamyl $C_7H_{13}N_3O_3S$	N,N'-Dimethyl-2-methyl carbamoyloxyimino-2-(methylthio)acetamide	0.051^a	−0.44	$280,000^a$	7
Pirimicarb $C_{11}H_{18}N_4O_2$	2-Dimethylamino-5, 6-dimethyl pyrimidin-4-yl dimethylcarbamate	0.4	1.7	3000	7–234

Sources: Data from Tomlin, C. (Ed.) in *The Pesticide Manual*, British Crop Protection Council, 2000; http://ec.europa.eu/food/plant/protection/evaluation/exist_subs_rep_en.htm; http://www.epa .gov/opprd001/factsheets/; Hornsby, A.G., Wauchope, R.D., and Herner, A.E. in *Pesticide Properties in the Environment*, Springer-Verlag, New York, 1996; De Liñan, C. in *Farmacología Vegetal*, Ediciones Agrotecnicas S.L., 1997.

[a] 25°C.

1.3.3 ORGANOCHLORINES

Endosulfan

p,p'-DDT

These insecticides are characterized by three kinds of chemicals: DDT analogs, benzene hexachloride (BHC) isomers, and cyclodiene compounds. DDT is one of the most persistent and durable of all contact insecticides because of its insolubility in water and very low vapor pressure. DDT has a wide spectrum of activity on different families of insects and related organisms. BHC isomers are active against a great variety of pests. Cyclodiene compounds are effective where contact action and long persistence are required. These compounds have a broad spectrum insecticide and have been used for the control of insect pests of fruits, vegetables, and cotton as soil insecticides and for seed treatment. Due to their persistence and toxicity, most of these organochlorine compounds have been banned or their use as pesticide has been restricted (Table 1.15).

1.3.4 ORGANOPHOSPHORUS

Organophosphorus insecticides are hydrocarbon compounds which contain one or more phosphorus atoms in their molecule. They are relatively short lived in biological systems.

Fenitrothion   Chlorpyrifos

The diversity of organophosphorus insecticide types makes them to form the most versatile group. There are compounds with nonresidual action and prolonged residual action, and compounds with a broad spectrum and very specific action that can have activity as systemic insecticides for plants, seed, and soil treatments, as well as for animals. In general, they are soluble in water and readily hydrolyzed and they dissipate from soil within a few weeks after application. Because of their low persistence and high effectiveness, these compounds are widely used as systemic insecticides for plants, animals, and soil treatments (Table 1.16).

1.3.5 PYRETHROIDS

Permethrin

Pyrethrins are natural insecticides obtained from pyrethrum, extracted from the flowers of certain species of chrysanthemum. The insecticide properties are due to five esters that are mostly present in the flowers. These esters have asymmetric carbon atoms and double bonds in both alcohol and acid moieties. The naturally occurring forms are esters from (+)-*trans* acids and (+)-*cis* alcohols. Synthetic pyrethrins, called

TABLE 1.15

Chemical Names and Properties of Organochlorine Insecticides

Common Name	IUPAC Name	Vapor Pressure mPa (20°C)	K_{ow} log P	Water Solubility mg/L (25°C)	Half-Life in Soil (Days)
Aldrin $C_{12}H_8Cl_6$	1,2,3,4,10,10-Hexachloro-1,4,4a,5,8,8a-hexahydro-exo-1,4-endo-5,8-dimethanonaphthalene	3.08	5.0–7.4	<0.05	365
p,p'-DDT $C_{14}H_9Cl_5$	1,1,1-Trichloro-2,2-bis(4-chlorophenyl)ethane	0.025	6.19–6.91	0.0055	2000
Dieldrin $C_{12}H_8Cl_6O$	1,2,3,4,10,10-Hexachloro-6,7-epoxy-1,4,4a,5,6,7,8,8a-octahydro-endo-1,4-exo-5,8-dimethanonaphthalene	0.4	4.32–5.40	0.186	1000
Dicofol $C_{14}H_9Cl_5O$	2,2,2-Trichloro-1,1-bis(4-chlorophenyl)ethanol	0.053	4.28	0.8	45
Endosulfan $C_9H_6Cl_6O_3S$	(1,4,5,6,7,7-Hexachloro-8,9,10-trinorborn-5-en-2,3-ylenebismethylene)sulfite	0.83	4.74	0.32[a]	30–70
γ-HCH $C_6H_6Cl_6$	1,2,3,4,5,6-Hexachlorocyclohexane	4.4[b]	3.5	8.5	400
Methoxychlor $C_{16}H_{15}Cl_3O_2$	1,1,1-Trichloro-2,2-bis(4-methoxyphenyl)ethane	<1	—	0.1	120
Tetradifon $C_{12}H_6Cl_4O_2S$	4-Chlorophenyl-2,4,5-trichlorophenylsulfone	3.2×10^{-5}	4.61	0.078[c]	—

Sources: Data from Tomlin, C. (Ed.) in *The Pesticide Manual*, British Crop Protection Council, 2000; http://ec.europa.eu/food/plant/protection/evaluation/exist _subs_rep_en.htm; http://www.epa.gov/opprd001/factsheets/; Hornsby, A.G., Wauchope, R.D., and Herner, A.E. in *Pesticide Properties in the Environment*, Springer-Verlag, New York, 1996; De Liñan, C. in *Farmacología Vegetal*, Ediciones Agrotecnicas S.L., 1997.

[a] 22°C.
[b] 24°C.
[c] 20°C.

TABLE 1.16
Chemical Names and Properties of Organophosphorus Insecticides

Common Name	IUPAC Name	Vapor Pressure mPa (25°C)	K_{ow} log P	Water Solubility mg/L (20°C)	Half-Life in Soil (Days)
Azinphos-methyl $C_{10}H_{12}N_3O_3PS_2$	S(3,4-Dihydro-4-oxobenzo [d]-[1,2,3]triazin-3-ylmethyl) O,O-dimethyl phosphorodithioate	5×10^{-4a}	2.96	28	10–40
Chlorfenvinphos $C_{12}H_{14}Cl_3O_4P$	2-Chloro-1-(2,4-dichlorophenyl)vinyl diethylphosphate	1	3.85	145^b	—
Chlorpyrifos $C_9H_{11}Cl_3NO_3PS$	O,O-Diethyl O-3,5,6-trichloro-2-pyridyl phosphorothioate	2.7	4.7	1.4^b	35–56
Chlorpyrifos-methyl $C_7H_7Cl_3NO_3PS$	O,O-Dimethyl O-3,5,6-trichloro-2-pyridyl phosphorothioate	3	4.24	2.6	1.5–33
Coumaphos $C_{14}H_{16}ClO_5PS$	O-3-Chloro-4-methyl-2-oxo-2H-chromen-7-yl O,O-diethyl phosphorothioate	0.013^a	4.13	1.5	—
Diazinon $C_{12}H_{21}N_2O_3PS$	O,O-Diethyl O-2-isopropyl-6-methylpyrimidin-4-yl phosphorothioate	12	3.30	60	11–21
Dichlorvos $C_4H_7Cl_2O_4P$	2,2-Dichlorovinyl dimethyl phosphate	2.1×10^3	1.9	18,000	0.5
Dimethoate $C_5H_{12}NO_3PS_2$	O,O-Dimethyl S-methyl carbamoylmethyl phosphorodithioate	0.25	0.704	23,800	2–4
Fenitrothion $C_9H_{12}NO_5PS$	O,O-Dimethyl O-4-nitro-m-tolyl phosphorothioate	18^a	3.43	14^b	12–28
Fenthion $C_{10}H_{15}O_3PS_2$	O,O-Dimethyl O-4-methylthio-m-tolyl phosphorothioate	1.4	4.84	4.2	34
Malathion $C_{10}H_{19}O_6PS_2$	S-1,2-Bis(ethoxycarbonyl) ethyl O,O-dimethyl phosphorodithioate	1.1^a	2.75	145^b	1
Methamidophos $C_2H_8NO_2PS$	O,S-Dimethyl phosphoramidothioate	2.3^a	−0.8	$>2 \times 10^5$	6
Methidathion $C_6H_{11}N_2O_4PS_3$	S-2,3-Dihydro-5-methoxy-2-oxo-1,3,4-thiadiazol-3-ylmethyl O,O-dimethyl phosphorodithioate	0.25^a	2.2	200^b	3–18
Oxydemeton-methyl $C_6H_{15}O_4PS_2$	S-2-Ethylsulfinylethyl O,O-dimethyl phosphorothioate	3.8^a	−0.74	1×10^6	2–20
Phosmet $C_{11}H_{12}NO_4PS_2$	O,O-Dimethyl S-phtalimidomethyl phosphorodithioate	0.065	2.95	25^b	4–20
Pirimiphos-methyl $C_{11}H_{20}N_3O_3PS$	O-2-Diethylamino-6-methylpyrimidin-4-yl O,O-dimethyl phosphorothioate	2^a	4.2	10	3.5–25
Profenofos $C_{11}H_{15}BrClO_3PS$	O-4-Bromo-2-chlorophenyl O-ethyl S-propyl phosphorothioate	0.12	4.44	28^b	7
Trichlorfon $C_4H_8Cl_3O_4P$	Dimethyl 2,2,2-trichloro-1-hydroxyethylphosphonate	0.5	0.43	1.2×10^5	5–30

Sources: Data from Tomlin, C. (Ed.) in *The Pesticide Manual*, British Crop Protection Council, 2000; http://ec.europa.eu/food/plant/protection/evaluation/exist_subs_rep_en.htm; http://www.epa.gov/opprd001/factsheets/; Hornsby, A.G., Wauchope, R.D., and Herner, A.E. in *Pesticide Properties in the Environment*, Springer-Verlag, New York, 1996; De Liñan, C. in *Farmacología Vegetal*, Ediciones Agrotecnicas S.L., 1997.

[a] 20°C.
[b] 25°C.

pyrethroids, present better activity for a larger spectrum of pests than natural ones. They show selective activity against insects and present low toxicity to mammals and birds. Pyrethroids are considered as contact poisons, affecting the insect nervous system and depolarizing the neuronal membranes. These compounds are degraded in soil and have no detectable effects on soil microflora. They have also been used in household to control flies and mosquitoes. Piperonyl butoxide ($C_{19}H_{30}O_5$) is used as a synergist for pyrethrins and related insecticides (Table 1.17).

1.4 FUNGICIDES

Fungicides used in agriculture to control plant diseases belong to various chemical classes. A wide variation of physicochemical properties of these substances can be observed, according to the different chemical structures of fungicides. Some fungicides are stereoisomers and they are normally commercialized as mixtures of these isomers. Fungicides can be applied pre- or postharvest for the protection of cereals, fruits, and vegetables from fungal diseases.

1.4.1 AZOLES

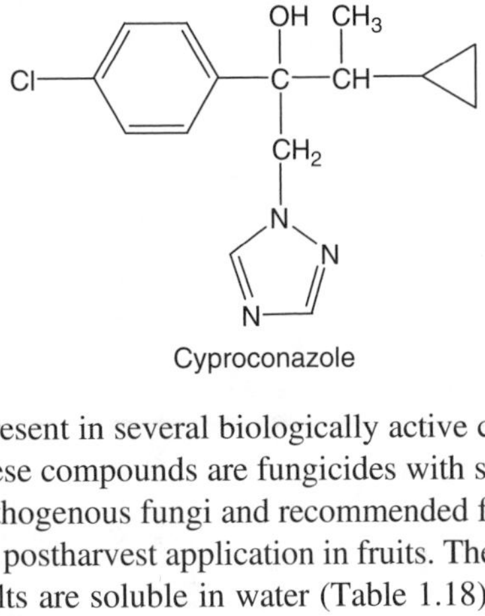

Cyproconazole

The imidazole ring is present in several biologically active compounds, while others have a triazole ring. These compounds are fungicides with systemic action, effective against several phytopathogenous fungi and recommended for seed dressing, as well as foliage fungicide and postharvest application in fruits. They are scarcely soluble in water, although their salts are soluble in water (Table 1.18).

1.4.2 BENZIMIDAZOLES

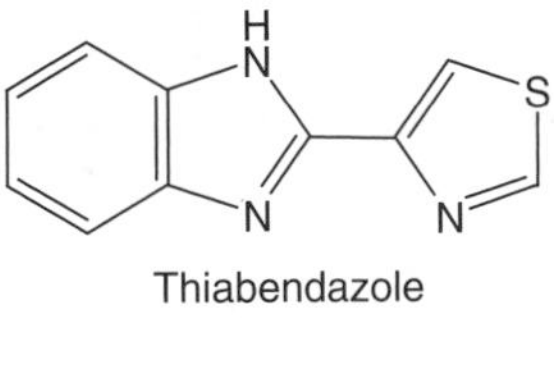

Thiabendazole

Fungicides of the benzimidazole type have a systemic action. Generally, they are taken up by the roots of the plants, and the active substances are then acropetally translocated through the xylem to the leaves. These compounds have been used in plant protection in the form of their insoluble salts. They are foliage and soil

TABLE 1.17

Chemical Names and Properties of Pyrethroid Insecticides

Common Name	IUPAC Name	Vapor Pressure mPa (20°C)	K_{ow} log P	Water Solubility mg/L (25°C)	Half-Life in Soil (Days)
Acrinathrin $C_{26}H_{21}F_6NO_5$	(S)-α-Cyano-3-phenoxybenzyl (Z)-(1R,3S)-2,2-dimethyl-3-[2-(2,2,2-trifluoro-1-trifluoromethyl ethoxycarbonyl)vinyl] cyclopropanecarboxylate	4.4×10^{-5}	5.6	≤0.02	5–100
Cyfluthrin $C_{22}H_{18}Cl_2FNO_3$	(RS)-α-Cyano-4-fluoro-3-phenoxybenzyl(1RS,3RS; 1RS,3SR)-3-(2,2-dichlorovinyl)-2,2-dimethyl cyclopropanecarboxylate	*Diastereoisomer* I: 9.6×10^{-4} II: 1.4×10^{-5} III: 2.1×10^{-5} IV: 8.5×10^{-5}	I: 6 II: 5.9 III: 6 IV:5.9	I: 2.2×10^{-3a} II: 1.9×10^{-3} III: 2.2×10^{-3} IV: 2.9×10^{-3}	56–63
Cypermethrin $C_{22}H_{19}Cl_2NO_3$	(RS)-α -Cyano-3-phenoxybenzyl(1RS,3RS; 1RS,3SR)-3-(2,2-dichlorovinyl)-2,2-dimethyl cyclopropanecarboxylate	2×10^{-4}	6.6	0.004	60
Deltamethrin $C_{22}H_{19}Br_2NO_3$	(S)-α-Cyano-3-phenoxybenzyl(1R,3R)-3-(2,2-dibromovinyl)-2,2-dimethyl cyclopropanecarboxylate	1.24×10^{-5b}	4.6	$<0.2 \times 10^{-3}$	23
Esfenvalerate $C_{25}H_{22}ClNO_3$	(S)-α-Cyano-3-phenoxybenzyl-(S)-2-(4-chlorophenyl)-3-methylbutyrate	2×10^{-4b}	6.22	0.002	35–88
Tau-fluvalinate $C_{26}H_{22}ClF_3N_2O_3$	(RS)-α-Cyano-3-phenoxybenzyl-N-(2-chloro-α,α,α-trifluoro-p-tolyl)-D-valinate	9×10^{-8}	4.26	0.001^a	12–92
Permethrin $C_{21}H_{20}Cl_2O_3$	3-Phenoxybenzyl-(1RS,3RS;1RS,3SR)-3-(2,2-dichlorovinyl)-2,2-dimethylcyclopropane carboxylate	*cis:* 0.0025 *trans:* 0.0015	6.1^a	6×10^{-3a}	<38

Sources: Data from Tomlin, C. (Ed.) in *The Pesticide Manual*, British Crop Protection Council, 2000; http://ec.europa.eu/food/plant/protection/evaluation/exist _subs_rep_en.htm; http://www.epa.gov/opprd001/factsheets/; Hornsby, A.G., Wauchope, R.D., and Hemer, A.E. in *Pesticide Properties in the Environment*, Springer-Verlag, New York, 1996; De Liñan, C. in *Farmacología Vegetal*, Ediciones Agrotecnicas S.L., 1997.

[a] 20°C.
[b] 25°C.

TABLE 1.18
Chemical Names and Properties of Azole Fungicides

Common Name	IUPAC Name	Vapor Pressure mPa (20°C)	K_{ow} log P	Water Solubility mg/L (20°C)	Half-Life in Soil (Days)
Cyproconazole $C_{15}H_{18}ClN_3O$	(2RS,3RS;2RS,3SR)-2-(4-Chlorophenyl)-3-cyclopropyl-1-(1H-1,2,4-triazol-1-yl)butan-2-ol	0.04	2.91	140[a]	90
Flusilazole $C_{16}H_{15}F_2N_3Si$	Bis(4-fluorophenyl)(methyl)(1H-1,2,4-triazol-1-ylmethyl)silane	0.04[a]	3.74	54	95
Hexaconazole $C_{14}H_{17}Cl_2N_3O$	(RS)-2-(2,4-Dichlorophenyl)-1-(1H-1,2,4-triazol-1-yl)hexan-2-ol	0.018	3.9	17	—
Imazalil $C_{14}H_{14}Cl_2N_2O$	(±)-1-(β-Allyloxy-2,4-dichlorophenylethyl)imidazole	0.158	3.82	180	150
Prochloraz $C_{15}H_{16}Cl_3N_3O_2$	N-Propyl-N-[2-(2,4,6-trichlorophenoxy)ethyl]imidazole-1-carboxamide	0.09	4.12	34.4[a]	120
Propiconazole $C_{15}H_{17}Cl_2N_3O_2$	(±)-1-[2-(2,4-Dichlorophenyl)-4-propyl-1,3-dioxolan-2-ylmethyl]-1H-1,2,4-triazole	0.03	3.72	100	110
Tebuconazole $C_{16}H_{22}ClN_3O$	(RS)-1-p-Chlorophenyl-4,4-dimethyl-3-(1H-1,2,4-triazol-1-ylmethyl)pentan-3-ol	0.002	3.7	36	—
Triadimefon $C_{14}H_{16}ClN_3O_2$	(1-(4-Chlorophenoxy)-3,3-dimethyl-1-(1H-1,2,4-triazol-1-yl)butan-2-one	0.02	3.11	64	6-60

Sources: Data from Tomlin, C. (Ed.) in *The Pesticide Manual*, British Crop Protection Council, 2000; http://ec.europa.eu/food/plant/protection/evaluation/exist _subs_rep_en.htm; http://www.epa.gov/opprd001/factsheets/; Hornsby, A.G., Wauchope, R.D., and Herner, A.E. in *Pesticide Properties in the Environment*, Springer-Verlag, New York, 1996; De Liñan, C. in *Farmacología Vegetal*, Ediciones Agrotecnicas S.L., 1997.

[a] 25°C.

TABLE 1.19

Chemical Names and Properties of Benzimidazole Fungicides

Common Name	IUPAC Name	Vapor Pressure mPa (25°C)	K_{ow} log P	Water Solubility mg/L (25°C)	Half-Life in Soil (Days)
Benomyl $C_{14}H_{18}N_4O_3$	Methyl 1-(butyl carbamoyl) benzimidazol-2-ylcarbamate	<0.005	1.37	0.003	67
Carbendazim $C_9H_9N_3O_2$	Methyl benzimidazol-2-yl carbamate	0.15	1.51	8	120
Thiabendazole $C_{10}H_7N_3S$	2-(Thiazol-4-yl)benzimidazole	4.6×10^{-4}	2.39	30[a]	33–120

Sources: Data from Tomlin, C. (Ed.) in *The Pesticide Manual*, British Crop Protection Council, 2000; http://ec.europa.eu/food/plant/protection/evaluation/exist_subs_rep_en.htm; http://www.epa .gov/opprd001/factsheets/; Hornsby, A.G., Wauchope, R.D., and Herner, A.E. in *Pesticide Properties in the Environment*, Springer-Verlag, New York, 1996; De Liñan, C. in *Farmacología Vegetal*, Ediciones Agrotecnicas S.L., 1997.

[a] 20°C.

fungicides with a specific and broad spectrum of action, also used for seed treatment and in postharvest (Table 1.19).

1.4.3 DITHIOCARBAMATES

Ethylenebisdithiocarbamates are prepared from ethylene diamine $H_2N–CH_2–CH_2–NH_2$.

$$[SC(S)NHCH_2CH_2NHCSSMn]_xZn_y$$
Mancozeb

These compounds are heavy metal salts of ethylenebisdithiocarbamate and these salts are unusually stable and suitable as fungicides. The dithiocarbamate fungicides are the most widely used organic fungicides and have a wide spectrum of activity as foliar sprays for fruits, vegetables, and ornamentals and as seed protectants (Table 1.20).

1.4.4 MORPHOLINES

$(H_3C)_3C—$[benzene ring]$—CH_2CHCH_2—N$ [morpholine ring with CH_3 groups]

Fenpropimorph

Morpholines are specific systemic fungicides against powdery mildew fungi and are used to control the disease in cereals, cucumbers, apples, and so on. These

TABLE 1.20
Chemical Names and Properties of Dithiocarbamate Fungicides

Common Name	IUPAC Name	Vapor Pressure mPa (20°C)	K_{ow} log P	Water Solubility mg/L (25°C)	Half-Life in Soil (Days)
Mancozeb $(C_4H_6 MnN_2S_4)_x(Zn)_y$	Manganese ethylenebis (dithiocarbamate) (polymeric) complex with zinc salt	<1	1.8	6.2	6–15
Maneb $C_4H_6MnN_2S_4$	Manganese ethylenebis (dithiocarbamate)	<0.01	−0.45	257	25
Metiram $(C_{16}H_{33}N_{11}S_{16}Zn_3)_x$	Zinc ammoniate ethylenebis (dithiocarbamate)- poly(ethylenethiuram disulfide)	<0.01	0.3	0.1	20
Nabam $C_4H_6N_2Na_2S_4$	Disodium ethylenebis (dithiocarbamate)	Negligible	—	2×10^5	—
Zineb $(C_4H_6N_2S_4Zn)_x$	Zinc ethylenebis (dithiocarbamate) (polymeric)	<0.01	≤1.3	10	23
Ziram $C_6H_{12}N_2S_4Zn$	Zinc bis (dimethyldithiocarbamate)	<0.001	1.23	1.58–18.3[a]	2

Sources: Data from Tomlin, C. (Ed.) in *The Pesticide Manual*, British Crop Protection Council, 2000; http://ec.europa.eu/food/plant/protection/evaluation/exist_subs_rep_en.htm; http://www.epa .gov/opprd001/factsheets/; Hornsby, A.G., Wauchope, R.D., and Herner, A.E. in *Pesticide Properties in the Environment*, Springer-Verlag, New York, 1996; De Liñan, C. in *Farmacología Vegetal*, Ediciones Agrotecnicas S.L., 1997.

[a] 20°C.

compounds are distributed in the plants by translocation from the root and foliage and protect the plants against infection by phytopathogenic fungi. They have a certain persistence in soil (Table 1.21).

1.4.5 MISCELLANEOUS

Captan   Chlorothalonil

TABLE 1.21

Chemical Names and Properties of Morpholine Fungicides

Common Name	IUPAC Name	Vapor Pressure mPa (20°C)	K_{ow} log P	Water Solubility mg/L (20°C)	Half-Life in Soil (Days)
Dodemorph $C_{18}H_{35}NO$	4-Cyclododecyl-2,6-dimethyl morpholine	0.48	4.14	<100	73
Fenpropimorph $C_{20}H_{33}NO$	(±)-*cis*-4-[3-(4-*tert*-Butylphenyl)-2-methylpropyl]-2,6-dimethylmorpholine	3.5	4.2	4.3	15–93
Tridemorph $C_{19}H_{39}NO$	2,6-Dimethyl-4-tridecyl morpholine	12	4.2	1.1	14–34

Sources: Data from Tomlin, C. (Ed.) in *The Pesticide Manual*, British Crop Protection Council, 2000; http://ec.europa.eu/food/plant/protection/evaluation/exist_subs_rep_en.htm; http://www.epa.gov/opprd001/factsheets/; Hornsby, A.G., Wauchope, R.D., and Herner, A.E. in *Pesticide Properties in the Environment*, Springer-Verlag, New York, 1996; De Liñan, C. in *Farmacología Vegetal*, Ediciones Agrotecnicas S.L., 1997.

Captan and folpet are fungicides used in foliar treatment of fruits, vegetables, and ornamentals, in soil and seed treatments, and in postharvest applications.

Procymidone is a dicarboximide-derived fungicide with moderate systemic action. It is rapidly absorbed not only through the roots but also through the stem or the leaves. It is used for the control of storage roots of fruits and vegetables and it is effective for seed dressing of cereals. Table 1.22 summarizes the properties of various frequently used fungicides belonging to different chemical classes.

1.5   MODE OF ACTION

The control of pests by pesticides depends on several factors like the mode of action of these compounds, the crop stage and the environmental conditions, moisture, soil type and temperature, among others, and numerous works have been published on these subjects [10–13]. The main modes of action of pesticides are summarized later.

1.5.1   HERBICIDES

1.5.1.1   Amino Acid Synthesis Inhibitors

Amino acid synthesis inhibitors act on a specific enzyme to prevent the production of certain amino acids, which are the key building blocks for normal plant growth and development.

One type of herbicide causes the inhibition of acetolactate synthase (ALS), the first common enzyme in the branched-chain amino acid biosynthetic pathway. ALS

TABLE 1.22
Chemical Names and Properties of Miscellaneous Fungicides

Common Name	IUPAC Name	Vapor Pressure mPa (25°C)	K_{ow} log P	Water Solubility mg/L (25°C)	Half-Life in Soil (Days)
Azoxystrobin $C_{22}H_{17}N_3O_5$	(E)-2-{2-[6-(2-Cyanophenoxy) pyrimidin-4-yloxy]phenyl}-3-methoxyacrylate	1.1×10^{-7a}	2.5	6[a]	7–56
Captan $C_9H_8Cl_3NO_2S$	N-(Trichloromethylthio) cyclohex-4-ene-1,2-dicarboximide	<1.3	2.8	3.3	1–10
Chlorothalonil $C_8Cl_4N_2$	Tetrachloroisophthalonitrile	0.076	2.92	0.81	5–36
Cyprodinil $C_{14}H_{15}N_3$	4-Cyclopropyl-6-methyl-N-phenylpyrimidin-2-amine	0.51	4	13	20–60
Fenhexamid $C_{14}H_{17}Cl_2NO_2$	N-(2,3-Dichloro-4-hydroxy phenyl)-1-methylcyclohexanecarboxamide	4.0×10^{-4a}	3.51	20[a]	≤1
Folpet $C_9H_4Cl_3NO_2S$	N-(Trichloromethylthio) phthalimide	2.1×10^{-2}	3.11	0.8	4.3
Iprodione $C_{13}H_{13}Cl_2N_3O_3$	3-(3,5-Dichlorophenyl)-N-isopropyl-2,4-dioxoimidazolidine-1-carboxamide	5×10^{-4}	3	13[a]	20–160
Metalaxyl-M $C_{15}H_{21}NO_4$	Methyl-N-(methoxy acetyl)-N-(2,6-xylyl)-D-alaninate	3.3	1.71	26,000	5–30
Ofurace $C_{14}H_{16}ClNO_3$	(±)-α-(2-Chloro-N-2,6-xylyl acetamido)-γ-butyrolactone	2×10^{-2a}	1.39	146[a]	26
Orthophenylphenol $C_{12}H_{10}O$	Biphenyl-2-ol	—	—	700	—
Procymidone $C_{13}H_{11}Cl_2NO_2$	N-(3,5-Dichlorophenyl)-1,2-dimethylcyclopropane-1,2-dicarboximide	18	3.14	4.5	7–21
Pyrimethanil $C_{12}H_{13}N_3$	N-(4,6-Dimethylpyrimidin-2-yl)aniline	2.2	2.84	121	7–54
Tolylfluanid $C_{10}H_{13}Cl_2FN_2O_2S_2$	N-Dichlorofluoromethylthio-N',N'-dimethyl-N-p-tolylsulfamide	0.2[a]	3.90	0.9[a]	2–11
Triforine $C_{10}H_{14}Cl_6N_4O_2$	N-N'-[Piperazine-1,4-diylbis [(trichloromethyl)methylene]] diformamide	80	2.2	9[a]	21

Sources: Data from Tomlin, C. (Ed.) in *The Pesticide Manual*, British Crop Protection Council, 2000; http://ec.europa.eu/food/plant/protection/evaluation/exist_subs_rep_en.htm; http://www.epa.gov/opprd001/factsheets/; Hornsby, A.G., Wauchope, R.D., and Herner, A.E. in *Pesticide Properties in the Environment*, Springer-Verlag, New York, 1996; De Liñan, C. in *Farmacología Vegetal*, Ediciones Agrotecnicas S.L., 1997.

[a] 20°C.

inhibitors include, among others, herbicides of the sulfonylurea family. These compounds vary greatly in selectivity; some of them remain extremely active.

The aromatic ring amino acids, tryptophan, phenylalanine, and tyrosine, are synthesized by plants through the shikimic acid pathway. Only one herbicide, glyphosate, inhibiting that pathway has been commercialized. The mode of action of glyphosate is the inhibition of the enzyme, 5-enolpyruvoyl-shikimate-3-phosphate synthase (EPSPS). This enzyme is present in plants, fungi, and bacteria, but absent in animals, which need to ingest those amino acids in the diet because they are not produced by them.

Another enzyme involved in amino acid synthesis used as a target for herbicides is the glutamine synthase (GS), which makes glutamine from glutamate and ammonia. This enzyme is present in plants, where it plays an important role in nitrogen assimilation, as well as in animals, as glutamate is a neurotransmitter that can be inactivated by GS. The mode of action of the herbicide glufosinate is the inhibition of the enzyme glutamine synthase.

1.5.1.2 Cell Division Inhibitors

This type of herbicide reacts with tubulin, a protein essential for building the intracellular skeleton in eukaryotic cells forming the wall of microtubules. These compounds disturb normal cell division by binding with tubulin.

Inhibitors of cell division are herbicides belonging to various chemical classes, such as dinitroanilines, benzoic acids, and pyridines.

1.5.1.3 Photosynthesis Inhibitors

Photosynthesis is a key process for plants and consequently is a main target for many herbicides. There are different mechanisms involved in the inhibition of photosynthesis, such as free radical generators, blockage of the electron transport system, and inhibition–destruction of protective pigments, but, in general, most herbicides interfere with the transfer of electrons to the plastoquinone pool by binding to a specific protein that regulates electron transfer.

The herbicides acting as photosynthesis inhibitors are all nitrogen-containing compounds with a diversity of chemical composition. These compounds, including phenyl ureas, triazines, pyridazines, phenyl carbamates, nitriles, and amides, are represented by various herbicide families, although some of these chemical classes also have specific herbicides that do not act as photosynthesis inhibitors.

1.5.2 Insecticides

1.5.2.1 Signal Interference in the Nervous System

Chemicals that disturb signal systems are frequently potent poisons. Pyrethroids and organochlorines are the most important insecticides in this category. Their mode of action is to inhibit the proper closing of the channels by acting at the voltage-gated sodium channels. Pyrethroids modify axonal conduction within the central nervous system of insects by altering the permeability of the nerve membrane to sodium and potassium ions. Organochlorines may interact with the pores of the lipoprotein structure of the insect nerve causing distortion and consequent excitation of nerve impulse transmission. The toxic properties of chlorinated cyclodiene insecticides, as lindane, reside in the blockade of the γ-aminobutyric acid-gated chlorine channels, inducing convulsions in insects.

1.5.2.2   Inhibitors of Cholinesterase

The target for many insecticides is an enzyme called acetylcholinesterase (AChE). This enzyme is an essential constituent of the nervous system and plays an important role in animals, but not in plants as they lack nervous system. AChE hydrolyzes acetylcholine, an ester released when nerve impulses are transmitted. Synapses, myoneural functions, and ganglia of the nervous system transmit neural impulses by the mediation of acetylcholine.

The organophosphorus insecticides have the capacity to phosphorylate the esteratic active site of the AChE. The phosphorylated enzyme is irreversibly inhibited and is not able to carry out its normal function of the rapid removal of acetylcholine (ACh). As a result, ACh accumulates and disrupts the normal functioning of the nervous system. Carbamates are also strong inhibitors of AChE and may also have a direct effect on acetylcholine receptors.

1.5.2.3   Inhibitors of Chitin Synthesis

Chitin is a very abundant polysaccharide in nature, although it is present in arthropods and fungi, but absent in plants and mammals.

Benzoylureas affect chitin synthesis in the insect cuticle by disrupting the process of connecting the N-acetylglucosamine units to the chitin chain, preventing in this way the normal moulting process of insects.

1.5.3   FUNGICIDES

1.5.3.1   Sulfhydryl Reagents

Sulfhydryl (SH) groups are important reactive groups often found in the active sites of many enzymes. Dithiocarbamate fungicides react with the SH-containing enzymes and coenzymes of fungal cells. Enzyme inhibition may also occur by complex formation of the active substance with the metal atoms of metal-containing enzymes.

The perhalogen mercaptans, captan and folpet, are good examples of pesticides that react with sulfhydryl groups in many enzymes. These fungicides affect the structure and functions of the cell membranes and inhibit the enzyme system causing tumors in the mitochondria.

1.5.3.2   Cell Division Inhibitors

Benzimidazole fungicides react with tubulin, a protein that is the building block of the intracellular skeleton in cells. The impairment of cell division is produced in most cases by inhibiting the formation of the microtubules. Benzimidazoles, such as benomyl, carbendazim, and thiabendazole, as well as other fungicide groups like carbamates, have this mode of action.

1.5.3.3 Inhibitors of Ergosterol Synthesis

The ergosterol inhibitor fungicides are active against many different fungi. Although they disturb sterol synthesis in higher plants, as well as the synthesis of gibberellins, their phytotoxicity is low. The synthesis of sterols is very complex and various groups of fungicides act on different targets of that synthesis. One large group of fungicides, called demethylase inhibitors, includes various compounds having a heterocyclic N-containing ring, such as azoles, morpholines, pyridines, and piperazines.

1.6   TOXICITY AND RISK ASSESSMENT

Pesticides are toxic compounds that may cause adverse effects on the human and the environment. Toxicity has been defined as the capacity of a substance to produce harmful effects, and other terms used in the risk assessment of chemicals are hazard, defined as the potential to cause harm, and risk, defined as the likelihood of harm. Risk characterization is the estimation of the incidence and severity of the adverse effects likely to occur in a human population, animals, or environmental compartments due to actual or predicted exposure to any active substance.

Humans can be exposed to pesticides by direct or indirect means. Direct or primary exposure normally occurs during the application of these compounds and indirect or secondary exposure can take place through the environment or the ingestion of food. Figure 1.4 summarizes the main routes of indirect exposure to pesticides.

A complete set of data is needed for the toxicological and ecotoxicological evaluation of pesticides.

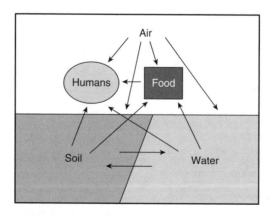

FIGURE 1.4 Routes of indirect exposure to pesticides.

Regarding the human health assessment, toxicological studies need to include the following tests:

A: Acute toxicity, which involves harmful effects in an organism through a single or short-term exposure, should be studied by:

   1. *Dermal toxicity test*: Rabbits are employed more often than any other species for studies of skin toxicity, although guinea pigs, rats, or mice are also used. The results are expressed in terms of LD_{50}, the dose under which the conditions stated will cause death in 50% of a group of test animals.

   2. *Mucus membrane and eye toxicity test*: The conjunctiva of the eye and the vaginal vault of experimental animals (rabbits and monkeys) have been employed in tests of the toxic or irritant effects of chemical substances on mucus membrane.

   3. *Inhalation toxicity test*: The procedures for the evaluation of potential hazards of gases, dusts, mists, or vapors via the inhalation route vary depending on the physical nature (solubility, particle size) of the pesticide.

   4. *Oral toxicity test*: The procedure normally employs the administration of compounds in the diet or intragastrically by gavage. The advantage is that it allows precise measurement of daily dosage to body weight. The oral toxicity generally takes place in three stages, acute (short-term), subacute (subchronic), and chronic (long-term).

B: Repeated dose test that comprises the adverse effects occurring in experimental animals as a result of repeated daily dosing with, or exposure to, a pesticide for a short part of their expected life span.

C: Reproduction and teratology that involves the endocrinological changes associated with the reproductive cycle in the female and the anabolic systems involved in embryologic and fetal growth constitute a challenging background against which to test the toxic potentiality of a pesticide where the enzymatic system plays an important role.

D: Carcinogenesis is to identify the carcinogenicity potential of pesticides in laboratory animals. The studies must be sufficient to establish the species and organ specificity of tumors induced and to determine the dose–response relationship. For nongenotoxic carcinogens, they identify doses that cause no adverse effects.

Estimation of a no observed adverse effect level (NOAEL), when possible, or of a lowest observable adverse effect level (LOAEL) is a critical step in the toxicological risk assessment of pesticides [14].

Concerning ecotoxicity, the estimation of hazard to wildlife involves the determination of the effects on different species. Toxicity data should be gathered for soil organisms, beneficial arthropods (as honeybees), aquatic species (fish, invertebrates, algae, and microorganisms), terrestrial vertebrates (mammals and birds), and plants. Several end points are of interest, depending on the species, namely acute toxicity, growth and activity inhibition, bioconcentration, and effects on reproduction. The poisoning of the species depends on the concentration of the pesticide in different environmental compartments (e.g., water, air, and soil), and consequently a predicted environmental concentration (PEC) has to be derived and compared with the corresponding predicted no effect concentration (PNEC).

Therefore, the toxicological and ecotoxicological effects as well as the concentration of pesticides in the environmental compartments and in food are required for the risk assessment of pesticides. The presence of pesticides in these matrices is normally referred to as pesticide residues, which are defined as any original or derived residue, including relevant metabolites, from a chemical. Analytical methods to determine pesticide residues in the mentioned matrices with adequate sensitivity and selectivity are then needed.

The analysis of pesticide residues in food and environmental samples, together with their monitoring in those matrices, will be described in the following chapters of this book.

REFERENCES

1. http://www.epa.gov/pesticides/about/
2. Carson, R.L. *Silent Spring*. The Riverside Press, Boston, MA, 1962.
3. http://www.fao.org/AG/AGP/AGPP/Pesticid/
4. http://www.codexalimentarius.net/mrls/pestdes/jsp/pest_q-e.jsp
5. Tomlin, C. (Ed.), *The Pesticide Manual*, British Crop Protection Council, 2000.
6. http://ec.europa.eu/food/plant/protection/evaluation/exist_subs_rep_en.htm
7. http://www.epa.gov/opprd001/factsheets/
8. Hornsby, A.G., Wauchope, R.D., and Herner, A.E., *Pesticide Properties in the Environment*, Springer-Verlag, New York, 1996.
9. De Liñan, C., *Farmacología Vegetal*, Ediciones Agrotecnicas S.L., 1997.
10. White-Stevens, R., *Pesticides in the Environment*, Marcel Dekker, New York, 1971.
11. Hutson, D.H. and Roberts, T.R., *Herbicides*, John Wiley & Sons, Chichester, UK, 1987.
12. Matolcsy, G., Nádasy, M., and Andriska, V., *Pesticide Chemistry*, Elsevier, Amsterdam, 1988.
13. Stenersen, J., *Chemical Pesticides. Mode of Action and Toxicology*, CRC Press, Boca Raton, FL, 2004.
14. http://ecb.jrc.it/tgd/

2 Sample Handling of Pesticides in Food and Environmental Samples

Esther Turiel and Antonio Martín-Esteban

CONTENTS

2.1   INTRODUCTION

The determination of pesticides in food and environmental samples at low concentrations is always a challenge. Ideally, the analyte to be determined would be already in solution and at a concentration level high enough to be detected and quantified by the selected final determination technique (i.e., HPLC or GC). Unfortunately, the reality is far from this ideal situation. Firstly, the restrictive legislations from European Union and World Health Organization devoted to prevent contamination of food and environmental compartments by pesticides make necessary the development of analytical methods suitable for detecting target analytes at very low concentration levels. Besides, from a practical point of view, even when the analyte is already in solution (i.e., water or juice), there are several difficulties related to the required sensitivity and selectivity of the selected determination technique that must be overcome, since the concentration of matrix-interfering compounds is much higher than that of the analyte of interest. Consequently, the development of an appropriate sample preparation procedure involving extraction, enrichment, and cleanup steps becomes mandatory to obtain a final extract concentrated on target analytes and as free as possible of matrix compounds.

In this chapter, the different sample treatment techniques currently available and most commonly used in analytical laboratories for the analysis of pesticides in food and environmental samples are described. Depending on the kind of sample (solid or liquid) and the specific application (type of pesticide, concentration level, multiresidue analysis), the final procedure might involve the use of only one or the combination of several of the different techniques described later.

2.2   SAMPLE PRETREATMENT

Generally, sampling techniques provide amounts of sample much higher (2–10 L of liquid samples and 1–2 kg of solid samples) than those needed for the final analysis (just few milligrams). Thus, it is always necessary to carry out some pretreatments to get a homogeneous and representative subsample. Even if the sample is apparently homogeneous, that is, an aqueous sample, it will be at least necessary to perform a filtration step to remove suspended particles, which could affect the final determination of target analytes. However, some hydrophobic analytes (i.e., organochlorine pesticides) could be adsorbed onto particles surface and thus, depending on the objective of the analysis, might be necessary to analyze such particles. This simple example demonstrates the necessity of establishing clearly the objective of the analysis, since it will determine the sample pretreatments to be carried out, and highlights the importance of this typically underrated analytical step.

Usually, environmental water samples just require filtration, whereas liquid food samples might be subjected to other kinds of pretreatments depending on the objective of the analysis. However, solid samples (both environmental and food samples) need to be more extensively pretreated to get a homogeneous subsample. The wide variety of solid samples prevents an exhaustive description of the different procedures in this chapter; however, some general common procedures will be described later.

2.2.1 DRYING

The presence of water or moisture in solid samples has to be taken into account since it might produce alterations (i.e., hydrolysis) of the matrix and/or analytes, which will obviously affect the final analytical results. Besides, water content varies depending on atmospheric conditions and thus, it is recommended to refer the content of target analytes to the mass of dry sample.

Sample drying uses to be carried out before crushing and sieving steps, although it is recommended drying again before final determination since rehydration process might occur. Typically, sample is dried inside an oven at temperatures about 100°C. It is important to stress that higher temperatures can be used to decrease the time devoted to this step but losses of volatile analytes might occur. In this sense, it is important to know a priori the physicochemical properties of target analytes to preserve the integrity of the sample. A more conservative approach, using low temperatures, can be followed but it will unnecessarily increase the drying time. Alternatively, lyophilization is recommended if a high risk of analytes losses exists and it is an appropriate procedure for food, biological material, and plant samples drying. However, even following this procedure, losses of analytes might occur depending on their physical properties (i.e., solubility, volatility).

The results are evident that it is not possible to establish a general rule on how to perform sample drying. Thus, studies on stability of target analytes in spiked samples should be carried out to guarantee the integrity of the sample before final determination of the analytes.

2.2.2 HOMOGENIZATION

As mentioned earlier, samples are heterogeneous in nature and thus, they must be treated to get a homogeneous distribution of target analytes.

Generally, soil samples are crushed, grinded, and sieved through 2 mm mesh. Grinding can be done manually or automatically using specially designed equipments (i.e., ball mills). It is important to stress that this procedure might provoke the local heating of the sample and thus, thermolabile or volatile compounds might be affected. In this sense, it is recommended to grind the sample at short time intervals to minimize sample heating. In addition, due to heating, water content may vary making necessary to recalculate sample moisture.

Food samples use to be cut down to small pieces with a laboratory knife before further homogenization with automatic instruments (i.e., blender). Sample freezing is a general practice to ease blending, especially recommended for samples with high fat content (i.e., cheese) and for soft samples with high risk of phase separation during blending (i.e., liver, citrus fruits).

Apart from these general guidelines, especially in food analysis, the determination of pesticides might be restricted to the edible part of the sample or to samples previously cooked and thus, sample pretreatments will vary depending on the objective of the analysis.

Finally, it is important to point out that, in most of the cases, samples need to be stored for certain periods of time before performing the analysis. In this sense, although sample storage cannot be considered a sample pretreatment, the addition of preservatives as well as the establishment of the right conditions of storage (i.e., at room temperature or in the fridge) to minimize analyte/sample degradation are typical procedures carried out at this stage of the analytical process and need to be taken into account to guarantee the accuracy of the final result.

2.3 EXTRACTION AND PURIFICATION

The main aim of any extraction process is the isolation of analytes of interest from the selected sample by using an appropriate extracting phase. Pesticides from liquid samples (i.e., environmental waters) are preferably extracted using solid phases by solid-phase extraction (SPE) or solid-phase microextraction (SPME) procedures, although for low volume samples, liquid–liquid extraction (LLE) can also be carried out. Extraction of pesticides from environmental or food solid samples is usually performed by mixing the sample with an appropriate extracting solution, where the mixture is subjected to some process (agitation, microwaves, etc.) to assist migration of analytes from sample matrix to the extracting solution. For certain applications, matrix solid-phase dispersion (MSPD) can also be a good alternative. In all cases, once a liquid extract has been obtained, it is subsequently subjected to a purification step (namely cleanup), which is usually performed by SPE or LLE. In some cases, extraction and cleanup procedures can be performed in a unique step (i.e., SPE with selective sorbents), which enormously simplifies the sample preparation procedure.

2.3.1 SOLID–LIQUID EXTRACTION

As mentioned earlier, solid–liquid extraction is probably the most widely used procedure in the analysis of pesticides in solid samples. Solid–liquid extraction includes various extraction techniques based on the contact of a certain amount of sample with an appropriate solvent. Figure 2.1 shows a scheme of the different steps

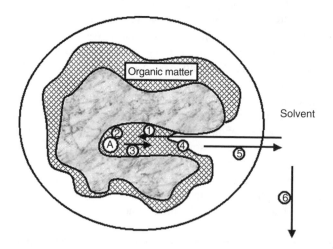

FIGURE 2.1 Scheme of the different steps involved in the extraction of a target analyte A from a solid particle.

that take place in a solid–liquid extraction procedure and will influence the final extraction efficiency. In the first stage (step 1), the solvent must penetrate inside the pores of the sample particulates to achieve desorption of the analytes bound to matrix active sites (step 2). Subsequently, analytes have to diffuse through the matrix (step 3) to be dissolved in the extracting solvent (step 4). Again, the analytes must diffuse through the solvent to leave the sample pores (step 5) and be finally swept away by the external solvent (step 6). Obviously, the proper selection of the solvent to be used is a key factor in a solid–liquid extraction procedure. However, other parameters such as pressure and temperature have an important influence on the extraction efficiency. Working at high pressure facilitates the solvent to penetrate sample pores (step 1) and, in general, increasing temperature increases solubility of the analytes on the solvent. Moreover, high temperatures increase diffusion coefficients (steps 3 and 5) and the capacity of the solvent to disrupt matrix–analyte interactions (step 2). Depending on the strength of the interaction between the analyte and the sample matrix, the extraction will be performed in soft, mild, or aggressive conditions. Table 2.1 shows a summary and a comparison of drawbacks and advantages of the different solid–liquid extraction techniques (which will be described later) most commonly employed in the analysis of pesticides in food and environmental samples.

2.3.1.1   Shaking

It is a very simple procedure to extract pesticides weakly bound to the sample and is very convenient for the extraction of pesticides from fruits and vegetables. It just involves shaking (manually or automatically) the sample in presence of an appropriate solvent for a certain period of time. The most commonly used solvents are acetone and acetonitrile due to their miscibility with water making ease the diffusion of analytes from the solid sample to the solution, although immiscible solvents such as dichloromethane or hexane can also be used for the extraction depending on the properties of target analytes. In a similar manner, the use of mixtures of solvents is a typical practice when analytes of different polarity are extracted in multiresidue analysis. Once analytes have been extracted, the mixture needs to be filtered before further treatments. Besides, since volume of organic solvents used following this procedure is relatively large, it is usually necessary to evaporate the solvent before final determination.

However, shaking might not be effective enough to extract analytes strongly bound to the sample. In order to achieve a more effective shaking, the use of ultrasound-assisted extraction is recommended. Ultrasound radiation provokes molecules vibration and eases the diffusion of the solvent to the sample, favoring the contact between both phases. Thanks to this improvement, both the time and the amount of solvents of the shaking process are considerable reduced.

An interesting and useful modification for reducing both the amount of sample and organic solvents is the so-called ultrasound-assisted extraction in small columns proposed by Sánchez-Brunete and coworkers [1,2] for the extraction of pesticides from soils. Briefly, this procedure just involves placing the sample (~5 g) in a glass column equipped with a polyethylene frit. Subsequently, samples are extracted with

TABLE 2.1

Solid–Liquid Extraction Techniques

Technique	Description	Advantages	Drawbacks
Shaking	Samples and solvent are placed in a glass vessel. Shaking can be done manually or mechanically	• Simple • Fast (15–30 min) • Low cost	• Filtration of the extract is necessary • Dependent of kind of matrix • Moderate solvent consumption (25–100 mL)
Soxhlet	Sample is placed in a porous cartridge and solvent recirculates continuously by distillation–condensation cycles	• Standard method • No further filtration of the extract necessary • Independent of kind of matrix • Low cost	• Time-consuming (12–48 h) • High solvent volumes (300–500 mL) • Solvent evaporation needed
USE	Samples and solvent are placed in a glass vessel and introduced in an ultrasonic bath	• Fast (15–30 min) • Low solvent consumption (5–30 mL) • Bath temperature can be adjusted • Low cost	• Filtration of the extract is necessary • Dependent of kind of matrix
MAE	Sample and solvent are placed in a reaction vessel. Microwave energy is used to heat the mixture	• Fast (~15 min) • Low solvent consumption (15–40 mL) • Easily programmable	• Filtration of the extract is necessary • Addition of a polar solvent is required • Moderate cost
PSE	Sample is placed in a cartridge and pressurized with a high temperature solvent	• Fast (20–30 min) • Low solvent consumption (30 mL) • Easy control of extraction parameters (temperature, pressure) • High temperatures achieved • High sample processing	• Initial high cost • Dependent on the kind of matrix

Note: USE, Ultrasound-assisted extraction; MAE, microwave-assisted extraction; PSE, pressurized solvent extraction.

around 5–10 mL of an appropriate organic solvent in an ultrasonic water bath. After extraction, columns are placed on a multiport vacuum manifold where the solvent is filtered and collected for further analysis.

2.3.1.2   Soxhlet Extraction

As indicated earlier, in some cases shaking is not enough for disrupting interactions between analytes and matrix components. In this regard, an increase of the temperature of the extraction is recommended. The more simple approach to isolate analytes bound to solid matrices at high temperatures is the Soxhlet extraction, introduced by Soxhlet in 1879, which is still the more used technique and of reference of the new techniques introduced during the last few years.

Sample is placed in an apparatus (Soxhlet extractor) and extraction of analytes is achieved by means of a hot condensate of a solvent distilling in a closed circuit. Distillation in a closed circuit allows the sample to be extracted many times with fresh portions of solvent, and exhaustive extraction can be performed. Its weak points are the long time required for the extraction and the large amount of organic solvents used.

In order to minimize the mentioned drawbacks, several attempts toward automation of the process have been proposed. Among them, Soxtec systems (Foss, Hillerød, Denmark) are the most extensively accepted and used in analytical laboratories and allow reducing the extraction times about five times compared with the classical Soxhlet extraction.

Table 2.2 shows a comparison of the recoveries obtained for several pesticides in soils after extraction using different techniques. In this case, it is clear that ultrasound-assisted extraction allows the isolation of target analytes, whereas the

---

TABLE 2.2

Recoveries (%) of Pesticides in Soils Obtained by Different Extraction Techniques

Pesticide	Concentration (mg/mL)	Ultrasound-Assisted Extraction	Soxhlet Extraction	Shaking
Atrazine	0.04	103.5 ± 2.8	201.9 ± 14.6	108.3 ± 6.2
Pyropham	0.05	79.7 ± 6.3	143.0 ± 18.6	65.1 ± 9.3
Chlorpropham	0.05	93.6 ± 7.9	155.6 ± 20.4	88.1 ± 10.0
α-Cypermethrin	0.12	97.2 ± 4.4	128.4 ± 16.4	90.1 ± 9.1
Tetrametrin	0.26	83.4 ± 4.2	64.3 ± 16.0	52.0 ± 8.3
Diflubenzuron	0.02	92.8 ± 4.0	182.5 ± 17.4	98.1 ± 8.9

Source: Reproduced from Babic, S., Petrovic, M., and Kastelan, M., *J. Chromatogr. A*, 823, 3, 1998. With permission from Elsevier.

Experimental conditions: 10 g of soil sample spiked at indicated concentration level. Ultrasound-assisted extraction: 20 mL of acetone, 15 min; Soxhlet extraction: 250 mL of acetone, 4 h; Shaking: 20 mL of acetone, 2 h.

simple shaking is not effective enough to extract the selected pesticides quantitatively. It is important to stress that recoveries after Soxhlet extraction were too high, which means that a large amount of matrix components were coextracted with target analytes. At this regard, it is clear that an exhaustive extraction is not always required and a balance between the recoveries obtained of target analytes and the amount of matrix components coextracted needs to be established.

2.3.1.3 Microwave-Assisted Extraction

Microwave-assisted extraction (MAE) has appeared during the last few years as a clear alternative to Soxhlet extraction due to the ability of microwave radiation of heating the sample–solvent mixture in a fast and efficient manner. Besides, the existence of several instruments commercially available able to perform the sequential extraction of several samples (up to 14 samples in some instruments), allowing extraction parameters (pressure, temperature, and power) to be perfectly controlled, has made MAE a very popular technique.

Microwave energy is absorbed by molecules with high dielectric constant. In this regard, hexane, a solvent with a very low dielectric constant, is transparent to microwave radiation whereas acetone will be heated in few seconds due to its high dielectric constant. However, solvents with low dielectric constant can be used if the compounds contained in the sample (i.e., water) absorb microwave energy.

A typical practice is the use of solvent mixtures (especially for the extraction of pesticides of different polarity) combining the ability of heating of one of the components (i.e., acetone) with the solubility of the more hydrophobic compounds in the other solvent of the mixture (i.e., hexane). As an example, a mixture of acetone:hexane (1:1) was used for the MAE of atrazine, parathion-methyl, chlorpyriphos, fenamiphos, and methidathion in orange peel with quantitative recoveries in <10 min [3].

As a summary, in general, the recoveries obtained are quite similar to those obtained by Soxhlet extraction but the important decrease of the extraction time (~15 min) and of the volume of organic solvents (25–50 mL) have made MAE to be extensively used in analytical laboratories.

2.3.1.4 Pressurized Solvent Extraction

Pressurized solvent extraction (PSE), also known as accelerated solvent extraction (ASE), pressurized liquid extraction (PLE), and pressurized fluid extraction (PFE), uses solvents at high temperatures and pressures to accelerate the extraction process. The higher temperature increases the extraction kinetics, whereas the elevated pressure keeps the solvent in liquid phase above its boiling point leading to rapid and safe extractions [4].

Figure 2.2 shows a scheme of the instrumentation and the procedure used in PSE. Experimentally, sample (~10 g) is placed in an extraction cell and filled up with an appropriate solvent (15–40 mL). Subsequently, the cell is heated in a furnace to the temperatures below 200°C, increasing the pressure of the system (up to a 20 Mpa) to perform the extraction. After a certain period of time (10–15 min),

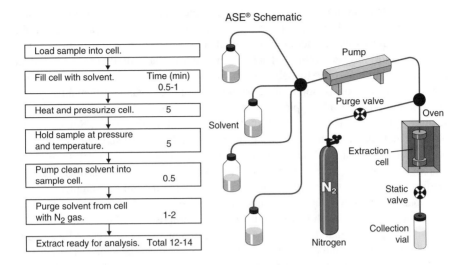

FIGURE 2.2 Pressurized solvent extraction equipment. (Courtesy of Dionex Corporation. With permission.)

the extract is directly transferred to a vial without the necessity of subsequent filtration of the obtained extract. Then, the sample is rinsed with a portion of pure solvent and finally, the remaining solvent is transferred to the vial with a stream of nitrogen. The whole process is automated and each step can be programmed, allowing the sequential unattended extraction of up to 24 samples.

This technique is easily applicable for the extraction of pesticides from any kind of sample and the high temperature used allows to perform very efficient extraction in a short time. In addition, the considerable reduction in the amount of organic solvents used makes PSE a very attractive technique for the extraction of pesticides. The main limitations of this technique are the high cost of the apparatus and the unavoidable necessity of purifying obtained extracts, which is common to other efficient extraction techniques based on the use of organic solvents as mentioned earlier.

2.3.2 SUPERCRITICAL FLUID EXTRACTION

Supercritical fluid extraction (SFE) has been widely used for the isolation of a great variety of organic compounds from almost any kind of solid samples. Supercritical fluids can be considered as a hybrid between liquids and gases, and possess ideal properties for the extraction of pesticides from solid samples. Supercritical fluids have in common with gases the ability to diffuse through the sample, which facilitates the extraction of analytes located in not easily accessible pores. In addition, the solvation power of supercritical fluids is similar to that of liquids, allowing the release of target analytes from the sample to the fluid.

Carbon dioxide has been widely used in SFE because it can be obtained with high purity, it is chemically inert, and its critical point (31.1°C and 71.8 atm) is easily accessible. Its main drawback is its apolar character, limiting its applicability to the extraction of hydrophobic compounds. In order to overcome, at least to a certain extent, this drawback, the addition of a small amount of an organic solvent modifier (i.e., methanol) has been proposed and permits varying the polarity of the fluid, thus increasing the range of extractable compounds. However, the role of the modifier during the extraction is not well understood. Figure 2.3 shows schematically the possible mechanisms taking place during the SFE of the herbicide diuron form soil samples using CO_2 as supercritical fluid modified with methanol [5]. Some authors propose that methanol molecules are able to establish hydrogen bonds with the phenolic moieties of the humic and fulvic acids present in soil samples and thus, diuron is displaced from active sites. However, other authors consider that the modifier is able to interact with target analyte releasing it from the sample.

Once target analytes are in the supercritical fluid phase, they have to be isolated for further analysis, which is accomplished by decompression of the fluid through a restrictor by getting analytes trapped on a liquid trap or a solid surface. With a liquid trap, the restrictor is immersed in a suitable liquid and thus, the analyte is gradually dissolved in the solvent while CO_2 is discharged into the atmosphere. In the solid surface method, analytes are trapped on a solid surface (i.e., glass vial, glass beads, solid-phase sorbents) cryogenically cooled directly by the expansion of the super-critical fluid or with the aid of liquid N_2. Alternatively, SFE can be directly coupled to gas chromatography or to supercritical fluid chromatography and is successful of such online coupling dependent of the interface used, which determines the quantitative transfer of target analytes to the analytical column [6].

As mentioned earlier, SFE has been widely used for the extraction of pesticides from solid samples; thanks to the effectiveness and selectivity of the extraction and to the possibility of online coupling to chromatographic techniques. However,

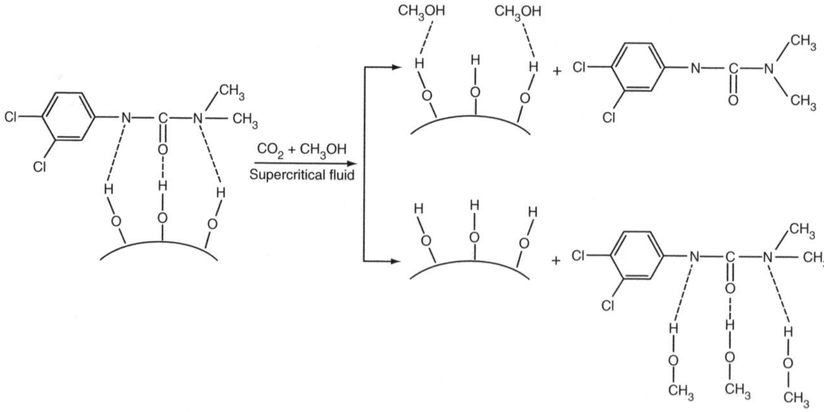

FIGURE 2.3 Mechanisms of the extraction of the herbicide diuron from sediments by SFE (CO_2 + methanol). (Reproduced from Martin-Esteban, A. and Fernandez-Hernando, P., *Toma y tratamiento de muestra*, Cámara, C., ed., Editorial Síntesis S.A., Madrid, 2002, Chap. 6. With permission from Editorial Síntesis.)

the costs of the instrumentation and the apparition in the market of new less sophisticated extraction instruments is making SFE to be displaced by other extraction techniques, especially by PSE.

2.3.3 LIQUID–LIQUID EXTRACTION

LLE has been widely used for the extraction of pesticides from aqueous liquid samples and, although to a lesser extent, for the purification of organic extracts. LLE is based on the partitioning of target analyte between two immiscible liquids. The efficiency of the process depends on the affinity of the analyte for the solvents, on the ratio of volumes of each phase, and on the number of successive extractions.

Most of the LLE applications deal with the extraction of pesticides from environmental waters. Hexane or cyclohexane are typical organic solvents used for extracting nonpolar compounds such as organochlorine and organophosphorus pesticides; and dichoromethane or chloroform for medium polarity organic compounds such as triazines or phenylurea herbicides. However, quantitative recoveries for relatively polar compounds by LLE are difficult to achieve. As an example, a recovery of 90% atrazine was obtained by LLE of 1 L water with dichloromethane, whereas the recoveries for its degradation products desisopropyl-, desethyl-, and hydroxyatrazine were 16%, 46%, and 46%, respectively [7].

In order to increase the efficiency and thus, the range of application, the partition coefficients may be increased by using mixtures of solvents, changing the pH (preventing ionization of acids or bases), or by adding salts ("salting-out" effect). At this regard, the recoveries for the atrazine degradation products of the previously mentioned example were 62%, 87%, and 63%, respectively, by carrying out the extraction with a mixture of dichloromethane and ethyl acetate with 0.2 M ammonium formate.

The high number of possible combinations of solvents and pHs makes ideally possible the isolation of any pesticide from water samples by LLE, which has been traditionally considered a great advantage of LLE. However, LLE is not exempt of important drawbacks. One of the most important drawbacks is the toxicity of the organic solvents used leading to a large amount of toxic residues. In this sense, the costs of the disposal of toxic solvents are rather high. However, it is important to mention that this problem is minimized when LLE is used for cleanup steps where low volumes are usually employed. Besides, the risk of exposure of the chemist to toxic solvents and vapors always exists. From a practical point of view, the formation of emulsions, which are sometimes difficult to break up, the handling of large water samples and the difficulties for automation of the whole process make LLE to be considered a tedious, time-consuming, and costly technique.

2.3.4 SOLID-PHASE EXTRACTION

SPE, as LLE, is based on the different affinity of target analytes for two different phases. In SPE, a liquid phase (liquid sample or liquid sample extracts obtained following the techniques mentioned earlier) is loaded onto a solid sorbent (polar, ion exchange, nonpolar, affinity), which is packed in disposable cartridges or enmeshed

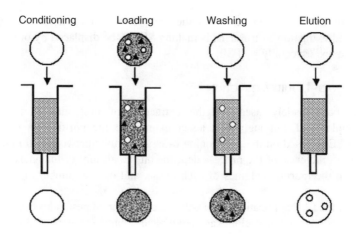

FIGURE 2.4 Solid-phase extraction steps.

in inert matrix of an extraction disk. Those compounds with higher affinity for the sorbent will be retained on it, whereas others will pass through it unaltered. Subsequently, if target analytes are retained, they can be eluted using a suitable solvent with a certain degree of selectivity.

The typical SPE sequence involving several steps is depicted in Figure 2.4. Firstly, the sorbent needs to be prepared by activation with a suitable solvent and by conditioning with same solvent in which analytes are dissolved. Then, the liquid sample or a liquid sample extract are loaded onto the cartridge. Usually, target analytes are retained together with other components of the sample matrix. Some of these compounds can be removed by application of a washing solvent. Finally, analytes are eluted with a small volume of an appropriate solvent. In this sense, by SPE, it is possible to obtain final sample extracts ideally free of coextractives; thanks to the cleanup performed, with high enrichment factors due to the low volume of solvent used for eluting target analytes. These aspects together with the simplicity of operation and the easy automation (see later) have made SPE a very popular technique widely used in the analysis of pesticides in a great variety of samples.

The success of a SPE procedure depends on the knowledge about the properties of target analytes and the kind of sample, which will help the proper selection of the sorbent to be used. Understanding the mechanism of interaction between the sorbent and the analyte is a key factor on the development of a SPE method, since it will ease choosing the right sorbent from the wide variety of them available in the market.

2.3.4.1 Polar Sorbents

The purification of organic sample extracts is usually performed by SPE onto polar sorbents. Within this group, the sorbent mostly used is silica, which possesses active silanol groups in its surface able to interact with target analytes. This interaction is stronger for pesticides with base properties due to the slightly acidic character of silanol groups. Other common polar sorbents are alumina (commercially available in its acid, neutral, and base form) and Florisil.

In the loading step, analytes compete with the solvent for the adsorption active sites of the sorbent, and elution is performed by displacing analytes from the active sites by an appropriate solvent. In this sense, the more polar the solvent is, the higher elution power it gets. The elution power is established by the eluotropic strength ($\varepsilon°$), which is a measure of the adsorption energy of a solvent in a given sorbent. The eluotropic series of different common solvents in alumina and silica are shown in Table 2.3. In this way, by a careful selection of solvents (or mixture of them), analytes (or interferences) will be retained on the sorbent by loading in a nonpolar solvent subsequently eluted using a second solvent with a higher eluotropic strength. Obviously, the selection of these solvents will be determined by the polarity of the analytes. Thus, after loading, hydrophobic pesticides such as pyrethroids can be eluted with a mixture of hexane:diethylether, whereas for eluting carbamates a more polar mixture such as hexane:acetone is necessary.

TABLE 2.3
Eluotropic Series

Solvent	$\varepsilon°$ Al_2O_3	$\varepsilon°$ SiOH
Pentane	0.00	0.00
Hexane	0.00–0.01	0.00–0.01
Iso-octane	0.01	0.01
Cyclohexane	0.04	0.03
Carbon tetrachloride	0.17–0.18	0.11
Xylene	0.26	—
Toluene	0.20–0.30	0.22
Chlorobenzene	0.30–0.31	0.23
Benzene	0.32	0.25
Ethyl ether	0.38	0.38–0.43
Dichloromethane	0.36–0.42	0.32
Chloroform	0.36–0.40	0.26
1,2-Dichloroethane	0.44–0.49	—
Methylethyl ketone	0.51	—
Acetone	0.56–0.58	0.47–0.53
Dioxane	0.56–0.61	0.49–0.51
Tetrahydrofuran	0.45–0.62	0.53
Methyl t-butyl ether	0.3–0.62	0.48
Ethyl acetate	0.58–0.62	0.38–0.48
Dimethyl sulfoxide	0.62–0.75	—
Acetonitrile	0.52–0.65	0.50–0.52
1-Butanol	0.7	—
n-Propyl alcohol	0.78–0.82	—
Isopropyl alcohol	0.78–0.82	0.6
Ethanol	0.88	—
Methanol	0.95	0.70–0.73

The number of developed methods based on SPE using polar sorbents for the determination of pesticides in food and environmental solid samples is huge, and thus, for specific examples, the interested reader should consult Chapters 6 through 8 of this book.

2.3.4.2 Nonpolar Sorbents

This kind of sorbent is appropriate for the trace-enrichment and cleanup of pesticides in polar liquid samples (i.e., environmental waters). Traditionally, *n*-alkyl-bonded silicas, mainly octyl- and octadecyl-silica, both in cartridges or disks, have been used due to its ability of retaining nonpolar and moderate polar pesticides from liquid samples. Retention mechanism is based on van der Waals forces and hydrophobic interactions, which allows handling large sample volumes and the subsequent elution of target analytes in a small volume of a suitable organic solvent (i.e., methanol, acetonitrile, ethyl acetate) getting high enrichment factors. However, for more polar pesticides, the strength of the interaction is not high enough and low recoveries are obtained due to the corresponding breakthrough volume is easily reached.

An easy manner of increasing breakthrough volumes is to increase the amount of sorbent used, which will increase the number of interactions that take place. A second option is the addition of salts to the sample, diminishing the solubility of target analytes (salting-out effect) and thus favoring their interactions with the sorbent. Table 2.4 shows the obtained recoveries of several triazines by the SPE of 1 L of water spiked at 1 μg/L concentration level of each analyte in different experimental conditions. It is clear that the combination of using two C_{18} disks and the addition of a 10% NaCl to the water sample allow the obtainment of quantitative recoveries for all the tested analytes including the polar degradation products of atrazine. However, these approaches do not always provide satisfactory results. In that case, the most direct way of increasing breakthrough volumes of most polar pesticides is the use of sorbents with higher affinity for target analytes. These sorbents include

---

TABLE 2.4
Recoveries (R%) and Relative Standard Deviations (RSD)
of Several Triazines Obtained by SPE of 1 L of LC Grade Water Spiked
with 1 μg/L of Each Triazine

| | 1 C_{18} Disk | | | | 2 C_{18} Disk | | | |
| | Without NaCl | | 10% NaCl | | Without NaCl | | 10% NaCl | |
Triazine	R%	RSD	R%	RSD	R%	RSD	R%	RSD
Desisopropylatrazine	21.5	18.6	42.3	13.6	35.8	18.2	89.2	8.7
Desethylatrazine	50.4	9.3	98.4	6.2	60.5	13.1	95.4	6.3
Simazine	100.2	6.1	93.5	7.8	96.4	8.7	91.7	6.2
Atrazine	94.6	8.7	98.3	4.9	104.3	4.6	97.0	4.1

Source: Adapted from Turiel, E., Fernández, P., Pérez-Conde, C., and Cámara, C., *J. Chromatogr. A*, 872, 299, 2000. With permission from Elsevier.

styrene–divinylbenzene-based polymers with a high specific surface (~1000 m^2/g), which are commercialized by several companies under different trademarks (i.e., Lichrolut, Oasis, Envichrom). The interaction of analytes with these sorbents is also based on hydrophobic interactions, but the presence of aromatic rings within the polymeric network leads to strong π–π^* interactions with the aromatic rings present in the chemical structure of many pesticides. Another alternative is the use of graphitized carbon cartridges or disks, which have a great capacity for the preconcentration of highly polar pesticides (acid, basic, and neutral) and transformation products such as oxamyl, aldicarb sulfoxide, and methomyl; thanks to the presence of various functional groups, including positively charged active centers on its surface.

2.3.4.3 Ion-Exchange Sorbents

Ionic or easily ionizable pesticides can be extracted by these sorbents. Sorption occurs at a pH in which the analyte is in its ionic form and then it is eluted by a change of the pH value with a suitable buffer. The mechanism involved provides a certain degree of selectivity. Phenoxy acid herbicides can be extracted by anion-exchangers and amines or *n*-heterocycles using cation-exchangers. However, its use is rather limited due to the presence of high amount of inorganic ions in the samples, which overload the capacity of the sorbent leading to low recoveries of target analytes.

2.3.4.4 Affinity Sorbents

The sorbents described earlier are able to extract successfully pesticides from a great variety of samples. However, the retention mechanisms (hydrophobic or ionic interactions) are not selective, leading to the simultaneous extraction of matrix compounds, which can negatively affect the subsequent chromatographic analysis. For instance, the determination of pesticides (especially polar pesticides) in soil and water samples by liquid chromatography using common detectors is affected by the presence of humic and fulvic acids. These compounds elute as a broad peak or as a hump in the chromatogram, hindering the presence of target analytes and thus making difficult in some cases to reach the required detection limits. Even using selective detectors (i.e., mass spectrometry) the presence of matrix compounds can suppress or enhance analyte ionization, hampering accurate quantification.

The use of antibodies immobilized on a suitable support, so-called immuno-sorbent (IS), for the selective extraction of pesticides from different samples appeared some years ago as a clear alternative to traditional sorbents [8,9]. In this approach, only the antigen which produced the immune response, or very closely related molecules, will be able to bind the antibody. Thus, theoretically, when the sample is run through the IS, the analytes are selectively retained and subsequently eluted free of coextractives. The great selectivity provided by immunosorbents has allowed the determination of several pesticides in different matrices such as carbofuran in potatoes, or triazines and phenylureas in environmental waters, sediments, and vegetables. However, this methodology is not free of important drawbacks. The obtainment of antibodies is time-consuming, expensive, and few antibodies for pesticides are commercially available. In addition, it is important to point out that after the antibodies have been obtained they have to be immobilized on an adequate

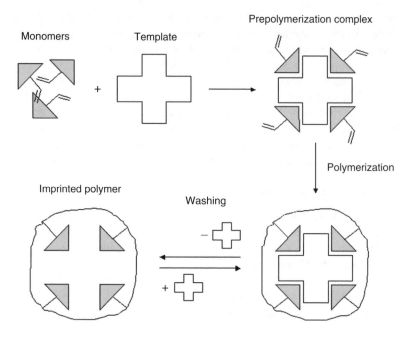

FIGURE 2.5 Scheme for the preparation of molecularly imprinted polymers.

support, which may result in poor antibody orientation or even complete denaturation. Because of these limitations, the preparation and use of molecularly imprinted
polymers (MIPs) has been proposed as a promising alternative.

MIPs are tailor-made macroporous materials with selective binding sites able to
recognize a particular molecule [10]. Their synthesis, depicted in Figure 2.5, is based
on the formation of defined (covalent or noncovalent) interactions between a
template molecule and functional monomers during a polymerization process in
the presence of a cross-linking agent. After polymerization the template molecule
is removed, cavities complementary in size and shape to the analyte are found. Thus,
theoretically, if a sample is loaded on it, in a SPE procedure, the analyte (the
template) or closely related compounds will be able to rebind selectively the polymer
subsequently eluted free of coextractives. This methodology, namely molecularly
imprinted SPE (MISPE), has been successfully employed in the determination of
pesticides such as triazines, phenylureas, and phenoxy acids herbicides, among
others, in environmental waters, soils, and vegetable samples. As an example of
the selectivity provided by MIPs, Figure 2.6 shows the chromatograms obtained
in the analysis of fenuron in potato sample extracts with and without MISPE onto a
fenuron-imprinted polymer. It is clear that the selectivity provided by the MIP
allowed the determination of fenuron at very low concentration levels [11].

Because of their easy preparation and excellent physical stability and chemical
characteristics (high affinity and selectivity for the target analyte), MIPs have
received special attention from the scientific community not only in pesticide residue
analysis but also in several fields. Besides, there are already MISPE cartridges

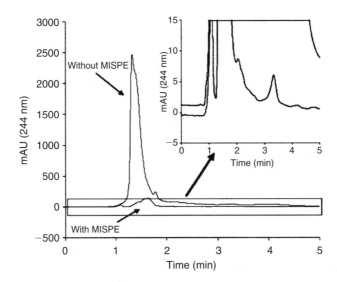

FIGURE 2.6 Chromatograms obtained at 244 nm with and without MISPE of potato sample extracts spiked with fenuron (100 ng/g). Graph insert shows the same chromatograms with different absorbance scale. (Reproduced from Tamayo, F.G., Casillas, J.L., and Martin-Esteban, A., *Anal. Chim. Acta*, 482, 165, 2003. With permission from Elsevier.)

commercially available for the extraction of certain analytes (i.e., triazines) and some companies offer custom synthesis of MIPs for SPE, which will ease the implementation of MISPE in analytical laboratories.

The wide variety of available sorbents as well as the reduced processing times and solvent savings have made SPE to be a clear alternative against LLE. Besides, automation is possible using special sample preparation units that sequentially extract the samples and clean them up for automatic injections. However, the typical drawbacks associated to off-line procedures, such as the injection in the chromatographic system of an aliquot of the final extract or the necessity of including a evaporation step remain, which affects the sensitivity of the whole analysis.

The use of SPE coupled online to liquid and gas chromatography can sort out the previously mentioned drawbacks. The coupling of SPE to liquid chromatography is especially simple to perform in any laboratory and has been extensively described for the online preconcentration of organic compounds in environmental water samples [12]. The simplest way of SPE–LC coupling is shown in Figure 2.7, where a precolumn (1–2 cm × 1–4.6 mm i.d.) filled with an appropriate sorbent is inserted in the loop of a six-port injection valve. After sorbent conditioning, the sample is loaded by a low-cost pump and the analytes are retained in the precolumn. Then, the precolumn is connected online to the analytical column by switching the valve, so that the mobile phase can desorb the analytes before their separation in the chromatographic column. Apart from a considerable reduction of sample manipulation, the main advantage is the fact that the complete sample is introduced in the analytical column. Besides, there are equipments commercially available for the whole automation of the process.

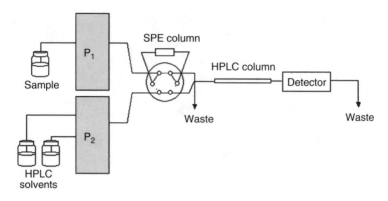

FIGURE 2.7 SPE–LC coupling setup.

Alkyl-bonded silicas (mainly C_{18}-silica) have been widely used as precolumn sorbent, although they are replaced by styrene–divinylbenzene copolymers, which offer higher affinity for polar analytes, so that permit the usage of larger sample volumes without exceeding the breakthrough volumes of analytes. Other materials successfully employed have been small extraction disks and graphitized carbons; and in order to provide selectivity to the extraction, precolumns packed with yeast cells immobilized on silica gel [13] or with immunosorbents have been proposed for the extraction of polar pesticides from environmental waters [14,15].

The coupling of SPE to GC is also possible, thanks to the ability of injecting large volumes into the gas chromatograph using a column of deactivated silica (retention gap) located between the injector and the analytical column. SPE–GC uses the same sorbents employed in SPE–LC but, in this case, after the preconcentration step, the analytes are desorbed with a small volume (50–100 μL) of an appropriate organic solvent, which is directly introduced into the chromatograph. In general, using only 10 mL of water sample, it is possible to reach detection limits at micrograms per liter level employing common detectors.

2.3.5 SOLID-PHASE MICROEXTRACTION

As it has been stated previously, SPE has demonstrated to be a very useful procedure for the extraction of a great variety of pesticides in food and environmental analysis. However, although in a lower extent than LLE, this technique still requires the use of toxic organic solvents and its applicability is restricted to liquid samples. With the aim of eliminating these drawbacks, Arthur and Pawliszyn introduced SPME in 1989 [16]. Its simplicity of operation, solventless nature, and the availability of commercial fibers have made SPME to be rapidly implemented in analytical laboratories.

As depicted in Figure 2.8, the SPME device is quite simple, and just consists of a silica fiber coated with a polymeric stationary phase similar to those used in gas chromatography columns. The fiber is located inside the needle (protecting needle) of a syringe specially designed to allow exposure of the fiber during sample analysis. As in any SPE procedure, SPME is based on the partitioning of target analytes

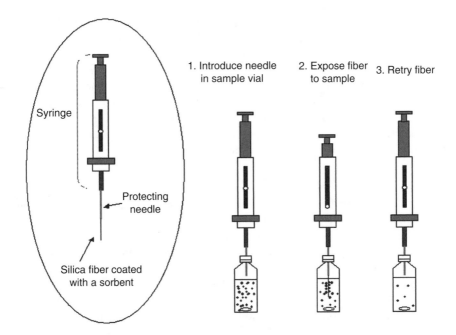

FIGURE 2.8 SPME device and typical mode of operation.

between the sample and the stationary phase and consists of two consecutive steps, extraction and desorption. An intermediate washing step can also be performed.

2.3.5.1   Extraction

The extraction step can be performed both by exposure of the fiber to the head-space (restricted to volatile compounds in liquid or solid samples) or by direct immersion of the fiber into the sample (aqueous-based liquid samples). As described in Figure 2.8, the experimental procedure is very simple. Firstly, the fiber is inside the protecting needle which is introduced into the sample vial. Then, the fiber is exposed to the sample to perform extraction by sorption of the analytes to the stationary phase. Finally, the fiber is retried inside the needle for further desorption and the whole device removed.

Obviously, a proper selection of the SPME sorbent is a key factor in the success of the analysis. In general, the polarity of the fiber should be as similar as possible to that of the analyte of interest. In this sense, there are nowadays a great variety of fibers commercially available that covers a wide range of polarities (i.e., carbowax/DVB for polar compounds or polydimethylsiloxane [PDMS] for hydrophobic compounds). In addition, both the fiber thickness and the porosity of the sorbent will influence the final extraction efficiency. Besides, other physical and chemical parameters such as temperature, exposition time, agitation, pH, or ionic strength (salting-out effect) of the sample can be optimized. As an example, it can be mentioned the extraction of dinoseb, an alquil-substituted dinitrophenol, in waters. The SPME of this compound can be favored by using a polyacrylate fiber and by adding 10% of NaCl at pH = 2 due to the produced salting-out effect and the lower ionization of dinoseb at low pH values.

Finally, concerning extraction, it is interesting to mention that from the mathematical model governing SPME, it can be concluded that when sample volume is much higher than the fiber volume, the extraction efficiency becomes independent of the sample volume. Although it is not applicable for laboratory samples (low volumes), this earlier fact makes SPME a very interesting tool for in-field sampling procedures, since the fiber can be exposed to the air or directly immersed into a lake or a river regardless of the sample volume.

2.3.5.2 Desorption

Desorption can be performed thermally in the injection port of a gas chromatograph, or by elution of the analytes by means of a suitable solvent. In the latter case, desorption can be carried out in a vial containing a small volume of the solvent to be further analyzed by chromatographic techniques or eluted with the mobile phase on an especially designed SPME–HPLC interface.

Thermal desorption of the analytes in the injector port of the GC instrument is based on the increase of the partition coefficient gas fiber with the increasing temperature. In addition, a constant flow of carrier gas inside the injector facilitates removal of the analytes from the fiber. The main advantage of the thermal desorption is the fact that the total amount of extracted analytes is introduced in the chromatographic system and analyzed, thus compensating the low recoveries usually obtained in the extraction step. However, unfortunately, thermal desorption cannot be used for nonvolatile or thermolabile compounds, thus necessary to use desorption with solvents. The procedure is similar to SPE elution but, in this case, the fiber is immersed in a small volume of elution solvent and agitated or heated to favor the transfer of the analytes to the solvent solution. A fraction of this extract or, for some applications, an evaporated and redissolved extract, is subsequently injected into the chromatographic system.

Recently, there are commercially available interfaces allowing the direct coupling of SPME to liquid chromatography. The coupling is similar to that described earlier in Figure 2.7 for SPE–HPLC but placing a specially designed little chamber instead of a precolumn in the loop of a six-port injection valve. This interface allows desorption of the analytes by the chromatographic mobile phase, where the total amount of compounds extracted introduced in the chromatographic system.

2.3.6 SOLID–SOLID EXTRACTION: MATRIX SOLID-PHASE DISPERSION

MSPD, introduced by Barker et al. in 1989 [17], is based on the complete disruption of the sample (liquid, viscous, semisolid, or solid), while the sample components are dispersed into a solid sorbent. Most methods use C_8- and C_{18}-bonded silica as solid support. Other sorbents such as Florisil and silica have also been used although to a lesser extent.

Experimentally, the sample is placed in a glass mortar and blended with the sorbent until a complete disruption and dispersion of the sample on the sorbent is obtained. Then, the mixture is directly packed into an empty cartridge as those used in SPE. Finally, analytes are eluted after a washing step for removing interfering compounds. The main difference between MSPD and SPE is that the sample is dispersed through the column instead of only onto the first layers of sorbent, which typically allows the obtainment of rather clean final extracts avoiding the necessity of performing a further cleanup.

MSPD has been successfully applied for the extraction of several pesticide families in fruit juices, honey, oranges, cereals, and soil, among others, and the achieved performance, compared with other classical extraction methods, has been found superior in most cases [18]. The main advantages of MSPD are the short extraction times needed, the small amount of sample, sorbent, and solvents required, and the possibility of performing extraction and cleanup in one single step.

2.3.7 OTHER TREATMENTS

2.3.7.1 Stir Bar Sorptive Extraction

Stir bar sorptive extraction (SBSE) is based on the partitioning of target analytes between the sample (mostly aqueous-based liquid samples) and a stationary phase-coated stir bar [19]. Until now, only PDMS-coated stir bars are commercially available, restricting the range of applications to the extraction of hydrophobic compounds (organochlorine and organophosphorus pesticides) due to the apolar character of PDMS.

The experimental procedure followed in SBSE is quite simple. The liquid sample and the PDMS-coated magnetic stir bar are placed in a container. Then, the sample is stirred for a certain period of time (30–240 min) until no additional recovery for target analytes is observed even when the extraction time is increased further. Finally, the stir bar is removed and placed in a specially designed unit in which thermal desorption and transfer of target analytes to the head of the GC column take place.

SBSE is usually compared and proposed as an alternative to SPME. The use of a PDMS-coated stir bar (10 mm length, 0.5 mm coating thickness) results in a significant increase in the volume of the extraction phase from ~0.5 μL for an SPME fiber (100 μm PDMS) to ~24 μL for a stir bar. Consequently, the yield of the extraction process is much greater when using a stir bar rather than an SPME fiber, both coated with PDMS. However, the greater coating area of magnetic stir bars is simultaneously its main drawback since the extraction kinetics are slower than for SPME fibers, and a high amount of interfering matrix compounds are coextracted with target analytes. Nevertheless, the simplicity of operation and its solventless nature make SBSE a very attractive technique, and the development of new stir bars coated with more polar and selective sorbents are expected in the near future.

2.3.7.2 Liquid Membrane Extraction Techniques

Liquid membrane extraction techniques (supported liquid membrane, SLME, and microporous membrane liquid–liquid, MMLLE, extractions) are based on the use a hydrophobic membrane, containing an organic solvent, which separates two immiscible phases. These extraction techniques are a combination of three simultaneous processes: extraction of analyte into organic phase, membrane transport, and reextraction in an acceptor phase. Chemical gradient existing between the two sides of the liquid membrane causes permeation of solutes. The compounds present in the donor phase diffuse across the organic liquid membrane to the acceptor phase, where they accumulate at a concentration generally greater than that in the donor phase. Depending on the sample volume, different membrane unit formats for liquid membrane extraction are applied [20]. The main advantages of liquid membrane extraction over the traditional separation methods are small amounts of organic phases used, mass transfer is performed in one step, and it is possible to achieve high separation and concentration factors.

The distinguishing factor of the use of SLMs or MMLLE is the possibility of connecting them online with an analytical system. MMLLE is easily interfaced to gas chromatography and normal-phase HPLC, whereas SLM is compatible with reversed-phase HPLC. These online connections result in an improvement of the overall reliability of analysis, since the number of steps involved in sample preparation is decreased and allows method automation. Additionally, significant reduction in analysis time is achieved. Till now, SLME and MMLLE have been successfully applied for enrichment of phenoxy acid, sulfonylurea, and triazine herbicides from environmental water samples. In those examples, similar or even better results were obtained in comparison with conventional sample preparation methods.

Thanks to their flexibility, SLME and MMLLE have proved to be interesting techniques to be combined with a second pretreatment technique (e.g., SPE). At this regard, detection limits as low as 30 µg/L have been achieved by combination of SLME and SPE for the determination of atrazine in fruit juices (orange, apple, blackcurrant, and grape) [21].

2.4 FUTURE TRENDS

In this chapter, a description of the different techniques developed during the last few years for the extraction and cleanup of pesticides from environmental and food samples has been made. It is evident that a great effort has been made to improve the techniques and procedures used for sample preparation. However, still nowadays, sample preparation is the limiting step of the analysis. Even using very powerful detection techniques such as LC–MS (MS), some sample preparation (including cleanup) is still necessary since otherwise interferences and signal suppression can occur.

Thus, since sample preparation cannot be avoided, further studies toward its simplification are expected in the near future. At this regard, environmental friendly, cost-effective, and selective procedures are required. In parallel, advances in miniaturization and automation will ease the integration of sample preparation and instrumental analysis leading to faster procedures with improved performance in terms of accuracy, precision, and traceability.

REFERENCES

1. Sánchez-Brunete, C., Pérez, R.A., Miguel, E., and Tadeo, J.L., Multiresidue herbicide analysis in soil samples by means of extraction in small columns and gas chromatography with nitrogen–phosphorus and mass spectrometric detection, *J. Chromatogr. A*, 823 (1–2), 17, 1998.

2. Castro, J., Sánchez-Brunete, C., and Tadeo, J.L., Multiresidue analysis of insecticides in soil by gas chromatography with electron-capture detection and confirmation by gas chromatography–mass spectrometry, *J. Chromatogr. A*, 918 (2), 371, 2001.
3. Bouaid, A., Martin-Esteban, A., Fernández, P., and Cámara, C., Microwave-assisted extraction method for the determination of atrazine and four organophosphorus pesticides in oranges by gas chromatography (GC), *Fresenius J. Anal. Chem.*, 367, 291, 2000.
4. Björklund, E., Nilsson, T., and Bøwadt, S., Pressurised liquid extraction of persistent organic pollutants in environmental analysis, *Trends Anal. Chem.*, 19 (7), 434, 2000.
5. Martin-Esteban, A. and Fernandez-Hernando, P., Preparación de la muestra para la determinación de analitos orgánicos, in *Toma y Tratamiento de Muestra*, Cámara, C., Ed., Editorial Síntesis S.A., Madrid, 2002, chap. 6.
6. Zougagh, M., Valcarcel, M., and Rios, A., Supercritical fluid extraction: a critical review of its analytical usefulness, *Trends Anal. Chem.*, 23 (5), 399, 2004.
7. Durand, G. and Barceló, D., Liquid-chromatographic analysis of chlorotriazine herbicides and its degradation products in water samples with photodiode array detection. 1. Evaluation of 2 liquid–liquid-extraction methods, *Toxicol. Environ. Chem.*, 25, 1, 1989.
8. Pichon, V., Chen, L., Hennion, M.-C., Daniel, R., Martel, A., Le Goffic, F., Abian, J., and Barceló, D., Preparation and evaluation of immunosorbents for selective trace enrichment of phenylurea and triazine herbicides in environmental waters, *Anal. Chem.*, 67, 2451, 1995.
9. Martín-Esteban, A., Fernández, P., and Cámara, C., Immunosorbents: a new tool for pesticide sample handling in environmental analysis, *Fresenius' J. Anal. Chem.*, 357, 927, 1997.
10. Sellergren, B., *Molecularly Imprinted Polymers: Man-Made Mimics of Antibodies and their Applications in Analytical Chemistry*, 1st edn, Elsevier Science BV, Amsterdam, 2001.
11. Tamayo, F.G., Casillas, J.L., and Martin-Esteban, A., Highly selective fenuron-imprinted polymer with a homogeneous binding site distribution prepared by precipitation polymerisation and its application to the clean-up of fenuron in plant samples, *Anal. Chim. Acta*, 482, 165, 2003.
12. Hennion, M.-C. and Scribe, P., Sample handling strategies for the analysis of organic compounds from environmental water samples, in *Environmental Analysis: Techniques, Applications and Quality Assurance*, Barceló, D., Ed., Elsevier Science Publishers BV, Amsterdam, 1993, chap. 2.
13. Martin-Esteban, A., Fernández, P., and Cámara, C., Baker's yeast biomass (*Saccharomyces cerevisae*) for selective on-line trace enrichment and liquid chromatography of polar pesticides in water, *Anal. Chem.*, 69, 3267, 1997.
14. Pichon, V., Chen, L., and Hennion, M.-C., On-line preconcentration and liquid chromatographic analysis of phenylurea pesticides in environmental water using a silica-based immunosorbent, *Anal. Chim. Acta*, 311, 429, 1995.
15. Martin-Esteban, A., Fernández, P., Stevenson, D., and Cámara, C., Mixed immunosorbent for selective on-line trace enrichment and liquid chromatography of phenylurea herbicides in environmental waters, *Analyst*, 122, 1113, 1997.
16. Arthur, C.L. and Pawliszyn, J., Solid phase microextraction with thermal desorption using fused silica optical fibers, *Anal. Chem.*, 62, 2145, 1990.
17. Barker, S.A., Long, A.R., and Short, C.R., Isolation of drug residues from tissues by solid phase dispersion, *J. Chromatogr.*, 475, 353, 1989.
18. Kristenson, E.M., Ramos, L., and Brinkman, U.A.Th., Recent advances in matrix solid-phase dispersion, *Trends Anal. Chem.*, 25, 96, 2006.

19. Baltussen, E., Sandra, P., David, F., and Cramers, C., Stir bar sorptive extraction (SBSE), a novel extraction technique for aqueous samples: theory and principles, *J. Microcolumn Sep.*, 11, 737, 1999.
20. Jönsson, J.A. and Mathiasson, L., Membrane-based techniques for sample enrichment, *J. Chromatogr. A*, 902, 205, 2000.
21. Khrolenko, M., Dzygiel, P., and Wieczorek, P., Combination of supported liquid membrane and solid-phase extraction for sample pretreatment of triazine herbicides in juice prior to capillary electrophoresis determination, *J. Chromatogr. A*, 975, 219, 2002.

3 Analysis of Pesticides by Chromatographic Techniques Coupled with Mass Spectrometry

Simon Hird

CONTENTS

3.1   INTRODUCTION

The Pesticide Manual lists 860 pesticides, most of which are still sold worldwide.[1] A considerable volume of registration data is submitted by the applicant for statutory approval, which relies on the determination of the concentration of the pesticide and associated metabolites and degradation products in a wide variety of matrices and their structural characterization. Another driver for the analysis of pesticide residues in food is to generate the monitoring data needed to back up the statutory approval process. Checks are carried out to ensure that no unexpected residues are occurring in crops and that residues do not exceed the statutory maximum residue levels (MRLs). Such surveillance is carried out as part of national and international programs and also by the food industry and their suppliers to demonstrate "due diligence" under food safety legislation. Following notification of an MRL violation, brand owners may choose to sample and analyze foodstuffs on a "positive release" basis to ensure that the materials are compliant before distribution. While laboratories undertaking the pesticide residue analysis for a survey might have up to 1 month to report their findings, a much more rapid approach to analysis is required for positive release situations (e.g., 24 h).

Residues of pesticides used for crop protection, on animals, for public hygiene use, in industry and in the home or garden are found in rivers and groundwater. In the United Kingdom, requirements for water analysis vary depending on whether it is for monitoring trends,[2] Environmental Quality Standards (EQSs),[3] tailored to local pesticide usage patterns,[4] or drinking water.[5] Analysis is also needed to investigate the concentration of pesticides and their metabolites in samples from humans; both by the chronic exposure of the general population to pesticides[6] and by occupational exposure for those working with pesticides.[7] Biological monitoring of exposure involves the measurement of a biomarker (normally the pesticide or its metabolite) in biological fluids.[8] Pesticides can poison wildlife, including beneficial insects and some pets. Cases relate to abuse of a product where the pesticide is used to deliberately and illegally poison animals.[9] Relevant tissues from casualties, including whole bees, are analyzed to help assess the probable cause of the incident and whether any pesticide residues found contributed to the death or illness of the animal.

Analyses must prove reliable, be capable of residue measurement at very low levels (sub ppb), and also provide unambiguous evidence of the identity and magnitude of any residues detected. More recently, additional emphasis has been on shortening analysis times to deal with high sample throughput. Depending on the purpose of the analysis, determination of pesticide residues may be termed target (compound) analysis or nontarget analysis. Checking food or surface and groundwater has been typically achieved by target compound analysis as the relevant analytes are fixed by the residue definition given in the various regulations, which may include significant metabolites or degradation products. In general, MRLs are in the range of 0.01–10 mg/kg but can be lower for infant food[10] where there is no approved use (Limit of Determination MRL, typically between 0.01 and 0.05 mg/kg). Although EU regulation of residues in drinking water does not contain detailed residue definitions, the high sensitivity of target compound analysis has been employed as a practical compromise to meet the 0.01 µg/L limit. For target compound analysis, characteristic ions for the analyte are selected before starting the analysis. An unexpected compound cannot be detected if its relevant ions are not selected and will be missed if present in the sample. Pesticide misuse can be missed due to incomplete target compound lists and strategic data regarding changing patterns in both legitimate and illegal use of pesticides cannot be captured. A nontarget analytical approach provides rapid and accurate screening of unknown substances in food and water and also when determining whether pesticide abuse caused the death of an animal. For nontarget screening, instruments must be able to generate sufficient information for elucidation of residues by providing either mass spectra for interpretation or accurate mass information from which empirical formulae can be deduced. This information must be generated while maintaining the high sensitivity required, for example, to detect violations of limit of detection (LOD)-based limits in food or water. When dealing with unknowns there is often a lack of reference standards, used in target compound analysis for unequivocal identification through the standard's characteristic chromatographic behavior and mass spectrum.

Over the past decades, approaches to trace level determination of pesticides have changed considerably, moving away from the use of GC with selective detectors to the sensitivity and selectivity offered by GC-MS. The commercialization of atmospheric pressure ionization with tandem mass spectrometers[11] enabled the determination of pesticides and their degradation products that are polar, relatively nonvolatile, and/or thermally labile, and, therefore not amenable to GC analysis.[12] Further developments in both detection and column technology enabled the scope for LC to be significantly enlarged and now LC-MS offers a similar breath of analysis to GC-MS (e.g., 171 pesticides and/or metabolites).[13] The use of alternative mass analyzers to the single quadrupole (Q) (i.e., various types of ion trap, triple quadrupoles, and time-of-flight [TOF]) and their various combinations (e.g., QTOF) has improved the capabilities of the instruments available. Table 3.1 provides an overview of the advantages and disadvantages of each analyzer for both GC-MS and LC-MS. The vast majority of pesticides sought is amenable to multiresidue approaches and can now be thoroughly isolated from water and complex food matrices without the large amounts of natural material coextracted with the pesticides interfering with the analysis. For example, out of ~400 pesticides routinely targeted using the QuEChERS method,[14] 217 are analyzed employing LC-MS/MS and 187 employing GC-MS and GC-TOF MS techniques. In a fascinating recent review,[15] Alder compared the scope and sensitivity of GC coupled with EI and single quadrupole MS with LC combined with tandem mass spectrometry for the analysis of 500 high-priority pesticides concluding that both techniques are still needed to cover the wide range of pesticides to be monitored. A number of compounds are not amenable to multiresidue analysis and so require separate, so-called, single residue methods.

TABLE 3.1

Capabilities of the Different Analyzers for Pesticide Residue Analysis

Analyzer	Advantages	Disadvantages
Quadrupole (Q): GC-MS, LC-MS	High sensitivity in SIM mode (0.1–1 pg), good dynamic range (five orders of magnitude), good selectivity in CI, low cost	Poor sensitivity in scan mode (50–500 pg), low selectivity for complex matrices, SIM needs preselection, unit mass resolution
Quadrupole ion trap (QIT): GC-MS, GC-MS/MS, LC-MS, LC-MS/MS	High/medium sensitivity in scan and product ion scan modes (0.1–10 pg), library-searchable EI and product ion spectra, good selectivity in CI, MS^n, fast acquisition rate, low cost	Low selectivity for complex matrices in MS mode, limit on number of ions that can be determined simultaneously, limited dynamic range (3–4 orders of magnitude), limited mass range in MS/MS, unit mass resolution
Triple Quadrupole (QqQ): GC-MS/MS, LC-MS/MS	Excellent sensitivity (10–100 fg) and selectivity in MRM mode, good dynamic range (five orders of magnitude), concurrent monitoring of many channels	The number of MRM channels that can be monitored at any one time is limited, MRM needs preselection, unit mass resolution, high cost
High-speed time-of-flight (TOF): GC-MS	High sensitivity (0.1–1 pg), library-searchable EI spectra, very fast acquisition rate	Low selectivity, limited dynamic range (four orders of magnitude), unit mass resolution, high cost
Enhanced resolution TOF: GC-MS, LC-MS	High sensitivity (0.1–1 pg), good selectivity, accurate mass, fast acquisition rate	Limited dynamic range (four orders of magnitude), not true "high resolution," high cost
Qq-linearIT (QqLIT): LC-MS/MS	Excellent sensitivity (10–100 fg) and selectivity in MRM mode, high sensitivity in product ion scan mode, MS^n	Unit mass resolution, high cost
QTOF: LC-MS/MS	High sensitivity (0.1–1 pg), good selectivity, accurate mass of both precursor and product ions, fast acquisition rate	Limited dynamic range (four orders of magnitude), not true high resolution, high cost

In the next sections, the use of GC and LC with selective detectors will be briefly discussed followed by an exploration of the three stages key to the successful application of both GC-MS and LC-MS: sample introduction, chromatography and subsequent ionization, and mass analysis (mass spectrometry).

3.2 GC AND LC WITH SELECTIVE DETECTORS

Original schemes for pesticide residue analysis comprised a number of different multiresidue methods for classes of compounds based on chemical composition. Traditionally, the vast majority of this work was achieved using GC with the so-called selective detectors; flame photometric detector (FPD) in phosphorus or sulfur mode for organophosphorus pesticides; electron capture detector (ECD) for organochlorine pesticides; and nitrogen phosphorus detector (NPD) for nitrogen and phosphorus compounds with a total coverage of about 300 compounds.[16] GC-MS was used purely for confirmation of identity. GC-FPD is still used in many laboratories for the determination of dithiocarbamate residues in fruits and vegetables by the determination of carbon disulfide (CS_2) generated by acid hydrolysis of dithiocarbamates.[17] Most of the pesticides that are not easily analyzed by GC can be separated using LC. Conventional LC detectors, such as the UV, diode array, or fluorescence, when used with extensive cleanup and/or with derivatization may exhibit sufficient selectivity and sensitivity[18] but many pesticides do not contain strong UV chromophores. The use of selective detectors for GC and LC, even in combination with different polarity columns, can only provide limited confirmatory evidence and cannot be used to identity unknowns. The use of mass spectrometry, with its information-rich content and unambiguous confirmation, is recommended for monitoring pesticide residues in the European Union.[19]

3.3 GAS CHROMATOGRAPHY-MASS SPECTROMETRY (GC-MS)

Originally, GC-MS analysis was restricted to the used packed columns coupled to a magnetic sector mass spectrometer via a jet separator so only a limited number of compounds could be detected in a single analysis. When fused capillary columns were coupled to affordable, benchtop mass spectrometers, GC-MS became an essential tool for pesticide residue analysis. A mixture of compounds to be analyzed is injected into the GC where the mixture is vaporized. The gas mixture travels through a GC column, where the compounds are separated as they interact with the column and then enter the mass spectrometer for ionization and mass analysis.

3.3.1 SAMPLE INTRODUCTION

The design of injection ports for GC has been constantly improved to achieve precise and accurate retention times and analyte response and also to handle large volume injections (LVI), either to lower detection limits or to simplify sample workup, and to allow direct coupling with sample preparation techniques.[20–22] An important issue when selecting an injector is the properties of the analyte, such as potential for chemical instability, thermal degradation, or discrimination of high-boiling-point compounds within the injector. A number of problematic pesticides are prone to degradation in the GC injector, including phthalimide fungicides (e.g., captan), organochlorines (e.g., DDT and chlorothalonil), organophosphorus pesticides (e.g., dimethoate), and pyrethroids.

The major source of inaccuracy in pesticide residue analysis by GC-MS, especially with food, is related to the injection of coextractives from the sample, the so-called "matrix effect." A buildup of coextractives in a GC inlet may lead to successive adverse changes in the performance of the chromatographic system such as the loss of analytes and peak tailing due to undesired interactions with active sites in the inlet and column. Analytes that give poor peak shapes or degrade have higher detection limits, are more difficult to identify and integrate, and are more prone to interferences than stable analytes that give narrow peaks. For susceptible analytes, significant improvements in peak quality are obtained when matrix components are present because they fill active sites, thus reducing analyte interactions. However, this can lead to problems with quantification. These matrix effects can produce an overestimation of the analyte concentration if calibration has been performed with standards in solvent. The presence of matrix effects should be evaluated for all tested analytes. There are a number of approaches for preventing, reducing, or compensating for the occurrence of matrix effects[23,24] including the use of matrix-matched calibrants,[25] which is recommended for the monitoring of pesticide residues within the European Union.[19]

3.3.1.1 Splitless Injection

Cold on-column (COC) injection is rarely used for food analysis due to contamination of the column inlet with nonvolatile materials.[26] The hot split/splitless injection technique is the most probably used for pesticide residue analysis by GC-MS. Split and splitless injection[27] are techniques that introduce the sample into a heated injection port as a liquid, and then rapidly and completely vaporize the sample solvent as well as all of the analytes in the sample. For most pesticide residue applications, the target analyte concentrations are so low that splitting the sample in the injection port will not allow an adequate signal from the detector; so the injector should be operated in the splitless injection mode. In splitless mode, the split outlet remains closed during the so-called splitless period so that sample vapors are transferred from the vaporizing chamber into the column. Flow through the split outlet is turned on again to purge the vaporizing chamber after most of the sample has been transferred. Transfer into the column is slow (e.g., 30–90 s), resulting in broad initial bands that must be focused by cold trapping or solvent effects. Although splitless injection is >30 years old, the vaporization process in the injector continues to be investigated and debated.[28,29] Splitless injection, however, is frequently performed incorrectly for a large number of reasons; vaporizing chambers can be too small, syringe needles too short, carrier gas supply systems poorly suited, sample volumes too large, needle technique inappropriate (cool versus hot) by slow instead of rapid injection with too low carrier gas flow rates, incorrect column temperature during the sample transfer, splitless periods that are too short, and liner packings at the wrong site. Some of these problems relate to a lack of understanding of the mechanisms involved (e.g., evaporation by "thermospray" (TSP) and "band formation").[30] Although compromises have to be made when dealing with multiresidue determinations, there is considerable benefit in evaluating each step of the injection process.

The limitations of splitless injections, small injection volumes (i.e., up to 2 μL), the potential to thermally degrade components, and incomplete transfer of compounds with high boiling points, can be overcome somewhat by using pressure-pulsed splitless injection.[31] The pulsed splitless technique uses high pressure (high column flow rate) during injection to sweep the sample out of the inlet rapidly.

After injection, the column flow rate is automatically reduced to normal values for chromatographic analysis. The pulsing effect maximizes sample introduction into the column while narrowing the sample bandwidth. Additionally, the sample has a very short residence time in the liner, thus minimizing the loss of active compounds.[32] Moreover, the pulsed splitless technique has been shown to enable an increase in the volume that can be injected[33] but this approach does not permit LVI (>10 µL) and compounds may still thermally degrade in the injector even when the injector temperature is lowered.

3.3.1.2 Programmed Temperature Vaporizing Injection

Temperature-programmed sample introduction was first described by Vogt[34] and based on this idea Poy[35] developed the programmed temperature vaporizing injector (PTV). Although the PTV injector closely resembles the classical split/splitless injector, the primary difference is temperature control. In PTV injectors, the vaporization chamber can be heated or cooled rapidly. Combining a cool injection step with a controlled vaporization eliminates a number of important disadvantages associated with the use of conventional hot sample inlets.[36] This type of injector is highly versatile and can be operated with a number of different configurations. PTV splitless (PTV SL) introduces the sample into a cold liner (temperature set below or near the solvent boiling point), the split exit is closed, and the chamber is rapidly heated. This technique offers more accurate and repeatable injection volumes, protection of heat sensitive materials, and more homogeneous evaporation for better analyte focusing. Some optimization of parameters and choice of liners are required for good performance.[37]

3.3.1.3 Large Volume Injection

Time-consuming and labor-intensive evaporation steps during sample preparation can be replaced by LVI in which the solvent is evaporated in the GC system, in a more rapid, automated, and controlled process. LVI can of course also be used to improve analyte detectability. If the sample extract is sufficiently clean and/or the detector selectivity is sufficiently high, the detection limits will improve proportionally with the volume injected. There are two main techniques by which injection volumes for GC can be increased: COC[38] and PTV.[39] Although COC techniques are very accurate, especially when thermally labile or volatile analytes are concerned, contamination of the column inlet with nonvolatile material is frequent and thus the number of samples that can be analyzed before disruption is limited. LVI using a PTV injector is based on selective evaporation of the sample solvent from the liner of the PTV injector while simultaneously trapping the less volatile components in the cold liner. During this stage of the sampling process, solvent vapors are discharged via the opened split exit of the injector. During solvent elimination, the split exit is closed and the components are transferred to the column in the splitless mode by rapid temperature-programmed heating of the injector. An advantage of using the PTV in the solvent vent mode is that it can also be used for the introduction of polar solvents, such as acetonitrile used for QuEChERS.[14] As the solvent is vented before introduction into the GC column, no band distortion occurs. The use of PTV injection also enables injection of large volumes of water to be directly injected into GC without any sample preparation.[40]

One more recent modification of the PTV inlet is the conversion of the inlet to allow for accommodation of a direct sample introduction device such as direct sample introduction (DSI)[41] or difficult matrix introduction (DMI).[42] In this approach, the standard PTV is converted into an intrainjector thermal desorption device where a microvial containing an extract volume up to 20 µL is inserted into a PTV injector liner using a holder or probe. For DMI an automated, robotic system is used to inject sample into a PTV liner holding the microvial, and then the PTV liner is robotically inserted into the injector. As in conventional PTV protocols, the start temperature is kept near the pressure-corrected boiling point of the solvent to allow evaporation and removal of solvent from the sample. The inlet is then heated rapidly to transfer volatile and semivolatile analytes to the column, leaving behind the nonvolatile components in the liner. After the separation the microvial or liner containing the microvial is removed, still containing nonvolatile matrix components, thus reducing build up of undesirable compounds in the PTV inlet or on the column.

Some modern microextraction techniques, such as solid-phase microextraction (SPME)[43] and stir bar sorptive extraction (SBSE),[44] can be directly coupled to GC-MS as sample introduction devices allowing the extraction and concentration steps to be focused into a single, solvent-free, automated step. Both provide high sensitivity because the whole extract can be introduced into the GC by thermal desorption rather than an aliquot of a liquid extract.

For laboratories faced with the determination of pesticides at levels significantly above the detection limits and where those pesticides are not thermally labile, injections of 1–2 µL via a splitless injector will probably suffice. For pesticides with high boiling points or that are thermally labile, a PTV inlet offers a robust solution for the injection of conventional volumes but with the additional capability of injecting large volumes to cope with the growing demand for lower detection limits in pesticide residue analysis and for coupling with microextraction devices. It will be interesting to see whether the degree of automation offered by devices such as DMI finds its way into routine use.

3.3.2 CHROMATOGRAPHY

3.3.2.1 Fast Gas Chromatography

One of the main goals in the development of an analytical method is to lift restrictions on limited sample throughput by increasing the speed of the determination step by means of "fast" gas chromatography.[45] With fast GC-MS, the reduced selectivity provided by the drop in separation efficiency sacrificed for speed on short capillary columns is compensated somewhat by the additional selectivity offered by the MS. Interest in high-speed GC-MS has resulted in many new techniques that have greatly reduced analysis times:[46] microbore columns, high-speed temperature programming, shorter columns, high carrier gas velocities, and column ensembles with adjustable column selectivity. Short columns or fast temperature programming can be combined with other techniques to reduce analysis time.

Although reducing the internal diameter of GC columns to 0.1, 0.15, or 0.18 mm i.d. ensures that separation efficiency is not compromised for speed of analysis, there are practical limitations to their use, as they require "specialist" equipment to both generate and deal with the narrower peaks (e.g., accommodation of higher inlet pressures and faster spectral acquisition rates). Although their low capacity combined with repeated injection of coextractives may result in loss of performance more quickly than conventional columns, 0.15 mm i.d. columns have been successfully used for the determination of pesticides.[47] Resistive heating[48] offers two major advantages over conventional ovens for fast temperature programming:[49] very rapid cool-down rates and excellent retention time repeatability.[50] In low-pressure gas chromatography (LP-GC),[51,52] lower column pressures lead to higher diffusivity of the solute in the gas phase, which shifts the optimum carrier gas velocity to a higher value, resulting in faster GC separations as compared with the use of the same column operated at atmospheric outlet pressures. The gain in speed becomes pronounced mainly for shorter and wider columns because they can be operated at low pressures along the entire column length. Conventional GC injection techniques are possible due to the addition of a short narrow restriction capillary at the inlet end. While shorter columns result in shorter analysis times, a complete separation is not always possible with complex mixtures.[53] In such cases, pressure-tunable selectivity strategies have been used to provide high-speed GC.[54]

3.3.2.2 Two-Dimensional Gas Chromatography

Isobaric interference from matrix coextractives, which coelute with the pesticides of interest, can hinder confirmation to identify using GC-MS. Especially problematic are those compounds whose EI spectra are dominated by low mass ions. This problem becomes even more critical when sensitivity is an issue or when seeking low reporting limits. One of the approaches employed to overcome these selectivity problems is to improve the chromatographic separation by using two-dimensional gas chromatography (GC × GC).[55,56] The sample is first separated on a normal-bore capillary column under programmed temperature conditions. The effluent of this column then enters a modulator, which traps each subsequent small portion of eluate, focuses these portions, and releases the compounds into a second column for further separation. The second separation is made to be fast enough to permit the continual introduction of subsequent, equally small fractions from the first column without mutual interference. The resulting chromatogram, generally presented as a contour plot, has two time axes (retention on each of the two columns) and a signal intensity corresponding to the peak height. Comparison between separation efficiency of one-dimensional and two-dimensional GC-MS for the determination of dichlorvos in peaches showed that this analyte could only be detected by GC × GC-MS as GC-MS could not separate dichlorvos from the coeluting, isobaric compound, 5-(hydroxymethyl)-2-furancarboxaldehyde.[57]

Unfortunately, while these innovative developments clearly have great potential to improve the speed of analysis and the separation of complex mixtures, they remain largely in the domain of research and development and have yet to transfer across for use in routine monitoring.

3.3.3 Mass Spectrometry

The ionization source and mass analyzer are fundamental parts of a mass spectrometer as production of gas-phase ions is necessary for manipulation by electric or magnetic fields. These ions are extracted into the analyzer region of the mass spectrometer where they are separated according to their mass-to-charge ratios (m/z).

3.3.3.1 Ionization

Although chemical ionization (CI) cannot be used for a primary multiresidue method, CI, and negative chemical ionization (NCI) in particular, gives better selectivity than EI for a limited number of specific compounds, which provides improvements in reporting limits.[58] NCI exhibits a high selectivity for "electron-trapping" compounds (e.g., halogen-containing and other heteroatomic compounds) and electron-deficient aromatic compounds. The sensitivity can be improved by two orders of magnitude compared with EI. Mass spectra produced by CI are usually dominated by ions, which, although optimum for enhancing sensitivity, offer less information in contrast with EI, which typically produces large number of characteristic fragments, which, when acquired under standardized condition (70 eV) can be compared with spectra of known pesticides in published libraries such as the NIST/EPA/NIH Mass Spectral Library and Wiley Registry of Mass Spectral Data, now available as a combined database[59] or used to generate bespoke libraries of known pesticides.[60] The identification process is based on search algorithms that compare the spectra acquired with those of the library. A spectral match and fit factor defines the certainty of the identification.

3.3.3.2 Single Quadrupole Analyzers

The quadrupole consists of four parallel metal rods. Each opposing rod pair is connected together electrically and a radio-frequency (RF) voltage is applied between one pair of rods and the other. A direct current (DC) voltage is then superimposed on the RF voltage. Ions travel down the quadrupole in between the rods. Only ions of a certain m/z will reach the detector for a given ratio of voltages: other ions have unstable trajectories and will collide with the rods. This allows selection of a particular ion, or scanning by varying the voltages. The quadrupole can be used in two modes: scan or single ion monitoring (SIM), also called single ion recording (SIR). In scan mode, the amplitude of the DC and RF voltages are ramped (while keeping a constant RF/DC ratio), to obtain the mass spectrum over the required mass range. Sensitivity is a function of the scanned mass range, the scan speed, and resolution. In SIM mode, the parameters (amplitude of the DC and RF voltages) are set to observe only a specific mass, or to "jump" between a selection of specific masses. Figure 3.1 shows a schematic of a single quadrupole instrument in SIM mode. This mode provides the highest sensitivity but users are restricted to acquiring specific ions, typically EI fragments, since more time, the dwell time, can be spent on each mass. A longer dwell time would result in better sensitivity but, the number of data points acquired across a single peak, and the total number of pesticides that could be analyzed in a single run, are reduced. Due to insufficient

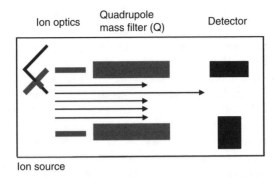

FIGURE 3.1 Schematic overview of a single quadrupole mass spectrometer.

sensitivity in scan mode, historically, quadrupoles have typically been operated in SIM for optimum sensitivity, limiting the amount of structural information that could be recorded. However, such information is critical to the successful confirmation of identity of target analytes. There have been numerous reports describing the development and implementation of methods for the simultaneous determination of anything up to typically a maximum of about 400 target pesticides by GC-MS using EI with SIM and the screening for 927 "pesticides and endocrine disrupters" was recently reported.[61]

3.3.3.3 Quadrupole Ion-Trap Analyzers

Three-dimensional quadrupole ion-trap analyzers (3D QIT),[62] also termed ion-trap detectors (ITDs),[63] have been used to carry out similar determinations.[64] Figure 3.2 shows a schematic of a 3D QIT instrument. The principle of the trap is to store the ions in a three-dimensional quadrupole field. Ions are removed one m/z value at a time by resonant ejection to obtain a scan recorded as a mass spectrum. Ions can be

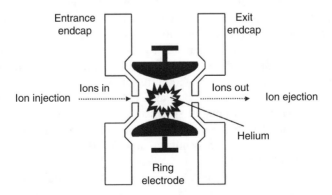

FIGURE 3.2 Schematic overview of a three-dimensional quadrupole ion-trap mass spectrometer.

formed internally and stored as they are formed or externally followed by injection and storage. The ability to selectively store ions provides a substantial improvement in sensitivity when compared with a quadrupole instrument when recording mass spectra and so permits the recording of complete mass spectra in concentration ranges in which quadrupole instruments have historically had to operate in SIM mode. This aspect is particularly advantageous for pesticide residue analysis when the components under investigation are present in the sample only in very small concentrations and an unambiguous identification is required. Moreover, no selection of characteristic ions is necessary during data acquisition for MS with a 3D QIT, permitting investigation of unknown samples, that is, screening of samples on the basis of complete mass spectra. Thus, substances that were not originally sought can be detected by revisiting the data. This is not possible in the case of the SIM technique, which operates on the principle of the selection of previously known substance and characteristic ions.

Disadvantages of GC-MS using original 3D QITs were related to space charge problems, leading to lower mass resolution and mass shifts[65] and ion/molecule reactions called "self-CI."[66] Although modern instruments have various techniques to prevent overfilling of the trap, this can still be a problem when analyzing pesticides at low levels in dirty matrices because the trap is filled with ions derived from matrix leaving little space for the small number of analyte ions. The limited storage of ions has also limited the dynamic range of the 3D QIT.

3.3.3.4 Tandem Mass Spectrometry Analyzers

As both single quadrupole and 3D QIT work at unit mass resolution, selectivity is limited, so these instruments can suffer from reduced sensitivity due to the contribution to the analyte signal from chemical noise. Although low reporting limits might be possible for simple matrices using GC-MS, these instruments can provide insufficient selectivity for complex food matrices. Tandem mass spectrometry (MS/MS),[67] in which mass-selected ions are subjected to a second mass spectrometric analysis, can provide increased selectivity, which reduces the contribution to the analyte signal from isobaric interference leading to improvements in sensitivity.[68] Hence, lower limits of detection become achievable when using GC-MS/MS for pesticide residue analysis in complex matrices. The same selectivity, achieved by monitoring the transition from one parent ion to a characteristic product ion, provides a greater degree of confidence for confirmation of identity than SIM, which can suffer from isobaric interferences. Based on the current EU quality control procedures for pesticide residue analysis,[19] if using GC-MS, four ions have to be detected and all ion ratios have to be within the specified tolerance intervals for identity to be confirmed. Additional legislation directed at residues of substances in live animals and animal products introduced an identification point (IP) system that was weighted to the selectivity of the method used.[69] When using the more selective MS/MS technique, monitoring and detection of two transitions exhibiting a ratio within tolerance is sufficient, as the precursor earns 1 point and each product ion earns 1.5 points, 4 points in total. The IP system has been applied to the determination of pesticide residues in animal products[25] and may find wider usage.

The capability of the 3D QIT to store ions of a single m/z value to the exclusion of ions of all other m/z values allows for MS/MS by means of collision-induced dissociation (CID) within a single mass analyzer.[70] An ion can be stored as a precursor, and that stored ion can then be manipulated to collide with the cooling gas molecules to produce product ions. By ramping the RF voltage, or by applying supplementary voltages on the end cap electrodes, or by combination of both, it is possible to keep only one ion in the trap, fragment it by inducing vibrations, and observe the fragments as they are sequentially ejected from the trap. The high efficiency for ion-trap MS/MS results from the parent and product ions remaining in a single ion trap and not transported from one chamber to another, eliminating transport losses. The application of wideband excitation (activation) and normalized collision energy leads to highly reproducible mass spectra. Hence, the main advantage of using a 3D QIT for GC-MS/MS is that full product ion spectra can be generated from trace amounts of pesticides for comparison with MS/MS libraries.[71]

The performance of tandem quadrupole GC-MS/MS has long been recognized,[72] but the price has been out of the reach of many laboratories involved in pesticide residue analysis that would benefit from this technology. With the recent introduction of GC-MS/MS instruments, based on the tandem quadrupole technology of existing LC-MS/MS platforms, the number of laboratories using it for pesticide residue analysis is growing. The analyzer of a triple quadrupole instrument consists in two quadrupoles, separated by a collision cell.[73] The first quadrupole is used in SIM mode to select a first ion (precursor), which is fragmented in the collision cell. This is typically achieved in the collision cell by accelerating the ions in the presence of a collision gas. The energy of the collision with the gas can be varied to allow different degrees of fragmentation. The resulting fragments are analyzed by the second quadrupole and also typically used in SIM mode to monitor a specific fragment (product), the process known as multiple reaction monitoring (MRM) (also called selected reaction monitoring, SRM). Figure 3.3 shows a schematic of a triple quadrupole instrument in MRM mode. As two analyzers increase the selectivity, the ion signal is reduced during the transmission, but the chemical noise, which is a major limitation for complex samples, is also largely decreased, leading to an improvement of the signal-to-noise ratio.

One limitation in GC-MS/MS, on either type of instrument, arises from the fragmentation provided by EI as often the total ion current is spread on many

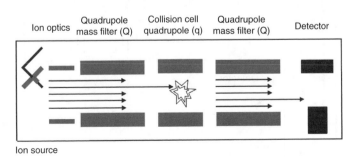

FIGURE 3.3 Schematic overview of a triple quadrupole mass spectrometer.

fragments, resulting in low intensity of ions that can be selected as parent ions for MS/MS experiments. The primary advantage of 3D QIT is that multiple MS/MS experiments can be performed quickly without having multiple analyzers. Hence, the introduction of MS/MS on a 3D QIT was a major breakthrough as it brought down the cost of tandem mass spectrometry.[74] Unlike MRM using a triple quadrupole, however, MS/MS using 3D QIT is restricted to concurrent acquisition of a limited number of precursor ions.

3.3.3.5   Time-of-Flight Analyzers

The use of TOF technology[75] provides an innovative approach to overcoming the drawbacks that limit the exploitation of mass spectrometers for detecting pesticides at trace levels while retaining full spectral information as a tool for confirmation of identity. The design of the orthogonal acceleration (oa) TOF,[76] into which pulses of ions are extracted orthogonally from a continuous ion beam, the availability of fast-recording electronics, together with improvements in signal deconvolution techniques, were major breakthroughs in the development of modern GC-TOF instruments.

Figure 3.4 shows a schematic of a TOF instrument. As the name implies, separation of ions in a TOF mass analyzer is accomplished by measuring their flight time in a field-free tube based on the fact that ion velocity is mass-dependent. The ions generated in an EI source are initially accelerated to get discrete packages with a constant kinetic energy, which are ejected into the mass analyzer using pulsed electric field gradient oriented orthogonally to the ion beam. Reflectrons (ion mirrors) are used to compensate for variations in initial energy distribution. Ions are reflected based on their forward kinetic energy. The more energetic the ion, the deeper it penetrates the retarding field of the reflection before getting reflected. This allows an energetic ion, traveling a longer flight path, to arrive at the detector at the

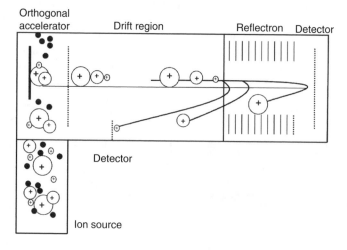

FIGURE 3.4 Schematic overview of a TOF mass spectrometer.

same time as the less energetic ions of the same mass. Ions are detected using a multichannel plate detector (MCP) and detection of "ion events" converted from an analog signal to a digital record. Flight times, which are proportional to the square root of the m/z value of an ion, are in the order of microseconds. Consequently, TOF MS can operate at very high repetition rates and between 20 and 500 spectra per second can be stored. The effort to exploit these unique features has resulted in the development of two types of commercial spectrometers differing in their basic characteristics: instruments using unit-resolution instruments that feature a high acquisition speed and elevated resolution analyzers with only moderate acquisition speed. The application potential of these approaches is obviously complementary.[77,78] Both approaches are characterized by high sensitivity due to improved mass analyzer efficiency and continuous acquisition of full range mass spectra. Mass analyzer efficiency of the TOF-MS instruments is as high as 25% in full spectra acquisition (quadrupole 0.1%). Generation of complete mass spectra from residues of pesticides even at trace levels enables searching against library reference spectra for identification. The fast repetition rate ensures that no changes in the ratios of analyte ions across the peak occur during the acquisition of the mass spectrum and, consequently, no spectral skew, which is commonly observed by scanning instruments, is encountered. The generation of full range mass spectra also provides the ability to review archived data for new compounds outside the scope of the initial analysis. For example, in response to an EU Rapid Alert in late December 2006, those laboratories within the EU using GC-TOF MS were able to look back again at historical data and identify residues of isofenphos-methyl even though no routine government testing program in any EU member state had included tests for isofenphos-methyl at that time.

High-speed TOFs, operating at unit mass resolution but with very fast scan rates (e.g., 500 spectra/s), can provide the data density necessary to accurately define narrow chromatographic peaks typical for fast and ultrafast GC[79] and GC × GC.[80] The high data acquisition rate and the absence of any spectral skew allow overlapping signals to be automatically deconvoluted based on their mass spectra.[81] The GC-TOF instruments with elevated resolution (typically 5000–7000 FWHM) and good mass accuracy (e.g., 5 ppm RMS) have so far had a more limited application to the analysis of pesticide residues.[82] The elevated resolution and good mass accuracy can be significantly used to reduce the contribution from isobaric interference by evaluating data with a narrow mass window (typically 50 mDa), which improves detectability of the analytes. Measurement of an accurate mass of a particular ion also provides additional information for confirmation of identity for target compound analysis and, more importantly, aids the assignment of unknown compounds based on calculation of their elemental composition. High mass accuracy is attainable by using a lock mass calibration procedure for which a reference compound is continuously supplied into the ion source during analysis. On the basis of a previously performed mass calibration over a given mass range and a defined exact mass of the reference ion (the lock mass), the values of all masses in the acquired spectra are automatic and continually corrected. Mass accuracy varies with ion intensity. Care is required when applying exact mass windows as if the window is set too narrow, peaks may be underestimated or missed altogether. It is possible to search for compounds of interest by measurement of exact mass and isotope pattern of fragment ions and compare with the accurate masses in the reference spectra. Using this approach, the identification of the analyte can be based not only on retention time and EI mass spectrum, but thanks to the exact mass measurement also on elemental composition. In addition, the measurement of exact mass may aid the identification of unknown compounds through the calculation of elemental composition. In practice, this is a very demanding task as EI spectra rarely exhibit a molecular ion so a good knowledge of the fragmentation mechanism is required.

The main disadvantage of the TOF as an analyzer for quantitative GC-MS is the limited linear dynamic range compared with conventional MS instrumentation. The analog signal from the ion detector is converted to a digital record by a fast analog-to-digital (ADC)-based continuous averager, also called an integrating transient recorder (ITR). With an ITR, the ion rates can be increased so that many ions of the same m/z value arrive at the ion detector simultaneously. The result is an analog voltage pulse whose amplitude is approximately proportional to the number of ions in the pulse. An ADC samples the output waveform of the ion detector and periodically converts the measured analog voltage to a digital representation, which is sequentially stored in a digital memory to form a single record. Although the linear dynamic range of the ITR is limited to two orders of magnitude, it has been expanded to approximately four orders of magnitude by application of continuing improvements in both hardware and software features.

While GC-MS, using EI with SIM on a single quadrupole, still provides the widest scope for pesticide residue analysis, GC-MS/MS and GC-TOF MS offer two unique solutions. However, both are still used by a few laboratories. Although initial purchase cost remains higher than conventional benchtop instruments, which may currently prohibit use within routine laboratories, both approaches offer considerable benefits and so usage is likely to grow.

3.4 LIQUID CHROMATOGRAPHY-MASS SPECTROMETRY (LC-MS)

Interest in the use of LC-MS for pesticide residue analysis has grown considerably during the last 10 years and, after ~30 years of continual and rapid development, the technique is in routine use in most laboratories. Analytes are introduced to a solvent flow path, carried through a column packed with specialized materials for component separation and then enter the mass spectrometer for ionization and mass analysis.

3.4.1 SAMPLE INTRODUCTION

By comparison with GC-MS, LC injections are simple and very precise. There are no inserts, no thermal degradation, and very few surface activity issues. Furthermore, these benefits hold true even when large-volume injection (LVI) is required for lower detection limits. As long as the solvent strength of the injected sample is properly matched to the initial mobile-phase solvent strength, or a precolumn is used, scaling up to LVI on LC-MS is equally simple.

Samples, prepared in a suitable submission solvent, are usually introduced by an automated syringe drawing samples from vials and injecting into the LC column via a port consisting of an injection valve (e.g., six-port rotary valves) and sample loop. Some consideration of injection technique (e.g., filled versus partial filling), loop dimensions, and dead volume is important as all influence analytical precision. In addition, the composition and volume of injection solvent used can influence the chromatography and peak shape in particular. A strong injection solvent, relative to the mobile-phase composition, can cause the sample to move quickly through the first portion of the column while it reequilibrates with the mobile phase, thereby broadening or even splitting peaks. Where possible, it is best to inject the sample in the starting mobile phase or a weaker solvent to avoid this type of problem. Alternatively, if injecting in a strong solvent is preferred, injection volumes must be kept small. Increasing the volume injected beyond conventional volumes is possible if a weak submission solvent is used, but LVI has rarely been applied to pesticide residue analysis of matrices using LC-MS. Although direct LVI of water samples has been shown to be a useful alternative to SPE of very polar pesticides,[83] online SPE systems appear to be the preferred approach for introduction of larger volumes of sample.

In spite of the high selectivity of the LC-MS technique, sensitivity is rarely high enough to directly determine the trace amounts of pesticides in food commodities or water at the levels required by legislation; so preconcentration is usually required. Fortunately, the majority of LC-MS is conducted on MS/MS instruments with enhanced selectivity; so extensive cleanup is less crucial when compared with analysis by GC-MS and there is a growing trend toward LC-MS analysis without cleanup.[84] Any cleanup of food extracts is still usually performed off-line before LC-MS, whereas the use of online trace enrichment has been successfully used for the determination of pesticide residue in water.[85] SPME, in-tube SPME, and SBSE have all been used as sample introduction devices for LC-MS.[86–88]

3.4.2 Chromatography

Chromatographic selectivity is a prerequisite for most applications of LC-MS, as mass selectivity does not completely eliminate isobaric interferences and matrix effects that may affect the relative response of analytes. Although separation of all analytes is not considered necessary for detection of pesticide residues due to the high selectivity provided by MS/MS, ion suppression is observed either from coelution with other analytes or, more likely, coeluting matrix components. The most important and widely used separation technique for pesticides is LC on reversed-phase (RP) columns. Separation is based on differences in hydrophobicity by partitioning between an apolar stationary phase and a polar mobile phase. Together with appropriate control of operational parameters such as solvent composition, pH, temperature, and flow rate, reversed phase can enable separations of many pesticides with a wide range of polarities and molecular weight. Columns with C_8 and C_{18} stationary phases on high purity silica are the most widely used.

3.4.2.1 Mobile Phases

The selection of mobile-phase constituents for LC-MS using RP is not an easy task, and the conditions described in the literature are clearly rarely optimized. The selection of mobile phase is important to obtain a good chromatographic separation, but it also affects the analyte ionization and the sensitivity of the mass spectrometer.[89] For example, analyte charge should be suppressed by manipulation of the mobile-phase pH for optimum retention but this can have a detrimental affect on MS response.[90] Contrary to conditions for RP LC retention, for optimized electrospray (ES) ionization, the pH should be adjusted to promote the charged state of the analyte over its neutral species as ionization takes place in the liquid phase. In contrast, for optimized atmospheric pressure chemical ionization (APCI), in which ionization takes place in the gas phase, formation of the neutral species is favored due to the higher volatility of the neutral versus charged species and hence better vaporization.[91] A compromise can be sought by experimentation but this is particularly difficult when a wide range of pesticides, with differing properties, have to be analyzed. The situation is further complicated if polarity switching is employed so that positive and negative ions are periodically sampled throughout the analytical run. Alternatively, the target pesticides are divided into anionic and cationic groups and analysis performed separately.

For routine application, even designs of orthogonal nebulizers for atmospheric pressure ionization interfaces are still restricted to the use of volatile buffers. The concentration of the buffer, or acid or base used to adjust/control the pH, should be as low as possible for ES. If not, competition between analyte and electrolyte ions for conversion to gas-phase ions decreases the analyte response. If a species is in large excess, it will cover the droplet surface and prevent other ions to access the surface, and thus to evaporate. A species in large excess will also catch all charges available and prevent the ionization of other molecules present at much lower concentration. Ammonium acetate and ammonium formate are generally applicable at pH 7 but concentrations should be kept to a minimum. RP LC separations are sometimes improved at acidic pH, using acetic acid or formic acid, as such or in combination with ammonium acetate or ammonium formate. The addition of reagents postcolumn can be used to generate pH conditions optimum for ionization without changing chromatographic separation but this approach is rarely implemented for routine analyses.

Methanol and/or acetonitrile are used as organic modifiers. Low surface tension and a low dielectric constant of the solvent promote ion evaporation, which favor the ionization process. The gas-phase basicity (proton affinity) and gas-phase acidity (electron affinity) are also important solvent properties in the positive and negative ionization modes, respectively. Those features encourage the use of methanol versus acetonitrile. Methanol is also preferred over acetonitrile when MS with ES is coupled to gradient LC because the lower eluotropic strength of methanol causes compounds to elute at a higher percentage of organic solvent, where ES sensitivity is increased. In most cases, more pesticides appear to elute in the middle of the analytical run. Given the extra demand on MS acquisition in terms of obtaining sufficient data points across a peak and a long enough dwell time for sensitivity, it is surprising that few authors have reported efforts to optimize the gradient conditions so that pesticides exhibit a wider elution profile.[92]

3.4.2.2 Ion Pair, Hydrophilic Interaction, and Ion Chromatography

Volatile ion pair reagents (e.g., heptafluorobutyric acid and tetrabutyl ammonium) have been added either to the sample vial[93] or to the mobile phase to improve the chromatography of ionic species, such as paraquat and diquat.[94] The introduction of an ion pair reagent into the mobile phase increases the interactions between the quaternary ammonium compounds and the C_{18} stationary phase, providing the necessary retention and resolution.[95] The type and quantity of ion pair reagent added has to be a compromise between improvement in separation and retention and minimizing the suppression observed in ES. Hydrophilic interaction chromatography (HILIC)[96] has been explored for the determination of paraquat and diquat by MS/MS without the need for ion-pairing reagents. HILIC separates compounds by passing a hydrophobic or mostly organic mobile phase across a neutral hydrophilic stationary phase, causing solutes to elute in order of increasing hydrophilicity. Although the chromatography behavior on HILIC is not as good as that observed using the ion pair systems, the MS sensitivity using the HILIC mobile phase was claimed to be significantly greater.[97]

Although ion chromatography (IC) allows separation of ionic compounds that have no retention on conventional reversed-phase LC columns, it is rarely used for the determination of pesticides by LC-MS due to the difficulties encountered by spraying the nonvolatile salts used in high ionic strength eluents. Exceptions include chlormequat,[98] as elution is possible with volatile buffers, and glyphosate,[99] when a suppressor is used between IC system and MS/MS to remove salts from the eluent to make coupling with the ES source possible.

3.4.2.3   Fast Liquid Chromatography

One of the primary parameters that influence LC separation is the particle size of the packing materials used to effect the separation. There has been a long trend of reducing particle size in LC (e.g., 10, 5, and 3 μm). Initially, the smallest particles were used in short columns, leading to fast analysis times but relatively modest gains in resolving power. Flow rates and column length were restricted by the back pressure generated. By using smaller particles, increases in speed of analysis, improved resolution, and sensitivity are possible.[100] According to the van Deemter equation, which describes the relationship between linear velocity (flow rate) and plate height (HETP or column efficiency), when the particle size decreases to <2.5 μm not only is there a significant gain in efficiency, but the efficiency does not diminish at increased flow rates or linear velocities. In order to take advantage of such small particle sizes, instrumentation capable of high-pressure operation with low system and dead volume is required; so, in 2004, Swartz and Murphy introduced the first LC system capable of operation up to 15,000 psi.[101] These systems have been termed ultraperformance liquid chromatography (UPLC) or ultra-HPLC (UHPLC) to differentiate them from HPLC.[102] With sub 2 μm particles, half-height peak widths of <1 s can be obtained, posing significant challenges for the MS. In order to accurately and reproducibly integrate an analyte peak, the MS must have a sufficient acquisition rate to capture enough data points across the peak (>10 points/peak), requiring very short dwell times and interchannel delays on a triple quadrupole instrument or the fast spectral acquisition of the TOF analyzer and the software tools to handle the increased number of results. Improved detection limits for LC-MS are achieved by narrower chromatographic peaks effectively increasing the concentration of analytes entering the MS source increasing signal intensity and the improved resolution may reduce ion suppression by separating species that may coelute in conventional LC.[103] In addition, shorter analytical run times are possible without compromising chromatographic resolution leading to an increase in sample throughput. This technology offers considerable benefits over conventional LC-MS and its application to the determination of pesticide residues[104,105] is growing rapidly.

3.4.3 MASS SPECTROMETRY

3.4.3.1 Ionization

The main prerequisite for MS is the introduction of ions in the gas-phase ions at reduced pressures. The main challenge is to generate gas-phase ions from analytes in the LC eluent, while removing the solvent and maintaining adequate vacuum level in the mass spectrometer. This is achieved by evaporation, pressure reduction, and ionization. Particle beam (PB) ionization[106] has the advantage of providing classical, library-searchable electron ionization spectra for compounds that were too thermally labile or nonvolatile to be analyzed by GC-MS. The robustness and sensitivity of PB was limited and so, with the introduction of alternatives, it had limited application for routine trace analysis. Currently, the most popular technologies create ions at atmospheric pressure and then sample ions through so-called atmospheric pressure (AP) interfaces. Blakely and Vestal[107] introduced the TSP ionization source, which was capable of producing ions from an aqueous solution sprayed directly into the mass spectrometer using conventional analytical LC flow rates. There was much debate over the models proposed for the ionization mechanism,[108] which later contributed to the construction of models for ionization taking place in other API technologies. The use of TSP for small molecule LC-MS applications also facilitated familiarity with the concept of generation of ions through cation attachment (e.g., $[M + NH_4]^+$) and the use of negative ion mode for acidic compounds. TSP found tremendous use as an interface for higher flow-rate LC/MS, including for pesticide residue analysis.[109] Although its usage lessened as other API technologies become more popular, existing instruments could be relatively easily upgraded with the new, more sensitive, and robust API devices. ES ionization[110] and APCI[111] are now among the most commonly used techniques for creating ions from pesticides in solution. The application of the more recent atmospheric pressure photoionization (APPI) to pesticide residue analysis is less common.[112] Although still at the development stage, there has been growing interest in EI for LC-MS[113,114] to obtain library-matchable, readily interpretable, mass spectra, to aid the confirmation of identity of targeted pesticides and, more importantly, for the characterization of unknowns such as metabolites or transformation products.

3.4.3.2 Electrospray Ionization

The basic aspects of ES ionization and APCI are described in detail elsewhere[115–120] but Willoughby et al. provide a good practical introduction to ionization used for LC-MS.[121] There have been considerable efforts directed at understanding the mechanisms involved in ion production for electrospray;[122,123] but as practitioners, we have come to accept that ES works and many might rather avoid considering the mechanisms involved. But, as Balogh argues, "understanding how ions are liberated from the liquid mobile phase in the gas-phase transition helps us understand and diagnose issues such as lack of expected sensitivity and ion suppression."[124] Two separate theories have been proposed, the charge residue mechanism and ion evaporation mechanism models, but Cole argues that both mechanisms might be working concurrently: the charge residue mechanism dominating at high mass and ion evaporation dominating for lower masses.[125] The ES probe, or device, is typically a conductive capillary, usually made of stainless steel, through which the eluent from the LC flows. A voltage is applied between the probe tip and the sampling cone. In most instruments, the voltage is applied on the capillary, while the sampling cone is held at low voltage. The capillary, contained within a larger bore tube, allows a concentric nitrogen flow applied to the aerosol at its exit point so that capillary acts as a nebulizer. While this variant was initially called "pneumatically assisted electrospray" or "ion spray,"[126] this terminology appears to be replaced by the more generic "electrospray." Aerosol droplet formation is enhanced by the added shear forces of the gas and heat transmitted from adjacent supplemental devices, direct heating of the gas itself or with the assistance of an additional heated desolvation gas. ES ionization takes place as a result of imparting the strong electrical field to the eluent flow as it emerges from the nebulizer, producing an aerosol of charged droplets. Due to the solvent evaporation, the size of the droplet reduces, and, consequently, the density of charges at the droplet surface increases. The repulsion forces between the charges increase until there is an explosion of the droplet. These coulombic fissions continue until droplets containing a single analyte ion remain. The charge residue model suggests that a gas-phase ion forms only when solvent from the last droplet evaporates. In the ion evaporation model, the electric field strength at the surface of the droplet is thought to be high enough for solvated ions to attain sufficient charge density to be ejected from the surface of the droplet and transfer directly into the gas phase without evaporation of all the solvent. Ensuring that the compound of interest is ionized in solution critical for ES ionization, so mobile phases should have a pH such that the analytes will be ionized. Charging is usually accomplished by adding or removing protons but cation or anion attachment generating adduct ions is also common.

3.4.3.3 Atmospheric Pressure Chemical Ionization

Although work that demonstrated APCI as an ionization technique for LC-MS was published by Carroll[127] some time before Whitehouse's work on electrospray,[128] it was not widely adopted until the latter was commercialized. The complimentary capability of APCI was marketed as enabling the analysis of compounds that resisted converting to gas-phase ions using ES, that is, the less polar and more volatile ones. In contrast to ES, APCI transfers neutral analytes into the gas phase by vaporizing the LC eluent contained within a nonconductive capillary inserted in a coaxial pneumatic nebulizer through which a gas is added to assist the ionization process. The mixture of gas and nebulized eluent passes through a heated zone that assists the solvent evaporation and the fine droplets are converted into desolvated molecules in the gas phase. The desolvated analyte molecules are then ionized via chemical ionization; the transfer of charged species between a reagent ion and a target molecule to produce a target ion that can be mass analyzed. The corona-discharge needle in the APCI source produces a stream of electrons that ionizes the atmosphere surrounding the corona electrode, which consists mainly of nebulizer and drying gases (typically nitrogen and/or air), the vapor generated from the HPLC eluent, and the analyte molecules. The process starts by ionizing nitrogen and finishes with protonated water, water clusters, and solvent clusters as possible reagent ions. For successful APCI, the analyte must be volatile and thermally stable and the mobile phase must be suitable for gas-phase acid–base reactions. For example, when working in positive ion mode, the proton affinity of the analyte must be higher than the proton affinity of the eluent: that is, the analyte can acquire a proton from the protonated solvent. Since water cluster ions are a major source of reagent ions, the proton affinity of these clusters relative to analyte ions will have a profound effect on sensitivity.[129] Similarly, the use of certain modifiers added to the mobile phase to enhance LC separation (e.g., triethylamine) can be the source of considerable ion suppression in APCI. A strong base will receive protons from the reactant ions to form their protonated forms. Subsequent proton transfer will occur only if the analyte is more basic than the modifier.[130]

Although the choice of the most appropriate interface as well as detection polarity are based on analyte polarity and LC operating conditions, many classes of compounds perform well using either technique and sometimes in both ion modes, whereas, for other compounds, the choice is more restricted.[131] Interfaces are selected based on individual preference derived from experience and available techniques as well as the magnitude of any matrix effects. Although there are a great number of examples of the use of APCI for pesticide residue analysis for both environmental and food applications, including some pioneering early work,[132] more recently the technique appears to be left in the wake of ES ionization's overwhelming popularity. This may be related to the increasing number and wider range of pesticides currently sought but perhaps also reflects the improvements in source and probe design for ES not yet paralleled with APCI. The choice between ES and APCI is irrelevant when using the recently introduced multimode sources, which deliver simultaneous ES ionization and APCI.[133]

3.4.3.4 Atmospheric Pressure Interfaces

Common to all API sources for mass spectrometers is an ion inlet orifice that forms an interface between the API region and the low-pressure region of the mass analyzer. This small orifice allows the vacuum system attached to the mass analyzer to maintain a satisfactory vacuum therein at a finite pumping speed. The API source of commercial LC-MS instruments is arranged orthogonally of the ion inlet orifice, providing improved tolerance to nonvolatile components in the LC eluent and a more stable ion signal than axial predecessors.[134] Consequently, cone orifices can be larger than in previous designs. The combination of larger orifices and noise reduction largely compensates for transmission losses due to the orthogonal geometry, giving a large gain in sensitivity. The sampling of ions from atmospheric pressure into the high vacuum region of the mass analyzer region requires significant pressure reduction. A gas stream introduced into a vacuum system expands and cools down. When this gas stream contains ions and solvent vapors, the formation of ion–solvent clusters is observed. To obtain good sensitivities and high-quality spectra, one of the key roles of the interface is to prevent cluster formation. Declustering of analyte ions may be achieved using one or a combination of the following approaches; using a countercurrent gas flow between interface and sampling plates, also known as "curtain gas," using a heated transfer capillary between the API region and the nozzle–skimmer region, and/or using a drift voltage between the nozzle and skimmer plates to promote intermolecular collisions between the analyte clusters and the background gas molecules. In APCI, nebulizer temperatures should be optimized to reduce cluster ion formation.

3.4.3.5   Characteristics of Atmospheric Pressure Ionization

ES and APCI are soft methods of ionization as very little residual energy is retained by the analyte on ionization. The major disadvantage of the techniques is that very little fragmentation is produced. The mass spectra generated by either technique are typically dominated by protonated or deprotonated molecules, $[M+H]^+$ or $[M-H]^-$, depending on the ion mode used and adducts (e.g., $[M+Na]^+$, $[M+NH_3]^+$, $[M+HCOO]^-$). This only provides information on molecular weight. This is very different from the information-rich spectra obtained with EI. For better selectivity through MS/MS or elucidation of structure, fragmentation is needed. Possible fragmentation techniques include "in-source" CID, CID in the collision cell of a tandem-type instrument and fragmentation in an ion trap.

One of the major problems encountered using LC-MS with ES is the presence of coeluting matrix compounds that alter the ionization of the target compounds, and which can reduce drastically the response affecting both quantification and detection of pesticide residues. This phenomenon is known as the matrix effect and, because it has an important impact on pesticide analysis, it has been the object of considerable study.[135,136] There are a number of ways matrix effects can be detected; the most straightforward way is the comparison of the response obtained from a standard solution with that from a standard solution prepared in a matrix extract. This approach can be extended to the comparison of calibration graphs obtained from the analysis of standards prepared in solution with those prepared in matrix extracts. A third approach is the postcolumn infusion system,[137] in which continuous postcolumn infusion of the analyte of interest is performed while blank extracts are injected into the LC column. This enables the evaluation of the absolute matrix effects on the analyte at different portions of the chromatogram, illustrating the need for change in the LC separation required to minimize the matrix effect. If matrix suppression cannot be eliminated by improved sample preparation or reoptimization of LC conditions, careful consideration of calibration strategy is needed to compensate as much as possible for matrix effects. Using matrix-matched calibrants, standard addition or stable isotope-labeled internal standards is recommended.

3.4.3.6 Tandem Mass Spectrometry Analyzers

A wide range of mass analyzers is used for pesticide residue LC-MS analysis.[138] Despite early successes,[139] single quadrupole instruments are rarely used now for pesticide residue analysis, as, when combined with API sources, the technique lacks the selectivity required for both detection in complex matrices and for confirmation of identity. Single quadrupole MS has been superseded by MS/MS. Originally, LC-MS/MS was mainly delivered on 3D QIT instruments, as they provided more cost-effective access to MS/MS than triple quadrupole instruments.[140] The application of wideband excitation (activation) and normalized collision energy leads to highly reproducible mass spectra without losses of sensitivity, which has enabled the publication of searchable libraries.[141] An additional key feature of the ion-trap instrument is that it provides multiple stages of MS/MS (MS^n). Product ions of a single m/z value produced by an MS/MS experiment are stored to the exclusion of product ions of all other m/z values to obtain a second iteration of MS/MS. This can be carried out for several more iterations. Although most commercial 3D QIT mass spectrometers allow for MS^n where n is 10 iterations, in practice the time it takes for a typical HPLC peak to elute limits the experiment to MS^3. Detailed studies of pesticide CID ion fragmentation processes and pathways have been reported using MS^n,[142] but few LC-MS^n methods for pesticide residue analysis have been developed making use of this extra selectivity.[143] Other restrictions of the commercial 3D QIT are its inability to trap product ions below m/z 50, and the existence of a upper limit on the ratio between the precursor mass and the lowest trapped fragment ion mass that is ~0.3 (dependent on the q/z value). The fragment ions with masses in the lower third of the mass range will not be detected. When Soler et al. compared the use of 3D QIT and triple quadrupole for the determination of pesticide residues, they found that product ion scanning with either MS/MS or MS^n could not provide sufficient sensitivity needed to monitor MRL compliance in oranges.[144] The future of ion-trap technology may lie with the new linear ion traps (LIT), which can be used either as ion accumulation devices in combination with quadrupole (Q), TOF, and Fourier transform ion cyclotron resonance (FT-ICR) devices or as commercially available, stand-alone mass spectrometers with MS^n capabilities.[145]

LC-MS/MS with a triple quadrupole instrument in MRM mode is currently the most widely used technique for the quantification of target pesticides[146–148] as it delivers the sensitivity required for monitoring compliance with the legislation. The use of other modes, such as neutral loss or precursor ion scanning, is limited by poor sensitivity and specificity. Generally, a triple quadrupole instrument used in MRM mode provides an order of magnitude better limit of quantification than product ion scanning on a 3D QIT instrument. Due to the enhanced selectivity, interfering peaks from other pesticides or matrix are rarely observed. For confirmation of identity, the ratio of at least two MRM transitions is required to match that of a reference standard. One of the current challenges for LC-MS/MS is to be able to acquire sufficient data points for quantification while acquiring an ever increasing number of MRM transitions, with very short dwell times (e.g., 5 ms), especially when coupled with UPLC. Reports of using LC-MS/MS for determination of 50 pesticides or more are becoming more common.[149]

The main limitation of triple quadrupoles in MRM mode is that confirmation of identity is based on the ratio of one or more MRM transition rather than full MS/MS product ion spectra. The replacement of Q3 in a QqQ instrument with a scanning LIT enhances its sensitivity in product ion scanning mode.[150] Additionally, the system has MS[3] capability and time-delayed fragmentation scans that aid structure elucidation. Quantitative (MRM) and qualitative (MS/MS or MS[3] product ion spectra) work can be performed concomitantly on the same instrument. Although reports of the use of the QqLIT instrument for pesticide residue analysis are currently limited to material from the vendor,[151,152] it is in routine use in some laboratories.

3.4.3.7 Time-of-Flight Analyzers

As the lists of targeted compounds get longer, setting up time-segmented MRM methods becomes more and more complicated. At the same time, interest in incorporating nontarget analysis into monitoring programs, especially for banned substances and unknown metabolites and transformation products, has grown considerably. Many of the new pesticides and their transformation products are readily analyzed by LC-MS. A screening and identification scheme for pesticides was reported, which employed searching a pesticides exact-mass library for the empirical formulas generated from accurate mass data acquired with high mass resolution.[153] High resolution and accurate mass are available from four types of instruments: double-focusing magnetic sector, LIT-Orbitrap, FT-ICR, and TOF mass spectrometers. Reports of coupling magnetic sector instrument with API interfaces are limited.[154] While both LIT-Orbitrap and FTICR instruments, which combine high trapping capacity, MS[n] capabilities with excellent mass accuracy and resolving power, have considerable potential for analysis of pesticide residues,[155,156] neither of them are yet used for routine monitoring. The current trend in pesticide residue analysis is to use LC-TOF MS or LC-QTOF MS systems because they are easier and less expensive to operate compared with the other three mass spectrometers.[157] Although TOF-MS can record accurate full spectral information with good sensitivity, this is of limited use unless resolution and mass accuracy are enhanced, as the spectra generated using the soft ionization API interfaces are usually characterized by a lack of fragment ions. LC-TOF instruments are capable of a resolving power of 10,000–20,000 (FWHM) and an accuracy of 2–5 ppm and the rapid acquisition rate combines well with fast chromatography systems. High mass resolution allows the reduction of the mass window when extracting a specific mass, leading to a substantial reduction in chemical noise, facilitating the detection of the analyte in the extracted ion chromatogram. If the mass window is set too narrow, mass errors either from drift in instrument calibration or derived from coelution of isobaric interferences from the matrix can lead to errors in quantification or, false negatives if the compound falls out of the mass window. As a compromise between enhanced selectivity and prevention of reporting false negatives, a 50 mDa mass window is recommended.[158] TOF was found to be around one order of magnitude less sensitive than a triple quadrupole instrument used in SRM mode[159] and so sensitivity of LC-TOF may not always be sufficient for the intended application. The introduction of a concentration step may be required to meet required reporting limits. Although acquisition rates have now improved (e.g., 20 spectra/s) and issues related to linear range still hamper the use of LC-TOF for quantification, some recent success has been reported.[160] None of the commercial benchtop TOF systems currently available meet the definition of high resolution given in Decision 2002/657/EC[69] and so no added weight should be conferred for an accurate mass measurement. Currently, at least four ions would have to be measured earning four IPs for confirmation, whereas if one considers LC-TOF to be a high-resolution technique, then only two ions need to be measured as each ion earns two IPs. In-source fragmentation can be used to increase the number of measured ions but the origin of the fragment ions may not be unequivocal and the technique is prone to isobaric interference. An alternative approach for confirmation of identity using LC-TOF is to measure the mass error from the accurate mass measurement of the suspect positive and reference standard but the number of compounds sharing the same empirical formula and therefore the exact mass can be surprisingly high which makes accurate mass measurements on the fragment ions necessary.[161] For elucidation of unknowns, the combination of accurate mass with accurate isotope composition and the use of database searches may lead to a reduced number of potential candidates but rarely a single answer. In the absence of a list of possible candidates, the complimentary use of other techniques is normally required to obtain a molecular formula and structure.[162]

The development of the hybrid quadrupole–time-of-flight (QTOF) instruments presents the analyst with all the advantages indicated for the LC-TOF with the additional capability of accurate mass product ion scans. Figure 3.5 shows a schematic of a QTOF instrument. QTOF allows the determination of elemental composition of all the product ions from an MS/MS experiment, a feature, which has been used for confirmation of identity for target compound analysis and elucidation of unknowns.[163] When confirming positive findings, both the exact

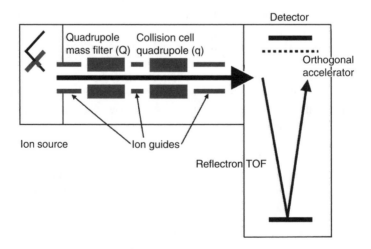

FIGURE 3.5 Schematic overview of a QTOF mass spectrometer.

masses and the relative intensities of all the product ions can be compared with those of the reference standard. The enhanced selectivity of MS/MS, when combined with high resolution and mass accuracy of the measurement of the product ions, provides low chemical background and hence improved quality of confirmation. Regardless of whether TOF is considered high resolution or not, the number of IPs possible with LC-QTOF is higher than for LC-TOF as the system is weighted for MS/MS. The LC-QTOF has been used for the elucidation of pesticide metabolites and transformation products, but the potential is more limited when no previous knowledge is available. In such cases, the most common approach is to search a database for the molecular formula. The accurate product ion mass spectrum provides additional structural information, which can be used for comparative purposes with reference standards or to distinguish between isomers. The application of the database approach is limited by the absence of many compounds, including pesticide transformation products, which are not included in commercial databases.

3.5 CONCLUSIONS

Recent innovations have made mass spectrometers increasingly more sensitive and selective. GC coupled to EI MS with single quadrupole analyzers and LC with ES ionization when combined with tandem MS are currently the most important instruments for targeted pesticide residue analysis, for environmental, food, and biological applications. Both techniques are influenced by matrix effects, which impact on analyte transmission from the GC injector to the column or suppression of response in ES ionization. The use of matrix-matched standards, although laborious, can reduce the problem. When using GC-MS, matrix components can also be the source of isobaric interference resulting in quantification errors and a reduction in the confidence of confirmation. The increase in the use of GC-MS/MS for pesticide residue analysis provides an increase in selectivity and hence a more secure confirmation of identity. The increase in sensitivity of LC-MS/MS instruments now available allows extracts to be diluted before analysis, reducing the matrix effect. The introduction of fast chromatography appears to be having immediate benefits for LC-MS, as UPLC-MS/MS is already used to generate residue data with considerable reduction in analytical run times.

Recent developments in TOF technology, both for GC-MS and LC-MS, enable the screening of sample extracts, both for targeted and nontargeted analysis, because full spectral information can be acquired with high sensitivity. Although this approach works well for EI spectra, high resolution and good accurate mass measurement are essential when using LC-MS due to the limited number of ions generated. In terms of quantification, TOF devices still present a common limitation, which derives from the low linear dynamic range. For confirmation of identity or elucidation of unknowns, until LIT-Orbitrap and FTICR become more readily available, the LC-QTOF, with its enhanced selectivity, is the instrument of choice. It should be stressed that the complimentary use of other techniques is usually needed for the elucidation of complete unknowns.

REFERENCES

1. Tomlin, C., *The Pesticide Manual: A World Compendium*, 14th edn, BCPC, Alton, Hampshire, 2006.
2. http://www.defra.gov.uk/environment/statistics/inlwater/iwpesticide.htm
3. EU, Directive 2000/60/EC of the European Parliament and of the Council of 23 October 2000 establishing a framework for Community action in the field of water policy, *Off. J. Eur. Commun.*, L327, 1–72, 2000.
4. Coggon, D., *A Guide to Pesticide Regulation in the UK and the Role of the Advisory Committee on Pesticides*, Defra/HSE, New York, 2003, p. 17.
5. EU, Council Directive 98/83/EC of 3 November 1998 on the quality of water intended for human consumption, *Off. J. Eur. Commun.*, L330, 32–54, 1998.
6. Margariti, M., Tsakalof, A., and Tsatsakis, A., Analytical methods of biological monitoring for exposure to pesticides: recent update, *Therap. Drug Monitor.*, 29 (2), 150–163, 2007.
7. Marin, A. et al., Assessment of potential (inhalation and dermal) and actual exposure to acetamiprid by greenhouse applicators using liquid chromatography-tandem mass spectrometry, *J. Chromatogr. B*, 804 (2), 269–275, 2004.
8. Barr, D. and Needham, L., Analytical methods for biological monitoring of exposure to pesticides: a review. *J. Chromatogr. B*, 778 (1–2), 5–29, 2002.
9. Barnett, E. et al., *Pesticide Poisoning of Animals in 2005: Investigations of Suspected Incidents in the United Kingdom*, Defra, New York, 2005.
10. EU, Commission Directive 2003/13/EC of 10 February 2003 amending Directive 96/5/EC on processed cereal-based foods and baby foods for infants and young children, *Off. J. Eur. Commun.*, L139, 29–31, 2003.
11. Niessen, W., Advances in instrumentation in liquid chromatography mass spectrometry and related liquid-introduction techniques, *J. Chromatogr. A*, 794 (1–2), 407–435, 1998.
12. Slobodnik, J., VanBaar, B., and Brinkman, U., Column liquid-chromatography mass-spectrometry—selected techniques in environmental applications for polar pesticides and related-compounds, *J. Chromatogr. A*, 703 (1–2), 81–121, 1995.
13. Hiemstra, M. and de Kok, A., Comprehensive multi-residue method for target analysis of pesticides in crops using liquid-chromatography-tandem mass spectrometry, *J. Chromatogr. A*, 1154, 3–25, 2007.
14. Anastassiades, M. et al., Fast and easy multiresidue method employing acetonitrile extraction/partitioning and "dispersive solid-phase extraction" for the determination of pesticide residues in produce, *J. AOAC Int.*, 86 (2), 412–431, 2003.
15. Alder, L. et al., Residue analysis of 500 priority pesticides: better by GC-MS or LC-MS/MS? *Mass Spectrom. Rev.*, 25 (6), 838–865, 2006.
16. van der Hoff, G. and van Zoonen, P., Trace analysis of pesticides by gas chromatography, *J. Chromatogr. A*, 843 (1–2), 301–322, 1999.
17. Malik, A. and Faubel, W., Methods of analysis of dithiocarbamate pesticides: a review, *Pestic. Sci.*, 55 (10), 965–970, 1999.
18. Michel, M. and Buszewski, B., HPLC determination of pesticide residue isolated from food matrices, *J. Liq. Chromatogr. Relat. Technol.*, 25 (13–15), 2293–2306, 2002.
19. EU, *Quality Control Procedures for Pesticide Residues Analysis, Document No. SANCO/10232/2006*, Brussels, 2006.
20. Grob, K., Injection techniques in capillary GC, *Anal. Chem.*, 66 (20), 1009A–1019A, 1994.
21. Gross, G., Reid, V., and Synovec, R., Recent advances in instrumentation for gas chromatography, *Curr. Anal. Chem.*, 1 (2), 135–147, 2005.

22. Bailey, R., Injectors for capillary gas chromatography and their application to environmental analysis, *J. Environ. Monit.*, 7, 1054–1058, 2005.

23. Hajslova, J. and Zrostlikova, J., Matrix effects in (ultra)trace analysis of pesticide residues in food and biotic matrices, *J. Chromatogr. A*, 1000 (1–2), 181–197, 2003.

24. Mastovska, K., Lehotay, S., and Anastassiades, M., Combination of analyte protectants to overcome matrix effects in routine GC analysis of pesticide residues in food matrixes, *Anal. Chem.*, 77 (24), 8129–8137, 2005.

25. Garrido-Frenich, A. et al., Characterization of recovery profiles using gas chromatography-triple quadrupole mass spectrometry for the determination of pesticide residues in meat samples, *J. Chromatogr. A*, 1133 (1–2), 315–321, 2006.

26. Muller, H. and Stan, H., Pesticide-residue analysis in food with CGC—study of long term stability by the use of different injection techniques, *J. High. Resolut. Chromatogr.*, 13 (10), 697–701, 1990.

27. Grob, K., *Split and Splitless Injection in Capillary GC*, Wiley-VCH, New York, 2001.

28. Grob, K., *Split and Splitless Injection for Quantitative Gas Chromatography: Concepts, Processes, Practical Guidelines, Sources of Error*, 4th edn, Wiley-VCH, New York, 2001.

29. Hinshaw, J., Setting realistic expectations for GC optimization, *LC-GC Eur.*, 20 (3), 136–142, 2007.

30. Grob, K. and Biedermann, M., The two options for sample evaporation in hot GC injectors: thermospray and band formation. Optimization of conditions and injector design, *Anal. Chem.*, 74 (1), 10–16, 2002.

31. Wylie, P. et al., Using electronic pressure programming to reduce the decomposition of labile compounds during splitless injection, *J. High. Resolut. Chromatogr.*, 15 (11), 763–768, 1992.

32. Wylie, P., Improved gas chromatographic analysis of organophosphorus pesticides with pulsed splitless injection, *J. AOAC Int.*, 79 (2), 571–577, 1996.

33. Godula, M., Hajlova, J., and Alterova, K., Pulsed splitless injection and the extent of matrix effects in the analysis of pesticides, *J. High Resolut. Chromatogr.*, 22 (7), 395–402, 1999.

34. Vogt, W. et al., Capillary gas chromatographic injection system for large sample volumes, *J. Chromatogr.*, 186, 197–205, 1979.

35. Poy, F., Visani, F., and Terrosi, F., Automatic injection in high-resolution gas-chromatography—a programmed temperature vaporizer as a general-purpose injection system, *J. Chromatogr.*, 217, 81–90, 1981.

36. Stan, H.-J. and Müller, H., Evaluation of automated and manual hot-splitless, cold-splitless (PTV), and on-column injection technique using capillary gas chromatography for the analysis of organophosphorus pesticides, *J. High Resolut. Chromatogr.*, 11 (1), 140–143, 1988.

37. Godula, M. et al., Optimization and application of the PTV injector for the analysis of pesticide residues, *J. Sep. Sci.*, 24 (5), 355–366, 2001.

38. Hikker, J., McCabe, T., and Morabito, P., Optimization and application of the large volume on-column introduction (LOCI) technique for capillary GC with preliminary online capillary solvent distillation concentration, *J. High Resolut. Chromatogr.*, 16 (1), 5–12, 1993.

39. Teske, J. and Engewald, W., Methods for, and applications of, large-volume injection in capillary gas chromatography, *TrAC, Trends Anal. Chem.*, 21 (9–10), 584–593, 2002.

40. Muller, S., Efer, J., and Engewald, W., Water pollution screening by large-volume injection of aqueous samples and application to GC/MS analysis of a river Elbe sample, *Fresen. J. Anal. Chem.*, 357 (5), 558–560, 1997.

41. Lehotay, S., Analysis of pesticide residues in mixed fruit and vegetable extracts by direct sample introduction/gas chromatography/tandem mass spectrometry, *J. AOAC Int.*, 83, 680–697, 2000.

42. Patel, K. et al., Analysis of pesticide residues in lettuce by large volume-difficult matrix introduction-gas chromatography-time of flight-mass spectrometry (LV-DMI-GC-TOF-MS), *Analyst*, 128 (10), 1228–1231, 2003.

43. Beceiro-Gonzalez, E. et al., Optimisation and validation of a solid-phase micro-extraction method for simultaneous determination of different types of pesticides in water by gas chromatography-mass spectrometry, *J. Chromatogr. A*, 1141 (2), 165–173, 2007.

44. Sandra, P., Tienpont, B., and David, F., Multi-residue screening of pesticides in vegetables, fruits and baby food by stir bar sorptive extraction-thermal desorption-capillary gas chromatography-mass spectrometry, *J. Chromatogr. A*, 1000 (1–2), 299–309, 2003.

45. Donato, P. et al., Rapid analysis of food products by means of high speed gas chromatography, *J. Sep. Sci.*, 30 (4), 508–526, 2007.

46. Mastovska, K. and Lehotay, S., Practical approaches to fast gas chromatography-mass spectrometry, *J. Chromatogr. A*, 1000 (1–2), 153–180, 2003.

47. Hercegova, A. et al., Fast gas chromatography with solid phase extraction clean-up for ultratrace analysis of pesticide residues in baby food, *J. Chromatogr. A*, 1084 (1–2), 46–53, 2005.

48. Dalluge, J. et al., Resistively heated gas chromatography coupled to quadrupole mass spectrometry, *J. Sep. Sci.*, 25 (9), 608–614, 2002.

49. McNair, H. and Reed, G., Fast gas chromatography: the effect of fast temperature programming, *J. Microcolumn Sep.*, 12 (6), 351–355, 2000.

50. Mastovska, K. et al., Fast temperature programming in routine analysis of multiple pesticide residues in food matrices, *J. Chromatogr. A*, 907 (1–2), 235–245, 2001.

51. Mastovska, K., Hajslova, J., and Lehotay, S., Ruggedness and other performance characteristics of low-pressure gas chromatography-mass spectrometry for the fast analysis of multiple pesticide residues in food crops, *J. Chromatogr. A*, 1054 (1–2), 335–349, 2004.

52. Walorczyk, S. and Gnusowski, B., Fast and sensitive determination of pesticide residues in vegetables using low-pressure gas chromatography with a triple quadrupole mass spectrometer, *J. Chromatogr. A*, 1128 (1–2), 236–243, 2006.

53. Snow, N., Fast gas chromatography with short columns: are speed and resolution mutually exclusive? *J. Liq. Chromatogr. Relat. Technol.*, 27 (7–9), 1317–1330, 2004.

54. Harynuk, J. and Marriott, P., Fast GC × GC with short primary columns, *Anal. Chem.*, 78 (6), 2028–2034, 2006.

55. Adahchour, M. et al., Recent developments in comprehensive two-dimensional gas chromatography (GC × GC) I. Introduction and instrumental set-up, *TrAC, Trends Anal. Chem.*, 25 (5), 438–454, 2006.

56. Adahchour, M. et al., Recent developments in comprehensive two-dimensional gas chromatography (GC × GC) II. Modulation and detection, *TrAC, Trends Anal. Chem.*, 25 (6), 540–553, 2006.

57. Zrostlikova, J., Hajslova, J., and Cajka, T., Evaluation of two-dimensional gas chromatography-time-of-flight mass spectrometry for the determination of multiple pesticide residues in fruit, *J. Chromatogr. A*, 1019 (1–2), 173–186, 2003.

58. Barreda, M. et al., Residue determination of captan and folpet in vegetable samples by gas chromatography/negative chemical ionization-mass spectrometry, *J. AOAC Int.*, 89 (4), 1080–1087, 2006.

59. Anon, *Wiley Registry, 8th Edition/NIST 2005 Mass Spectral Library*, John Wiley & Sons, Hoboken, 2006.

60. Klaffenbach, P. and Stan, H., Automated screening of food and water samples for pesticide-residues by means of capillary GC-MSD using a library of pesticide spectra and macro programming, *J. High Resolut. Chromatogr.*, 14 (11), 754–756, 1981.

61. Wylie, P., Szelewski, M., and Meng, C.-K., Screening for 927 pesticides and endocrine disrupters in a single GCMS run using deconvolution reporting software with SIM and Scan databases, presented at PittCon 2007, Chicago, February 26–March 01, 2007.

62. March, R. and Todd, J., *Quadrupole Ion Trap Mass Spectrometry*, 2nd edn, John Wiley & Sons, Hoboken, 2005.

63. Hubschmann, H., The ion-trap detector—technology and applications, *Fresen. Z. Anal. Chem.*, 320 (7), 693–694, 1985.

64. Cairns, T. et al., Multiresidue pesticide analysis by ion-trap mass-spectrometry, *Rapid Commun. Mass Spectrom.*, 7 (11), 971–988, 1993.

65. Cleven, C. et al., Mass shifts due to ion–ion interactions in a quadrupole ion-trap mass spectrometer, *Rapid Commun. Mass Spectrom.*, 8 (6), 451–454, 1994.

66. Eichelberger, J., Budde, W., and Slivon, L., Existence of self chemical ionization in the ion trap detector, *Anal. Chem.*, 59 (22), 2730–2732, 1987.

67. Busch, K., Glish, G., and McLuckey, S., *Mass Spectrometry/Mass Spectrometry*, Wiley-VCH, New York, 1988.

68. Kotretsou, S. and Koutsodimou, A., Overview of the applications of tandem mass spectrometry (MS/MS) in food analysis of nutritionally harmful compounds, *Food Rev. Int.*, 22 (2), 125–172, 2006.

69. EU, Commission Decision of 12 August 2002 implementing Council Directive 96/23/EC concerning the performance of analytical methods and the interpretation of results (2002/657/EC), *Off. J. Eur. Commun.*, L221, 8–36, 2002.

70. Louris, J. et al., Instrumentation, applications, and energy deposition in quadrupole ion-trap tandem mass spectrometry, *Anal. Chem.*, 59 (13), 1677–1685, 1987.

71. Beguin, S. et al., Protocols for optimizing MS/MS parameters with an ion-trap GC-MS instrument, *J. Mass Spectrom.*, 41 (10), 1304–1314, 2006.

72. Johnson, J. and Yost, R., Tandem mass-spectrometry for trace analysis, *Anal. Chem.*, 57 (7), 758A–768A, 1985.

73. Dawson, P. et al., The use of triple quadrupoles for sequential mass-spectrometry. 1. The instrument parameters, *Org. Mass Spectrom.*, 17 (5), 205–211, 1982.

74. Sheehan, T., GC-MS-MS: the next logical step in benchtop GC-MS, *Am. Lab.*, 28 (17), V28, 1996.

75. Guilhaus, M., Mlynski, V., and Selby, D., Perfect timing: time-of-flight mass spectrometry, *Rapid Commun. Mass Spectrom.*, 11 (9), 951–962, 1997.

76. Dawson, J. and Guilhaus, M., Orthogonal-acceleration time-of-flight mass spectrometer, *Rapid Commun. Mass Spectrom.*, 3 (5), 155–159, 1989.

77. Cajka, T. and Hajslova, J., Gas chromatography-high-resolution time-of-flight mass spectrometry in pesticide residue analysis: advantages and limitations, *J. Chromatogr. A*, 1058 (1–2), 251–261, 2004.

78. Cajka, T. and Hajslova, J., Gas chromatography-time-of-flight mass spectrometry in food analysis, *LC-GC Eur.*, 20 (1), 25–30, 2007.

79. Davis, S., Makarov, A., and Hughes, J., Ultrafast gas chromatography using time-of-flight mass spectrometry, *Rapid Commun. Mass Spectrom.*, 13 (4), 237–241, 1999.

80. Zrostlikova, J., Hajslova, J., and Cajka, T., Evaluation of two-dimensional gas chromatography-time-of-flight mass spectrometry for the determination of multiple pesticide residues in fruit, *J. Chromatogr. A*, 1019 (1–2), 173–186, 2003.

81. Veriotti, T. and Sacks, R., Characterization and quantitative analysis with GC/TOFMS comparing enhanced separation with tandem-column stop-flow GC and spectral deconvolution of overlapping peaks, *Anal. Chem.*, 75 (16), 4211–4216, 2003.

82. Dalluge, J., Roose, P., and Brinkman, U., Evaluation of a high-resolution time-of-flight mass spectrometer for the gas chromatographic determination of selected environmental contaminants, *J. Chromatogr. A*, 970 (1–2), 213–223, 2002.

83. Ingelse, B. et al., Determination of polar organophosphorus pesticides in aqueous samples by direct injection using liquid chromatography-tandem mass spectrometry, *J. Chromatogr. A*, 918 (1), 67–78, 2001.

84. Pizzutti, I. et al., Method validation for the analysis of 169 pesticides in soya grain, without clean up, by liquid chromatography-tandem mass spectrometry using positive and negative electrospray ionization, *J. Chromatogr. A*, 1142 (2), 123–136, 2007.

85. Hernandez, F. et al., Rapid direct determination of pesticides and metabolites in environmental water samples at sub-mu g/l level by on-line solid-phase extraction-liquid chromatography-electrospray tandem mass spectrometry, *J. Chromatogr. A*, 939 (1–2), 1–11, 2001.

86. Aulakh, J. et al., A review on solid phase micro extraction-high performance liquid chromatography (SPME-HPLC) analysis of pesticides, *Crit. Rev. Anal. Chem.*, 35 (1), 71–85, 2005.

87. Wu, J. et al., Analysis of polar pesticides in water and wine samples by automated in-tube solid-phase microextraction coupled with high-performance liquid chromatography-mass spectrometry, *J. Chromatogr. A*, 976 (1–2), 357–367, 2002.

88. Blasco, C. et al., Comparison of solid-phase microextraction and stir bar sorptive extraction for determining six organophosphorus insecticides in honey by liquid chromatography-mass spectrometry, *J. Chromatogr. A*, 1030 (1–2), 77–85, 2004.

89. Gao, S., Zhang, Z., and Karnes, H., Sensitivity enhancement in liquid chromatography/atmospheric pressure ionization mass spectrometry using derivatization and mobile phase additives, *J. Chromatogr. B*, 825 (2), 98–110, 2005.

90. Schaefer, W. and Dixon, F., Effect of high-performance liquid chromatography mobile phase components on sensitivity in negative atmospheric pressure chemical ionization liquid chromatography mass spectrometry, *J. Am. Soc. Mass. Spectrom.*, 7 (10), 1059–1069, 1996.

91. Willoughby, R., Sheehnan, E., and Mitrovich, S., *A Global View of LC/MS, How to Solve Your Most Challenging Analytical Problems*, 2nd edn, Global View Publishing, Pittsburgh, 2002, p. 417.

92. Kovalczuk, T. et al., Ultra-performance liquid chromatography-tandem mass spectrometry: a novel challenge in multiresidue pesticide analysis in food. *Anal. Chim. Acta*, 577 (1), 8–17, 2006.

93. Hernandez, F. et al., Rapid determination of fosetyl-aluminum residues in lettuce by liquid chromatography/electrospray tandem mass spectrometry, *J. AOAC Int.*, 86 (4), 832–838, 2003.

94. Marr, J. and King, J., A simple high performance liquid chromatography ionspray tandem mass spectrometry method for the direct determination of paraquat and diquat in water, *Rapid Commun. Mass Spectrom.*, 11, 479–483, 1997.

95. Bluhm, L. and Li, T., The role of analogue ions in the ion-pair reversed-phase chromatography of quaternary ammonium compounds, *J. Chromatogr. Sci.*, 41 (1), 6–9, 2003.

96. Hemstrom, P. and Irgum, K., Hydrophilic interaction chromatography, *J. Sep. Sci.*, 29 (12), 1784–1821, 2006.

97. Young, M. and Jenkins, K., Oasis® WCX: a novel mixed-mode SPE sorbent for LC–MS. Determination of paraquat and other quaternary ammonium compounds. Recent applications in LC–MS, *LC-GC Eur.*, 17 (11a), 51–52, 2004.

98. Startin, J. et al., Determination of residues of the plant growth regulator chlormequat in pears by ion-exchange high performance liquid chromatography-electrospray mass spectrometry, *Analyst*, 124 (7), 1011–1015, 1999.

99. Granby, K., Johannesen, S., and Vahl, M., Analysis of glyphosate residues in cereals using liquid chromatography-mass spectrometry (LC-MS/MS), *Food Addit. Contam.*, 20 (8), 692–698, 2003.

100. Nguyen, D. et al., Chromatographic behaviour and comparison of column packed with sub-2 mu m stationary phases in liquid chromatography, *J. Chromatogr. A*, 1128 (1–2), 105–113, 2006.

101. Swartz, M. and Murphy, B., Ultra performance liquid chromatography: tomorrow's HPLC technology today, *LabPlus Int.*, 18 (3), 6–9, 2004.

102. Wu, N. and Thompson, R., Fast and efficient separations using reversed phase liquid chromatography, *J. Liq. Chromatogr. Relat. Technol.*, 29 (7–8), 949–988, 2006.

103. Churchwell, M. et al., Improving LC-MS sensitivity through increases in chromatographic performance: comparisons of UPLC-ES/MS/MS to HPLC-ES/MS/MS, *J. Chromatogr. B*, 825 (2), 134–143, 2005.

104. Mezcua, M. et al., Application of ultra performance liquid chromatography-tandem mass spectrometry to the analysis of priority pesticides in groundwater, *J. Chromatogr. A*, 1109 (2), 222–227, 2006.

105. Leandro, C. et al., Ultra-performance liquid chromatography for the determination of pesticide residues in foods by tandem quadrupole mass spectrometry with polarity switching, *J. Chromatogr. A*, 1144 (2), 161–169, 2007.

106. Creaser, C. and Stygall, J., Particle beam liquid chromatography-mass spectrometry: instrumentation and applications, *Analyst*, 118 (12), 1467–1480, 1993.

107. Blakely, C. and Vestal, M., Thermospray interface for liquid chromatography/mass spectrometry, *Anal. Chem.*, 55 (4), 750–755, 1983.

108. Yergey, A. et al., *Liquid Chromatography/Mass Spectrometry Techniques and Applications*, Plenum Publ. Co., New York, 1989.

109. Abian, J., Duran, G., and Barcelo, D., Analysis of chloroatrazines and their degradation products in environmental samples by selected various operation modes in thermospray HPLC/MS/MS, *J. Agric. Food Chem.*, 41, 1264–1273, 1993.

110. Yamshita, A. and Fenn, J., Electrospray ion-source. Another variation on the free-jet theme, *J. Phys. Chem.*, 88 (20), 4451–4459, 1984.

111. Horning, E. et al., New picogram detection system based on a mass spectrometer with an external ionization source at atmospheric pressure, *Anal. Chem.*, 45 (6), 936–943, 1973.

112. Bos, S., van Leeuwen, S., and Karst, U., From fundamentals to applications: recent developments in atmospheric pressure photoionization mass spectrometry, *Anal. Bioanal. Chem.*, 384 (1), 85–99, 2006.

113. Cappiello, A. and Palma, P., Electron ionization in LC-MS: a technical overview of the direct EI interface, in *Advances in LC-MS Instrumentation, J. Chromatogr.*, Library, Volume 72, Cappiello, A., Ed., Elsevier, Oxford, 2007, chap. 3.

114. Granot, O. and Amirav, A., Electron ionization LC-MS with supersonic molecular beams, in *Advances in LC-MS Instrumentation, J. Chromatogr.*, Library, Volume 72, Cappiello, A., Ed., Elsevier, Oxford, 2007, chap. 4.

115. Cech, N. and Enke, C., Practical implications of some recent studies in electrospray ionization fundamentals, *Mass Spectrom. Rev.*, 20 (6), 362–387, 2001.

116. Manisali, I., Chen, D., and Schneider, B., Electrospray ionization source geometry for mass spectrometry: past, present, and future, *TrAC, Trends Anal. Chem.*, 25 (3), 243–256, 2006.

117. Cole, R., *Electrospray Ionization Mass Spectrometry, Fundamentals, Instrumentation & Applications*, Wiley-VCH, New York, 1997.

118. Bruins, A., Mass-spectrometry with ion sources operating at atmospheric pressure, *Mass Spectrom. Rev.*, 10 (1), 53–77, 1991.

119. Munson, B., Development of chemical ionization mass spectrometry, *Int. J. Mass Spectrom.*, 200 (1–3), 243–251, 2000.

120. Carroll, D. et al., Atmospheric-pressure ionization mass-spectrometry, *Appl. Spectrosc. Rev.*, 17 (3), 337–406, 1981.

121. Willoughby, R., Sheehan, E., and Mitrovich, S., *A Global View of LC/MS, How to Solve Your Most Challenging Analytical Problems*, 2nd edn, Global View Publishing, Pittsburgh, 2002, p. 66.

122. Kebarle, P., A brief overview of the present status of the mechanisms involved in electrospray mass spectrometry, *J. Mass Spectrom.*, 35 (7), 804–817, 2000.

123. Rohner, T., Lion, N., and Girault, H., Electrochemical and theoretical aspects of electrospray ionisation, *Phys. Chem. Chem. Phys.*, 6 (12), 3056–3068, 2004.

124. Balogh, M., Ionization revisited, *LC-GC N. Am.*, 24 (12), 1284–1288, 2006.

125. Cole, R., Some tenets pertaining to electrospray ionization mass spectrometry, *J. Mass Spectrom.*, 35 (7), 763–772, 2000.

126. Bruins, A., Covey, T., and Henion, J., Ion spray interface for combined liquid chromatography/atmospheric pressure ionisation mass spectrometry, *Anal. Chem.*, 59 (22), 2642–2646, 1987.

127. Carroll, D. et al., Atmospheric pressure ionization mass spectrometry. Corona discharge ion source for use in a liquid chromatograph-mass spectrometer-computer analytical system, *Anal. Chem.*, 47 (14), 2369–2373, 1975.

128. Whitehouse, C. et al., Electrospray interface for liquid chromatographs and mass spectrometers, *Anal. Chem.*, 57 (3), 675–679, 1985.

129. Sunner, J., Nicol, G., and Kebarle, P., Factors determining relative sensitivity of analytes in positive mode atmospheric-pressure ionization mass-spectrometry, *Anal. Chem.*, 60 (13), 1300–1307, 1998.

130. Raffaelli, A., Atmospheric pressure chemical ionization (APCI): new avenues for an old friend, in *Advances in LC-MS Instrumentation, J. Chromatogr.*, Library, Volume 72, in *Advances in LC-MS Instrumentation, J. Chromatogr.*, Library, Volume 72, Cappiello, A., Ed., Elsevier, Oxford, 2007, chap. 2.

131. Thurman, E., Ferrer, I., and Barcelo, D., Choosing between atmospheric pressure chemical ionization and electrospray ionization interfaces for the HPLC/MS analysis of pesticides, *Anal. Chem.*, 73 (22), 5441–5449, 2001.

132. Doerge, D. and Bajic, S., Analysis of pesticides using liquid-chromatography atmospheric-pressure chemical ionization mass-spectrometry, *Rapid Commun. Mass Spectrom.*, 6 (11), 663–666, 1992.

133. Balogh, M., A case for congruent multiple ionization modes in atmospheric pressure ionization mass spectrometry, in *Advances in LC-MS Instrumentation, J. Chromatogr.*, Library, Volume 72, Cappiello, A., Ed., Elsevier, Oxford, 2007, chap. 5.

134. Bruins, A., Covey, T., and Henion, J., Ion spray interface for combined liquid chromatography/atmospheric pressure ionisation mass spectrometry, *Anal. Chem.*, 59 (22), 2642–2646, 1987.

135. Hajslova, J. and Zrostlikova, J., Matrix effects in (ultra)trace analysis of pesticide residues in food and biotic matrices, *J. Chromatogr. A*, 1000 (1–2), 181–197, 2003.

136. Niessen, W., Manini, P., and Andreoli, R., Matrix effects in quantitative pesticide analysis using liquid chromatography-mass spectrometry, *Mass Spectrom. Rev.*, 25 (6), 881–899, 2006.

137. King, R. et al., Mechanistic investigation of ionization suppression in electrospray ionization, *J. Am. Soc. Mass Spectrom.*, 11 (11), 942–950, 2000.

138. Pico, Y., Blasco, C., and Font, G., Environmental and food applications of LC-tandem mass spectrometry in pesticide-residue analysis: an overview, *Mass Spectrom. Rev.*, 23 (1), 45–85, 2004.

139. Kawasaki, S. et al., Simple, rapid and simultaneous measurement of 8 different types of carbamate pesticides in serum using liquid-chromatography atmospheric-pressure chemical-ionization mass-spectrometry, *J. Chromatogr. Biomed.*, 620 (1), 61–71, 1993.

140. Andreu, V. and Pico, Y., Liquid chromatography-ion trap-mass spectrometry and its application to determine organic contaminants in the environment and food, *Curr. Anal. Chem.*, 1 (3), 241–265, 2005.

141. Bristow, A. et al., Reproducible product-ion tandem mass spectra on various liquid chromatography/mass spectrometry instruments for the development of spectral libraries, *Rapid Commun. Mass Spectrom.*, 18 (13), 1447–1454, 2004.

142. Larsen, B., Ion-trap multiple mass spectrometry in pesticide analysis, *Analysis*, 28 (10), 941–946, 2000.

143. Blasco, C., Font, G., and Pico, Y., Multiple-stage mass spectrometric analysis of six pesticides in oranges by liquid chromatography-atmospheric pressure chemical ionization-ion trap mass spectrometry, *J. Chromatogr. A*, 1043 (2), 231–238, 2004.

144. Soler, C., Manes, J., and Pico, Y., Comparison of liquid chromatography using triple quadrupole and quadrupole ion trap mass analyzers to determine pesticide residues in oranges, *J. Chromatogr. A*, 1067, 115–125, 2005.

145. Douglas, D., Frank, A., and Mao, D., Linear ion traps in mass spectrometry, *Mass Spectrom. Rev.*, 24 (1), 1–29, 2005.

146. Kuster, M., de Alda, M.L., and Barcelo, D., Analysis of pesticides in water by liquid chromatography-tandem mass spectrometric techniques, *Mass Spectrom. Rev.*, 25 (6), 900–916, 2006.

147. Pico, Y. et al., Control of pesticide residues by liquid chromatography-mass spectrometry to ensure food safety, *Mass Spectrom. Rev.*, 25 (6), 917–960, 2006.

148. Hernandez, F., Sancho, J., and Pozo, O., Critical review of the application of liquid chromatography/mass spectrometry to the determination of pesticide residues in biological samples, *Anal. Bioanal. Chem.*, 382 (4), 934–946, 2005.

149. Soler, C. and Pico, Y., Recent trends in liquid chromatography-tandem mass spectrometry to determine pesticides and their metabolites in food, *TrAC, Trends Anal. Chem.*, 26 (2), 103–115, 2007.

150. Hager, J. and Yves Le Blanc, J., Product ion scanning using a Q-q-Q linear ion trap (Q TRAP) mass spectrometer, *Rapid Commun. Mass Spectrom.*, 17 (10), 1056–1064, 2003.

151. Dahlmann, J., Galvin, B., and Kuracina, M., Advances in linear ion trap technology for food contaminant analysis, presented at EPRW, Corfu, May 21–25, 2006.

152. Blake, D. et al., Detection and confirmation of unknown contaminants in untreated tap water using a hybrid triple quadrupole linear ion trap LC/MS/MS system, presented at IMSC, Prague, August 27–September 1, 2006.

153. Ferrer, I. et al., Exact-mass library for pesticides using a molecular-feature database, *Rapid Commun. Mass Spectrom.*, 20 (24), 3659–3668, 2006.

154. Startin, J., Hird, S., and Sykes, M., Determination of ethylenethiourea (ETU) and propylenethiourea (PTU) in foods by high performance liquid chromatography-atmospheric pressure chemical ionisation-medium-resolution mass spectrometry, *Food Addit. Contam.*, 22 (3), 245–250, 2005.

155. Hollender, J., Singer, H., and Fenner, K., Opportunities and limits of the combination of linear ion trap with Orbitrap analyzer to detect and identify contaminants in environmental water samples, presented at SGMS Annual Meeting, Interlaken, November 2–3, 2006.

156. Mol, H. et al., Comprehensive pesticide residue analysis: taking advantage of advanced mass spectrometric detection, presented at EPRW, Corfu, May 21–25, 2006.

157. Lacorte, S. and Fernandez-Albaz, A., Time of flight mass spectrometry applied to the liquid chromatographic analysis of pesticides in water and food, *Mass Spectrom. Rev.*, 25 (6), 866–880, 2006.

158. Sancho, J. et al., Potential of liquid chromatography/time-of-flight mass spectrometry for the determination of pesticides and transformation products in water, *Anal. Bioanal. Chem.*, 386 (4), 987–997, 2006.

159. Hernandez, F. et al., Strategies for quantification and confirmation of multi-class polar pesticides and transformation products in water by LC-MS2 using triple quadrupole and hybrid quadrupole time-of-flight analyzers, *TrAC, Trends Anal. Chem.*, 24 (7), 596–612, 2005.

160. Ferrer, I. et al., Multi-residue pesticide analysis in fruits and vegetables by liquid chromatography-time-of-flight mass spectrometry, *J. Chromatogr. A*, 1082 (1), 81–90, 2005.

161. Thurman, E. et al., Feasibility of LC/TOFMS and elemental database searching as a spectral library for pesticides in food, *Food Addit. Contam.*, 23 (11), 1169–1178, 2006.

162. Thurman, E. et al., Discovering metabolites of post-harvest fungicides in citrus with liquid chromatography/time-of-flight mass spectrometry and ion trap tandem mass spectrometry, *J. Chromatogr. A*, 1082 (1), 71–80, 2005.

163. Petrovic, M. and Barcelo, D., Application of liquid chromatography/quadrupole time-of-flight mass spectrometry (LC-QqTOF-MS) in the environmental analysis, *J. Mass Spectrom.*, 41 (10), 1259–1267, 2006.

4 Immunoassays and Biosensors

Jeanette M. Van Emon, Jane C. Chuang,
Kilian Dill, and Guohua Xiong

CONTENTS

4.1 INTRODUCTION

Monitoring and exposure data are critical to accurately determine the impact of pesticides and environmental contaminants on human health [1]. This is especially true for infants and young children, as well as the elderly and those with compromised immune systems. Uncertainties in the assessment of human exposures to exogenous compounds may be reduced using data obtained from dietary and environmental

Notice: The U.S. Environmental Protection Agency (EPA), through its Office of Research and Development, funded and collaborated in the research described here under Contracts 68-D99-011 and EP-D04-068 to Battelle. It has been subjected to agency review and approved for publication. Mention of trade names or commercial products does not constitute endorsement or recommendation for use.

monitoring measurement studies. Faster and more cost-effective analytical methods can facilitate the collection of data concerning particular target analytes that may impact human health and the environment. Immunoassays and biosensors can provide fast, reliable, and cost-effective monitoring and measurement methods [2].

In 1993, the United States National Academy of Sciences (NAS) issued a major report on pesticides in the diet of children. The report, "Pesticides in the Diets of Infants and Children" [3] recommended that U.S. pesticide laws be revised to make foods safer for children. The Food Quality Protection Act [4] of 1996 was passed in response to the Academy's report. The FQPA is predicated on the need to reduce exposure to pesticides in foods particularly for vulnerable groups. The purpose of the FQPA is to eliminate high-risk pesticide uses, not to eliminate pesticide use entirely. The Academy report recommended that pesticide residue monitoring programs target foods often consumed by children, and that analytical testing methods be standardized, validated, and subjected to strict quality control and quality assurance programs [3].

The FQPA requires the U.S. Environmental Protection Agency (EPA) to look at all routes and sources (i.e., food, air, water, pets, indoor environments) when setting limits on the amount of pesticides that can remain in food. Based on these requirements and the recommendations in the Academy's report, there are major analytical challenges to fully implement the FQPA. Dietary and nondietary exposures must now be considered in an integrated manner. This aggregate exposure approach clearly requires cost-effective analytical methods for a variety of analytes in different matrices.

Immunoassay detection methods were initially developed for clinical applications where their sensitivity and selectivity provided improvements in diagnostic capabilities. Clinical chemists developed highly successful methods for medical and health-care applications by leveraging the sensitivity and selectivity of the specific antibody interaction with large target analytes such as drugs, hormones, bacteria, and toxins. Pesticide residue chemists recognized the potential of immunochemical technology for small molecule detection in the 1970s [5]. Since that time, immunoassays have been successfully adapted for the analysis of a wide range of pesticides [6] and other potential environmental contaminants including PCBs, PAHs, dioxins, and metals [7–10].

Immunoassay methods range from high sample throughput methods, providing cost-effective analytical detection for large-scale monitoring studies [11], to self-contained rapid testing formats. Immunoassays can provide rapid screening information or quantitative data to fulfill stringent data quality requirements. These methods have been used for the selective analyses of many compounds of environmental and human health concern. For water-soluble pesticides or compounds with low volatility, immunoassays can be faster, less expensive, and significantly more sensitive and reproducible than many other analytical procedures.

Biosensor technology also had its genesis in clinical applications. Medical diagnostic sensors designed for point-of-care use are small, portable devices, easy-to-use, and give rapid, quantitative results. These attributes are also important for unattended remote sensing of environmental contaminants and for monitoring pesticides and pesticide biomarkers [12]. Several pesticide biosensors have been reported for various monitoring situations [13–17].

4.2 IMMUNOASSAYS

All immunochemical methods are based on selective antibodies combining with a particular target analyte or analyte group. The selective binding between an antibody and a pesticide analyte has been used to analyze a variety of sample matrices for pesticide residues. Methods range from the determination of pesticide dislodgeable foliar residues on crops to monitoring dietary consumption, dust and soil exposures, and determining pesticide biomarkers in urine [18,19].

4.2.1 GENERAL OVERVIEW FOR IMMUNOASSAYS

Immunoassays have been routinely used in medical and clinical settings for the quantitative determination of proteins, hormones, and drugs with a molecular mass of several thousand Daltons (Da). Immunoassay techniques including the enzyme-linked immunosorbent assay (ELISA) have also proven useful for environmental monitoring and human observational monitoring studies [6,19]. Common environmental pollutants (i.e., pesticides) are typically small molecules with a molecular mass of <1000 Da. This small size will not elicit antibody production. Small molecules (haptens) can be used for antibody production when conjugated to carrier molecules such as proteins. The small molecule of interest is usually modified to introduce a chemical moiety capable of covalent binding. The small molecule, or hapten, is then converted to an immunogenic substance through conjugation to the carrier molecule for antibody production. The design of a hapten greatly affects the selectivity and sensitivity of the resulting antibody. The distinguishing features of the small molecule must be preserved while introducing an additional chemical group (i.e., –COOH, –OH, –SH, –NH_2) and linker chain or spacer arm for binding [5]. Hapten design, hapten synthesis, and antibody production are among the critical initial steps in developing immunoassays for small environmental pollutants.

A stepwise diagram for an ELISA is shown in Figure 4.1. This format is based on the immobilization of an antigen (i.e., the target analyte hapten conjugated to a

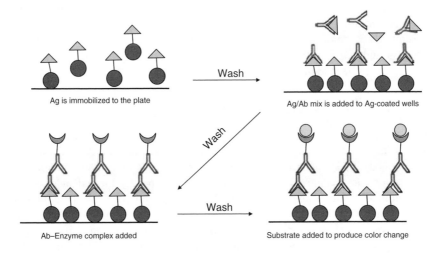

FIGURE 4.1 Indirect competitive ELISA.

protein) to a solid-phase support such as a test tube or a 96-well microtiter plate [20]. The sample extract for a microplate format (in a water-soluble solvent) and a solution of specific antibody (typically in phosphate-buffered saline [PBS] pH 7.4 containing 0.5% Tween 20) are added to the antigen-sensitized wells. The target analyte in solution and the immobilized antigen compete for binding sites on the specific antibody. The wells are rinsed with buffer to remove antibody not bound to the solid-phase antigen. The amount of antibody that can bind to the immobilized antigen on the plate is inversely related to the amount of analyte in the sample. A secondary antibody (species-specific that binds to the primary antibody) labeled with an enzyme (antibody-enzyme conjugate) is added to help visualize the presence of the bound primary antibody. Alkaline phosphatase and horseradish peroxidase are two commonly used enzyme labels. Another buffer rinse removes unbound excess enzyme-labeled secondary antibody. The addition of a chromogenic substrate produces a colored end product that can be measured spectrophotometrically or kinetically for quantitation of analyte. This indirect competitive format is useful to support large observational studies due to its high sample throughput, adaptation to automation, availability of commercial labels and substrates, and the high-performance level that can be achieved. For extremely high sample throughput capability, microtiter plates containing 384 microwells can be used. In-depth details on how to develop antibodies and immunoassays, as well as data analysis are presented by Van Emon [2].

There are several permutations to the basic indirect competitive ELISA. Figure 4.2 depicts an immunoassay format using immobilized antibody and an enzyme-labeled tracer [21]. Analyte in the sample competes with a known amount of enzyme-labeled analyte for binding sites on the immobilized antibody. In the initial step, the antianalyte antibody is adsorbed to the side of a test tube or microtiter plate well. The analyte and an enzyme-labeled analyte are next added to the antibody-coated wells and competition for antibody binding occurs. After an incubation step, all unbound reagents are rinsed from the wells. Substrate is added for color development that is inversely related to the concentration of analyte present in the sample. This particular format is commonly used in immunoassay testing kits as a few procedural steps are eliminated. However, this format does not have the convenience of commercially available reagents (i.e., enzyme-labeled secondary antibody) and requires the synthesis or labeling of either the analyte or hapten which may not be straightforward.

4.2.2 METHOD DEVELOPMENT

The development of an immunoassay method closely parallels the steps necessary for an instrumental analysis. A critical step is presenting the analyte to the detector (e.g., antibody, mass spectrometer, electron capture, flame ionization) in a form that the detector can recognize. A major difference is typically the extent of sample preparation required for an immunoassay. Frequently, immunoassays do not require the same amount of sample cleanup as an instrumental method, providing savings in time and costs. Many methods have reported simply using a dilution series to remove interfering matrix substances [22,23]. Solid-phase extraction (SPE) can be used for

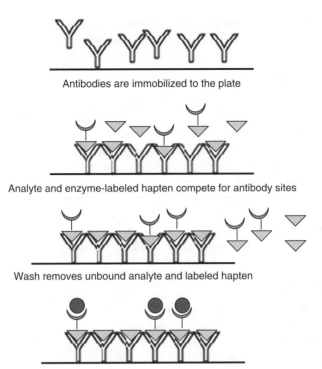

Antibodies are immobilized to the plate

Analyte and enzyme-labeled hapten compete for antibody sites

Wash removes unbound analyte and labeled hapten

Substrate is added for color detection

FIGURE 4.2 Direct competitive ELISA.

either unprocessed samples or in tandem with accelerated solvent extraction (ASE) methods [24–28]. Key to successful methods development is presenting the analyte to the antibody in a manner that is compatible with antibody function. As antibodies prefer an aqueous medium, the sample extract must be soluble in the buffer in which the immunoassay is performed.

Organic solvents, insoluble or miscible in water, can be used for the initial extraction, provided extracts are exchanged into a compatible solvent such as methanol or acetonitrile prior to ELISA. Methanol is one commonly used extraction solvent for ELISA detection. Other organic solvents such as acetone, acetonitrile, dichloromethane (DCM), or hexane can be used as an extraction solvent; however, a solvent-exchange step into an assay-friendly solvent is necessary. The tolerance of organic solvents must be determined in each specific method as it is dependent on the immunoreagents employed. For complex sample matrices such as soil, sediment, and fatty foods, extraction techniques and cleanup procedures may be required before ELISA detection. The extraction techniques employed in instrumental methods including shaking, sonication, supercritical fluid extraction (SFE), ASE, or SPE have also been used for ELISA methods. The shaking method is common for field applications. However, the shaking method may not provide adequate extraction efficiency depending on the shaking time, analyte, and sample matrix [29]. The efficiency and reproducibility should be evaluated and documented for any extraction techniques before application to field samples. This can be accomplished through recoveries of target analytes from fortified samples.

4.2.3 ELISA METHODS FOR PESTICIDES

ELISA is a common format that has been reported in the literature for determining pesticides and their metabolites in foods, as well as environmental and biological sample matrices [2,5,23,28,30–49]. These pesticides include organochlorine (OC) and organophosphorus (OP) compounds, carbamates, sulfonylurea pyrethroids, and many herbicides. Depending on the specificity of the antibody and the design of the hapten, ELISA methods can be very selective for a specific target pesticide and used for quantitative measurements. Other methods employing less selective antibodies, having a high cross-reactivity for structurally similar pesticides, can be used as qualitative monitoring tools or to develop exposure equivalency indices. Tables 4.1 and 4.2 summarize some of the ELISA methods developed for foods as well as environmental and biological samples.

Assay performance must be demonstrated before applying the ELISA method to field or study samples. For laboratory-based ELISA methods, immunoreagents such as antibodies and coating antigens may only be available from the source laboratories while enzyme conjugates and substrates are commercially available. Generally, the protocols provided by the source laboratories should be used as starting points for determining optimal concentrations of immunoreagents for the particular analysis. Checkerboard titrations can be performed to determine the optimal concentrations of the antibodies and coating antigens. Whenever new lots of immunoreagents are used, they should be examined for their performance with previously used reagents. Protocols provided with commercial testing kits should be followed in the specified manner and reagents used within the expiration date. Most ELISA methods can offer comparable or better analytical precision (e.g., within ±20%) and accuracy (e.g., greater than ±80% of expected value) as conventional instrument methods for analyzing pesticides. Calibration curves based on standard solutions must reflect the composition of the sample extract. Standards should be prepared in the same buffer/solvent solution as the samples. Ideally, the standards should also include the same amount of matrix as the samples. This is particularly important when sample dilution is used as the cleanup. For example, if a food extract contains 20% orange juice the standards should also contain 20% orange juice (analyte-free before spiking). When assay performance is extremely well-documented as to the extent of the matrix effect, the matrix may be omitted and a conversion factor applied to the buffer standard curve to account for the matrix in the sample.

Recently, a laboratory-based ELISA method was adapted to determine 3-phenoxy benzoic acid (3-PBA) in human urine samples collected in subsets from two observational field studies. 3-PBA is a common urinary metabolite for several pyrethroid pesticides (cypermethrin, cyfluthrin, deltamethrin, esfenvalerate, permethrin) that contain the phenoxybenzyl group. The anti-PBA antibody had negligible cross-reactivity toward the parent pyrethroids but also recognized and reacted with 4-fluoro-3-PBA (FPBA). The cross-reactivity to the structurally similar FPBA was 72%

TABLE 4.1

Examples of ELISA Methods for Determining Pesticides and Metabolites in Foods

Analyte	Food Matrix	Assay Format	LOD	References
2,4-D	Apple, grape, potato, orange, peach	Magnetic particle, DC ELISA	5 ppb	[34]
Acephate	Analyte-fortified tap water, mulberry leaves, lettuce	IC ELISA	2 ng/mL	[39]
Acetamiprid	Fruits, vegetables	DC ELISA	0.053 ng/g	[46]
Alachlor, carbofuran, atrazine, benomyl, 2,4-D	Beef liver, beef	Magnetic particle DC ELISA (per each analyte)	1–14 ppb	[33]
Atrazine	Extra virgin olive oil	Plate DC and DC sensor ELISA	0.7 ng/mL	[50]
Azoxystrobin	Grape extract	ELISA, FPIA, TR-FIA	3 pg/mL (ELISA) 36 pg/mL (PFIA) 28 pg/mL (TR-FIA)	[51]
Carbaryl (1-naphthyl methyl carbamate)	Apple, Chinese cabbage, rice, barley	Test tube, ELISA	0.7 ng/g	[15]
Carbaryl, endosulfan	Rice, oat, carrot, green pepper	Flow-through and lateral-flow, membrane-based gold particles	10–100 ng/mL	[52]
Chlorpyrifos	Fruits and vegetables	DC ELISA	0.32 ng/mL	[45]
Chlorpyrifos	Olive oil	Microtiter plate IC ELISA	0.3 ng/mL	[42]
DDT and metabolites	Drinking water, various foods	ELISA-CL	0.06 ng/mL (DDT) 0.04 ng/mL (metabolites)	[37]

(continued)

TABLE 4.1 (continued)
Examples of ELISA Methods for Determining Pesticides and Metabolites in Foods

Analyte	Food Matrix	Assay Format	LOD	References
Difenzoquat	Beer, cereal, bread	IC ELISA	0.8 ng/mL (beer); 16.0 ng/g (cereals)	[35]
Fenazaquin	Apple and pear	IC ELISA	8 ng/mL	[40]
Fenitrothion	Apple and peach	DC ELISA microtiter plate	20.0 ng/g	[47]
Fenthion	Vegetable samples	Microtiter plate DC ELISA and dipstick ELISA	0.1 ng/mL (plate); 0.5 ng/mL (dipstick)	[53]
Imidacloprid	Fortified water samples	Microtiter plate IC ELISA	0.5 ng/mL	[54]
Imidacloprid	Fruit juices	Microtiter DC ELISA	5–20 ng/mL	[49]
Iprodione	Apple, cucumber, eggplant	Microtiter plate DC ELISA	0.3 ng/g	[48]
Isofenphos	Fortified rice and lettuce	IC ELISA	5.8 ng/mL	[55]
Methyl parathion and parathion	Water and several food matrices	DC ELISA	0.05 ng/mL (methyl parathion); 0.5 ng/mL (parathion)	[56]
Methyl parathion	Vegetable, fruit	IC and DC ELISA; FPIA	IC: 0.08 ng/mL; DC: 0.5 ng/mL; FPIA: 15 ng/mL	[41]
Pirimiphos-methyl	Spiked grains	IC ELISA	0.07 ng/mL	[57]
Tebufenozide	Red and white wine	DC ELISA	10 ng/mL	[58]

CL, Chemiluminescence; DC, direct competitive; IC, indirect competitive; PFIA, fluorescence polarization immunoassay; TR-FIA, time-resolved fluorescence immunoassay; ELISA, enzyme-linked immunosorbent assay.

TABLE 4.2
Examples of ELISA Methods for Determining Pesticides and Metabolites in Biological and Environmental Samples

Analyte	Sample Matrix	Assay Format	LOD	References
2,4-D	Urine	Microtiter plate IC ELISA	30 ng/mL in urine	[23]
3,5,6-TCP	Urine	Microtiter plate IC ELISA	1 ng/mL in urine	[38]
3,5,6-TCP	Dust, soil	Magnetic particle DC ELISA	0.25 ng/mL in assay buffer	[38]
4-Nitrophenol parathion	Soil	Microtiter plate IC ELISA	0.2–1 ng/mL buffer	[25]
Atrazine mercapturic acid	Urine	Microtiter plate IC ELISA	0.05–0.3 ng/mL in urine	[22,28]
DDE	Soil	Microtiter plate IC ELISA	IC_{50} = 20 ng/mL	[59]
Glycine conjugate of cis/trans-DCCA	Urine	Microtiter plate IC ELISA	1 ng/mL in urine	[27]
Glyphosate, atrazine, metolachlor mercapturate	Water, urine	Multiplexed fluorescence microbead immunoassay	0.03–0.11 ng/mL	[60]
Methyl parathion	Soil	Microtiter plate IC and DC ELISA and FPIA	0.08 ng/mL (IC) 0.5 ng/mL (DC) 15 ng/mL (FPIA)	[41]
Triazine herbicides	Surface water, groundwater	Test tube DC ELISA	0.2–2 ng/mL in water	[24]

FPIA, Fluorescence polarization immunoassay; IC, indirect competitive; DC, direct competitive; ELISA, enzyme-linked immunosorbent assay.

as reported by the source laboratory [61]. FPBA is the metabolite for cyfluthrin (a pyrethroid pesticide containing a fluorophenoxybenzyl group). This high cross-reactivity is advantageous as this 3-PBA ELISA can be used as a monitoring tool for determining a broad exposure to pyrethroids. For assay development, the anti-PBA antibody, coating antigen, and initial assay protocol were provided by the source laboratory. Checkerboard titration experiments were performed to determine the optimal concentrations of anti-PBA antibody, coating antigen, and a commercial enzyme-conjugated secondary antibody. The optimal conditions established for the 3-PBA ELISA were 0.5 ng/mL of coating antigen, a 1:4000 dilution of anti-PBA antibody, and a dilution of 1:6000 of the commercial enzyme-labeled secondary antibody conjugate (goat anti-rabbit labeled with horseradish peroxidase). The assay procedures were modified by preparing the standard solutions in a 10% methanol extract of 10% hydrolyzed drug-free urine in PBS. Calibration curves (Figure 4.3) for 3-PBA were generated based on 10 concentration levels ranging from 0.00256 to 500 ng/mL (1:5 dilution series). The percent relative standard deviation (%RSD) values of the triplicate analyses were <20% for the standard solutions. Day-to-day variation for the quality control (QC) standard solution (1.0 ng/mL) was within 13.1% (1.2 ± 0.16 ng/mL) over a period of 4 months. The estimated assay detection limit was 0.2 ng/mL. Quantitative recoveries of 3-PBA were achieved by ELISA (92% ± 18%) in the fortified urine samples. Approximately 100 human urine samples were prepared and analyzed by the ELISA method. Different aliquots of the urine samples were also analyzed by gas chromatography/mass spectrometry (GC/MS). The GC/MS results indicated that 3-PBA was detected in 95% of the samples, whereas FPBA was only detected in 8.4% (10 out of 119 samples) of the samples. Similar results suggesting that FPBA was detected at much lower rate than 3-PBA in human urine samples collected from residential settings was also

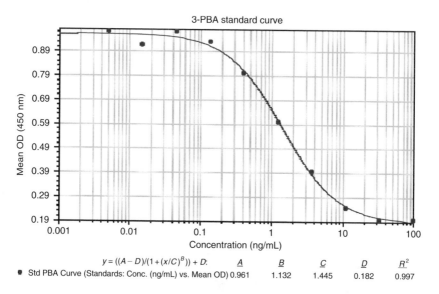

FIGURE 4.3 Calibration curve for 3-PBA immunoassay.

reported in the CDC third National Report on Human Exposure to Environmental Chemicals [62]. The ELISA-derived 3-PBA concentrations correlated well with the GC/MS results. The Pearson correlation coefficient between the 3-PBA concentrations of the two methods was 0.952, which was statistically significant ($p < 0.0001$). A nonsignificance outcome ($p = 0.756$) was also observed from the paired t-test indicating that there was no significant difference in measurements between the two analytical methods (ELISA vs. GC/MS) for a given sample. This study demonstrated that the ELISA method could be used as a monitoring tool for the urinary biomarker, 3-PBA in human urine samples, for assessing human exposure to pyrethroids.

As most fruit and vegetable baby food preparations generally contain a significant amount (>80%) of water, ELISA methods have the advantage over instrumental methods in determining pesticides in this aqueous sample matrix. We investigated various sample preparation methods for determining pesticides in baby foods using either GC/MS or ELISA methods [26]. A streamlined direct ELISA method consisting of dilution, filtration, and ELISA was evaluated on spiked baby foods at 1, 2, 5, 10, or 20 ppb. Quantitative recoveries (90%–140%) were achieved for atrazine in the nonfat baby foods (i.e., pear, apple sauce, carrot, banana/tapioca, green bean). The performance of other ELISA testing kits was not as good as the atrazine-ELISA testing kit. Over-recoveries were observed for carbofuran and metolachlor testing kits in banana/tapioca and green bean. This was probably due to a sample matrix interference that was not completely removed by dilution. An off-line coupling of enhanced solvent extraction (ESE) with ELISA was developed to determine atrazine in a more complex sample matrix of fatty baby foods. The results indicated that the extraction temperature was an important factor to recover atrazine. The ESE-ELISA method consisted of extracting the food at 150°C and 2000 psi with water and performing ELISA on the aqueous extract.

In an on-going study, different sample preparation procedures are being investigated for a magnetic particle ELISA analysis for permethrin. Quantitative recoveries (>90%) were obtained when the fortified soil samples were extracted with sonication using DCM, methyl-t-butyl ether (MTBE) or 10% ethyl ether (EE) in hexane. Recoveries were <50% from the fortified soil samples when the shaking method was employed (shaking with methanol for 1 h). A longer shaking time (16 h, overnight) was evaluated, using methanol, yielding recoveries of over 200% by ELISA. The longer shaking time extracted substances that interfered with the ELISA detection. This interference was also detected in the GC/MS analysis and persisted even after the SPE cleanup. Satisfactory recovery data (>90%) for post-spiked dust samples and a spiked dust sample were obtained. DCM was selected as the extraction solvent, as it was easily evaporated, facilitating the solvent-exchange step. The collected field samples were extracted with DCM using sonication. The DCM extract was concentrated and solvent exchanged into methanol. The methanol extract was diluted with reagent water (1:1) before ELISA.

Interferences caused by sample matrix components are a concern for both conventional instrument methods and ELISA methods. In immunoassays, sample matrix effects may result from nonspecific binding of the analyte to the matrix as well as the matrix to the antibody or enzyme or denaturation of the antibody or enzyme. The matrix interferences can often be removed by a series of dilutions if a practical detection limit can still be achieved [23]. Alternatively, cleanup methods for instrumental methods (e.g., SPE or column chromatographic separation) can also be performed before ELISA detection. Another effective cleanup method is immunoaffinity column chromatography that can be applied for the purification of sample extracts for either instrumental or ELISA detection [2,63].

In a recent study [64], an effective bioanalytical method for atrazine in complex sample media (soil, sediment, and duplicate-diet food samples) was developed. The method consisted of an ASE procedure with DCM, followed by immunoaffinity column cleanup with detection by a magnetic particle ELISA. Quantitative recoveries were achieved in fortified soil and sediment (93% ± 17%) as well as in food (100% ± 15%) samples. The ELISA data were in good agreement with the GC/MS data for these samples (the Pearson correlation coefficient was 0.994 for soil and sediment and 0.948 for food). However, the ELISA values were slightly higher than those obtained by GC/MS. This was probably the result of the solvent-exchange step required for the GC/MS but not the ELISA. This bioanalytical approach is more streamlined than the GC/MS analysis and could be applied to future large-scale environmental monitoring and human exposure studies.

4.2.4 DATA ANALYSIS

Calculations of sample analyte concentrations in ELISA methods are similar to those used in instrumental methods. A set of standard solutions covering the working range of the method is used to generate the calibration curve, and the concentration of target analyte is calculated according to the calibration data. For the 96-microwell format, it is easy to include a standard curve on each plate along with the samples. Thus, a calibration curve can be generated in the same 96-microwell plate along with the samples. For test tube formats, a standard curve series can be interspersed among the samples. Many mathematical models have been used to construct ELISA calibration curves including four-parameter logistic-log, log–log transforms, logistic-log transforms, and other models. The four-parameter logistic-log model is commonly used for 96-microwell plate assays and is built into commercial data analysis software [65]. The four-parameter logistic-log model is described as follows: $y = (A-D)/(1 + (x/C)^B) + D$ where x is the concentration of the analyte and y is the absorbance for colorimetric end point determinations.

Specifications are determined from each calibration curve for an expected midpoint on the curve at 50% inhibition (IC_{50}), a maximum absorbance for the lower asymptote (A), and a minimum absorbance for the upper asymptote (D). An established ELISA method usually has well-documented historical data for the specifications of the curve-fit constants, such as the slope of the curve (B), and central point of the linear portion of the curve (C). The specific curve-fit constants may vary from day to day and the accepted ranges of such variations must be determined and documented. Triplicate analyses of each standard, control, and sample are generally performed for 96-microwell plate assays. The %RSD of measured concentrations from triplicate analysis is usually within ±30% and can be as low as ±10%, depending on the specific assay and required data quality objectives. Recoveries of positive controls and back-calculated standard solutions typically range from 70% to 130% or better. If the results of the samples are outside the calibration range, the sample is diluted and reanalyzed. Effects of the sample matrix can be determined by analyzing a number of samples at different dilutions. Typically, results from different dilutions should be within ±30%. Larger variations in the data suggest a matrix interference problem, indicating cleanup procedures may be necessary.

When a commercial ELISA testing kit is used as a quantitative ELISA method, similar assay performance is expected as those previously described for laboratory-based 96-microwell plate assays. The samples need to be diluted and reanalyzed if the results of the samples are outside the calibration range. However, some of the commercial magnetic particle ELISA testing kits have a small dynamic optical density range (i.e., 1.0–0.35 OD) and small changes in OD correlate to large changes in derived concentrations. The differences between absorbance values and duplicate assays are generally small, and are well within the acceptance requirement (<10%) for the calibration standard solutions. However, the percent difference (%D) of the derived concentrations of the standard solution from duplicate assays sometimes may exceed 30%. The greater %D values obtained for some of the measured concentrations for the standards and samples may be due to a small volume of standard or sample retained in the pipette tip during the transfer step [8]. If the ELISA testing kit is to be used as a quantitative method, extreme care should be taken when transferring each aliquot of standard or sample. A trace amount of aliquot not delivered may result in a large variation in the data from duplicate analyses. The analyst should be alert in following the protocol when performing the assay.

To ensure the quality of the ELISA data, analytical quality control (QC) measures need to be integrated into the overall ELISA method. The QC samples may include: (1) negative and positive control standard solutions, (2) calibration standard solutions, (3) laboratory and field method blank, (4) fortified matrix samples, and (5) duplicate field samples. The assay performance can be monitored by characterization of the calibration curve and the data generated from the QC samples. The QC results will provide critical information such as assay precision, accuracy, detection limit, as well as overall method precision (including sample preparation and/or cleanup), accuracy, and detection limit when evaluating and interpreting the ELISA data.

Before applying an ELISA method for field application, the ELISA method needs to be evaluated and validated for its performance. The data generated from the ELISA method are usually compared with the data generated by a conventional instrument method (e.g., GC/MS). Various types of statistical analyses have been employed to compare the results between ELISA and GC/MS. For example, the Pearson correlation coefficient, commonly used, measures the extent of a general linear association between the ELISA and GC/MS data, and a parametric statistical test is performed to determine whether the calculated value of this correlation coefficient was significantly positive [66]. The slope of the established linear regression equation can also be used as guidance to determine if a 1:1 relationship exists for the ELISA and GC/MS data. The paired t-test [67] can be used to determine whether the measured ELISA and GC/MS concentrations differ significantly for a given sample at a 0.05 or 0.01 level of significance. Other nonparametric tests, namely, the Wilcoxon signed-rank test and the sign test, can also be performed on the sample-specific differences between ELISA and GC/MS data. These nonparametric tests can be used to determine if the median difference between the ELISA and GC/MS measurements among the samples is significantly different from zero [68]. The Wilcoxon signed-rank test is applied to differences between log-transformed measurements, as this test assumes that the differences have a symmetric distribution. In contrast, the sign test does not make this assumption and therefore does not require log transformations of the data. The McNemar's test of association can also be performed to determine whether there is any significant difference between the two methods in the proportion of samples having measurable levels that were at or above a specified threshold. The false-negative and false-positive rates can then be obtained at the specified concentration level.

4.3 BIOSENSORS

Biosensors are analytical probes composed of two components: a biological recognition element such as a selective antibody, enzyme, receptor, DNA, microorganism, or cell, and a transducer that converts the biological recognition event into a measurable physical signal to quantitate the amount of analyte present. Biosensors must rapidly regenerate to provide continuous monitoring data, yielding a response in real time. Analytical considerations such as sample preparation, matrix effects, and quality control measures must also be addressed in biosensor development. Matrix effects and the effect of sample on the recognition element are key issues for unattended sensors. Sensors that are easily fouled have limited reliability and application for environmental monitoring. Since biosensors use a biological recognition element, they may provide information on the effects of toxic substances as well as analytical measurements. Sensors for biochemical responses may assist in toxicity studies or human exposure assessments. Several pesticide biosensors have been reported for detecting various pesticides. Table 4.3 illustrates the application of biosensor technology to pesticide monitoring.

4.3.1 GENERAL DESCRIPTIONS

Biosensors can provide rapid and continuous in situ, measurements for on-site or remote monitoring. Several different transducer types such as optical, electrochemical, piezoelectric, and thermometric can be employed. Immunosensors contain specific antibodies for biological recognition and a transducer that converts the binding event of antibody to antigen to a physical signal.

Antibodies may be immobilized on membranes, magnetic beads, optical fibers; or embedded in polymers, or placed on metallic surfaces. In some types of sensors, such as those employing surface plasmon resonance (SPR), evanescent waves, or piezoelectric crystals, the binding of antigen and antibody can be detected directly. With other transducers, an indicator molecule (either a labeled antigen or labeled secondary antibody) is required. An indicator may be fluorescent or it may be an enzyme that alters a colorimetric or fluorescent signal or produces a change in pH affecting the electrochemical parameters.

Optical biosensors may measure fluorescence, fluorescence transfer, fluorescence lifetime, time-resolved fluorescence, color (either by absorbance or reflectance), evanescent waves, or an SPR response. Optical immunosensors are very rapid as they detect the antigen/antibody binding directly without requiring labeled reagents. Data in real time can be generated with devices applied to continuous

TABLE 4.3
Examples of Biosensors for Determining Pesticides and Metabolites in Biological and Environmental Samples

Analyte	Sensor Type	Matrix	Range or LOD	References
Atrazine	Electrochemical immunosensor	Orange juice	0.03 nmol/L	[17]
Atrazine	Electrochemical magnetoimmunosensor	Orange juice	0.027 nmol/L	[69]
Carbaryl, paraoxon	Disposable screen-printed thick-film electrode	Milk	20 µg/L (carbaryl) 1 µg/L (paraoxon)	[70]
Carbofuran	Flow-injection electrochemical biosensor	Fruits, vegetables, dairy products	1–100 nmol	[71]
Dichlorvos	Flow-injection calorimetric biosensor	Water	1 mg/L	[72]
Dichlorvos	Electrochemical biosensor	Wheat	0.02 µg/g	[73]
Fenthion	Dipstick electrochemical immunosensor	Water	0.01–1000 µg/L	[74]
Malathion, dimethoate	Amperometric biosensor	Vegetables	Malathion: 0.01–0.59 µM Dimethoate: 8.6–520 µM	[14]
OP pesticides	Fluorescence-based fiber-optic sensor	Buffer	1–800 µM (paraoxon) 2–400 µM (DFP[a])	[75]
OP pesticides and nerve agents	Electrochemical sensor using nanoparticles (ZrO$_2$) as selective sorbents	Water	1–3 ng/mL	[76]
OP pesticides and nerve agents	Flow-injection amperometric biosensor using carbon nanotube-modified glassy carbon electrode	Water	0.4 pM	[77]
Thiabendazole	Fluorescence-based optical sensor	Citrus fruits	0.09 mg/kg	[16]

[a] Diisopropyl phosphorofluoridate (a nerve agent).

monitoring situations such as effluent or runoff measurements from hazardous or agricultural waste streams. Optical immunosensors based on SPR employ immobilized specific antibody on a metal layer. When antigen binds, there is a minute change in the refractive index that is measured as a shift in the angle of total absorption of light incident on the metal layer. This technique was used to develop an SPR sensor to detect atrazine at 0.05 ppb in drinking water [78].

Fiber optic biosensors are based on the transmission of light along silica glass or plastic fibers. The advantages of fiber optic sensors are numerous: they are not subjected to electrical interference; a reference electrode is not needed; immobilized reagent does not have to be in contact with the optical fiber; they can be miniaturized; and they are highly stable. A major advantage of these sensors is that they can respond simultaneously to more than one analyte and are useful for remotely monitoring hazardous environments or municipal water supplies.

Electrochemical biosensors offer the advantages of being effective with colored or opaque matrices and do not contain light-sensitive components. In an immunosensor format, the binding of antigen to antibody is visualized as an electrical signal. The response may be coupled to signal amplification systems such as an enzyme-conjugated secondary antibody, conferring very low detection limits. Amperometric sensors measure current when an electroactive species is oxidized or reduced at the electrode. Potentiometric sensors detect the change in charge of an antibody when it binds to an antigen. Organophosphorus pesticides may be detected in a number of ways including potentiometric or amperometric methods. In both of these cases, enzymes such as organophosphorus hydrolase or urease may be employed. Dependent on the structure of the analyte, the release of hydrogen ions can either be measured via a pH change or a p-nitrophenol (PNP) group may be produced to give a redox compound for an electron shuttle.

Piezoelectric crystals are nonmetallic minerals (usually quartz), which conduct electricity and which develop a surface charge when stretched or compressed along an axis. The crystals vibrate when placed in an alternating electric field. The frequency of the vibration is a function of the mass of the crystal. Antibodies can be immobilized to the surface of piezoelectric crystals and the new vibrational frequency determined as a baseline measurement. The binding of analyte to the immobilized antibody alters the mass and vibrational frequency of the antibody–crystal system. This change in vibration can be measured to determine the amount of analyte detected.

Electroconductive polymer sensors have a specific antibody embedded in a conducting polymer matrix such as polypyrrole. When an analyte binds to the antibody, the ions in the matrix are less free to move, which decreases the ability of the polymer to conduct current. A reagentless electrochemical DNA biosensor has been reported using an Au–Ag nanocomposite material adsorbed to a conducting polymeric polypyrrole [79]. The detection limit was 5.0×10^{-10} M of target oligonucleotides with a response time of 3 s. The integration of nanotechnology and sensor development will provide new analytical platforms and formats. Although new designs may first appear for clinical applications, these advancements will favorably impact the development of sensors for environmental measurements. Table 4.3 summarizes several pesticide biosensors that have been reported for various monitoring situations [14,16,17,69–77].

4.3.2 MICROARRAYS

Microarrays contain minute amounts of materials (DNA, proteins, aptamers, etc.) that are placed onto a matrix in an array format. The matrix is a solid support onto which a biological or organic material is placed. The solid support material can be plastic, glass, complimentary metal oxide semiconductor (CMOS), gold, platinum, membranes, or other substance on which the reagents can be attached and still maintain function. The method of attachment can be covalent, hydrophobic, or through some tight-affinity reagent, such as a biotin/streptavidin couple [80].

A microarray can be defined in terms of the number of spots (or electrodes) per chip/slide. By this definition, a low-density array may contain as little as 16 spots or as many as 96 spots. High-density arrays may have >500,000 spots. Lower density arrays are considered to be sensors, as microsensor detection is typically at the lower end of array density. Based on these classifications, there are several companies that produce lower density microsensor arrays (Antara Biosciences, and Osmetech International, among others).

There are numerous methods used for array production. Arrays may result from "spotting" onto activated surfaces using robots to produce high-density arrays. Proteins or DNA are spotted onto activated surfaces (aldehydes, amines, etc.) so that either a chemical bond is formed or proteins can adhere through hydrophobic interaction. Another means of producing arrays is by photolithography using masks or lasers. This method has been used to produce in situ DNA- or peptide-based arrays. In this specific case, a photolabile group is used on the $5'$-nucleotide end or photolabile groups are used as amino protection groups (peptides). The use of lasers or masks removes the labile group from a specific electrode or spot, promoting peptide bond or oligonucleotide bond formation. Conversely, this can also be accomplished using acid that is generated at a specific electrode. DNA and peptides can also be synthesized in this manner. The protecting groups are removed only at specific electrodes that generate acid resulting in an elongated nucleotide or peptide. The oligomers or peptides can be used as aptamers to capture specific molecules, such as pesticides, heavy metals, or other environmental contaminants. The method can also be extended to any synthesis procedure, providing an acid- or base-labile group is present. Products from Antara Biosciences and Osmetech traditionally use cyclic voltammetry (CV). In this mode, a redox active species is used in conjunction with the assay. In arrays sold by CombiMatrix, the electrochemical amplification is enzyme-based and relies on a charge build up at a capacitor near that electrode. The capacitor is discharged and the quantity of charge is converted to nanoamps. As the current is determined by the charge buildup over time, this is an indirect measurement for the current developed.

In the early developmental stages of either a microarray or a large sensor technique, the starting point is typically one or two electrodes. Much of the recorded electrochemical sensor data are based on just a few electrodes, as a particular technique may or may not be converted to a microarray. The decision to convert to a high-density array is dependent on many parameters such as reading times and hardwire issues. Detection methods in microarrays employ various techniques including fluorescence, luminescence, visible, electrochemical, Raman scattering,

SPR, and electrochemiluminescence, among others. The detection method used depends on the matrix and if the chip is hardwired. Typically, the light-based method can accommodate almost any matrix and production method. However, a laser scanner or CCD camera is required, which tends to be very expensive increasing start-up costs (which may exceed $50K). Electrochemical methods require chip hardwire in tandem with various detection methods. Amperometric detection, cyclic voltammetry, and the evaluation of a charge build up on the electrode surface have all been employed.

4.3.3 BIOSENSORS METHODS FOR PESTICIDES

Several types of biosensors have been developed for measuring pesticides in various sample media. However, the use of biosensors for obtaining environmental measurements is not as common as for immunoassay. This section presents the application of biosensor techniques for detecting pesticides and illustrates the potential of various sensor designs for environmental monitoring.

4.3.3.1 Potentiometric, Light Addressable Potentiometric Sensor, and Amperometric Detection

Molecular devices employ the use of a "Light Addressable Potentiometric Sensor" (LAPS) for detection on large arrays. The samples are captured on membranes via vacuum filtration into discreet spots on a membrane [81]. The detection is pH-based using a sensitive LAPS method that can detect the urease enzyme conversion of urea in a pH-sensitive manner (potentiometric readings). This technique has been applied to the herbicide atrazine. As atrazine is a small molecule, a competitive assay format was developed. Fluorescein-labeled anti-atrazine antibodies and atrazine covalently linked to biotin-DNP were used as reagents. When the fluorescein-labeled antibody is bound to the biotinylated atrazine, the complex will bind to the streptavidin-coated membrane. If nonbiotinylated atrazine (from the sample) is added to the mix, any antibody bound to this species will be washed away. Thus, in this competitive assay format, the fluorescein-labeled anti-atrazine antibody can either bind to the nonlabeled or biotin-labeled atrazine. A species-specific secondary antibody labeled with urease reacts with the bound anti-atrazine antibody to generate a pH flux, providing the signal for the LAPS sensor. In this mode, there is an inverse relationship between signal and amount of nonlabeled analyte found in solution. The largest signal output is seen when there is no atrazine present and the lowest signal is observed when a large quantity of nonlabeled atrazine is present. Thus, if there is a large amount of environmental atrazine measured, the signal will be low. The result is a sigmoidal curve similar to the one shown in Figure 4.3 for the ELISA to detect 3-PBA. Note that the detection range tends to be narrow using this format (due to the sigmoidal curve) and the sensitivity can be limited. This assay would be classified as a biosensor as eight simultaneous assays can be performed using this system.

In addition to using a fluorogenic substrate for detection, other means may be used to detect the presence of pesticide analytes in environmental samples. One of the simplest techniques is a potentiometric sensor based on pH changes. In this case, a simple biosensor that is sensitive to changes in pH would be adequate. The enzyme organophosphorus hydrolase needs only to be attached to the electrode, encompassed in a polymer and attached to a bioresin over the electrode for OP detection. Organophosphorus hydrolase catalyzes the hydrolysis of a wide range of OP pesticides (e.g., coumaphos, diazinon, dursban, ethyl parathion, methyl parathion, and paraoxon). The attached or trapped hydrolase then acts on the OP compound to produce an alcohol and an acid. The resulting acid compound is monitored as a pH change at the electrode. This is a very simple system to use and is similar to LAPS detection.

Mulchandani et al. [82] developed an assay where organophosphorus hydrolase was placed onto an electrode. The phosphate hydrolysis product was monitored by measuring the current produced at the electrode. The output of the amperometric sensor could be correlated to the concentration of pesticide in sample solutions of soil and vegetation. This detection method can be incorporated into large arrays, such as the one used by CombiMatrix on electroactive electrode arrays.

Another biosensor method is applicable to other OP compounds that produce PNP as a releasing compound. These compounds include ethyl parathion, methyl parathion, paraoxon, fenithrothion, and O-ethyl O-(4-nitrophenyl) phenylphosphonothioate (EPN). The released PNP is oxidized at the anode to insert a hydroxyl group that is *ortho* to the nitro group. In this case, the oxidation current is measured amperometrically at a fixed potential. The signal is linear to the concentration of PNP present. The analysis relies on the OP compound to be trapped or conjugated to material over the electrode.

4.3.3.2 Piezoelectric Measurements

Many pesticides (e.g., organophosphates and carbamates) or their metabolites are cholinesterase inhibitors. This phenomenon can be used to develop sensors for the detection of these types of compounds. Using a piezoelectric sensor format, paraoxon was bound to an electrode (gold on a piezo/quartz surface) as the recognition element [83]. The analysis was performed by allowing a cholinesterase to interact with the modified electrode surface and with free paraoxon in a standard or sample. An oscillation change can be observed in terms of hertz or an electronic occurrence. A competitive assay was developed that allowed competition for cholinesterase between a cholinesterase inhibiting pesticide in solution and the inhibitor bound to the electrode surface. The ability of cholinesterase to bind to the paraoxon immobilized on the electrode is minimized or prevented in the presence of free inhibitor (analyte) in solution. In this case, the cholinesterase remains in solution bound to the pesticide in the sample. The sensing surface can be regenerated for reuse. The format can be used to develop better inhibitors and to quantitate OP compounds in solutions of environmental samples.

4.3.3.3 Surface Plasmon Resonance

SPR technology has been used in the biosensor field for some time and many sensors of this type are commercially available. The technique depends on the change in the reflectance angle (Plasmon) due to mass changes at the surface. Binding of proteins and small materials change the mass number at the surface and the reflectance angle is altered [84,85]. SPR detection has demonstrated the usage of many types of compounds. Initially, the technique was applied only to large molecules but as the technology has matured so has its potential for monitoring various pesticides, including photosynthetic inhibitors.

The crux of the system is a gold film on a glass surface. Attached to the gold film are self-assembled monolayers (SAMs) and capture reagents. These capture reagents may be antibodies, receptors, enzymes, ssDNA, streptavidin, and protein A or G (dependent on the type of antibody used) as well as other reagents. As the specific species is captured, the mass on the chip surface increases and changes the specific reflection angle. In this technique, a herbicide such as atrazine may be detected in several modes. The simplest mode would be to attach an anti-atrazine antibody (as a whole or in parts) to the chip surface. If the solution under test shows the presence of atrazine, a signal response on the chip would be detected.

Another option would be to attach the photosynthetic reaction center (RC) from a purple bacterium to the sensing chip. This can be accomplished in a number of ways, but literature evidence suggests that histidine (His) tags can be conveniently used. The system can easily be reused as the RC can be removed and the chip regenerated once the assay is completed. Samples of atrazine are introduced and the signal is monitored. A positive response can be quantitated and the chip can be reactivated for the next sample.

4.3.3.4   Conductive Polymers

One way to increase the use of electrochemical detection methods is to use conductive polymers [86]. The concept is that the interference from sample components is limited and many conductive polymers can be formed in situ directly over the electrode. Most of the polymers that have been used are electrochemically derived (synthesized in situ), formed by a host of starting materials. Additionally, many can be tethered to electrochemical conducting wires or even be encapsulated in a biopolymer matrix such as microgels [86–91]. A sensor using an electrodeposited conductive layer was able to detect the herbicide diruon [92] and could be applied to other substituted urea compounds.

For this technique to function, an enzymatic system is often used, such as glucose oxidase. Other enzymes may be employed, dependent on the nature of the biosensor developed and the anticipated monitoring applications. One application that appears to dominate for commercial development is that of a glucose sensor. Glucose is converted to gluconic acid and amperometric signals are observed based on the production of hydrogen peroxide. The polymer may encapsulate the electrode or be placed on the electrode using microparticle slurries.

Another polymer that can be used is a water-soluble Os-poly(vinyl imidazole) redox hydrogel. Again, the electron transfer is very efficient and necessitates a redox enzyme placed in the gel. A polypyrrole film has also been used in conjunction with $NADH^+$ ferro-/ferricyanide redox chemistries. An enzyme is required whose function is to use $NADP^+$ in conjunction with an enzymatic substrate to release a product and the cofactor, NADPH. The ferricyanide is present to efficiently shuttle the electrons.

There are also reports on the use of PVPOs(bpy) polymer and poly(mercapto-*p*-benzoquinone) on gold electrodes or within conducting hydrogels. For these systems, the redox enzyme horseradish peroxidase is used or the CV of the substrate, sulfo-*p*-benzoquinone (SBQ) is monitored. The types of solid supports and electro-chemical methods are almost limitless.

4.4 CURRENT DEVELOPMENTS

Immunochemical methods can either be performed independently or coupled with other analytical techniques to produce powerful tandem methods for pesticide analysis. Currently, our laboratory is investigating immunoaffinity separation techniques coupled to immunoassay and instrumental methods to support environmental monitoring studies including:

- Immunoaffinity chromatographic separation of a group of structurally similar pesticides. This may be accomplished by using either the high cross-reactivity of an antibody to a certain group of pesticides or using mixed antibodies that possess a combined affinity to a pesticide group.
- Hybrid affinity separation of multiple pesticides based on the integration of immunoaffinity chromatography and surface imprinting techniques. Hybrid affinity columns can be prepared by mixing one or more antibodies with one or more types of molecularly imprinted polymers.

Other methods this laboratory is investigating are the online combination of immunoaffinity separation with liquid chromatography-mass spectrometry (LC-MS) to provide rapid separation and detection of pesticides with a high degree of selectivity and sensitivity. Similar combinations can also be performed between immunoaffinity separation and flow-injection analysis. The online combination of immunoassay and sample preparation techniques such as SPE, or the online integration of SPE and immunoaffinity cleanup can provide efficient analytical methods.

4.5 FUTURE TRENDS

Immunoassay is a mature analytical technology with broad application to pesticide analysis. Extensive fundamental investigations as well as technical improvements will make immunoassay methods more powerful tools for the identification and determination of a variety of pesticides. New breakthroughs in the development and application of immunoassays will result from the integration of future state-of-the-art research in several key areas including antibody production, new platforms and detection systems, and nanotechnology.

Future research that may enhance the use of immunoassays and immunosensors for pesticide analysis is the development of novel antibodies for individual pesticide compounds. This includes the design and synthesis of new haptens using the latest concepts and techniques, better understanding and control of the combination of hapten molecules and macromolecular carriers, and improving the efficiency of existing laboratory procedures to increase the yield of antibodies having the desired characteristics.

Molecularly imprinted polymers (MIPs) and aptamers are emerging as possible reagents (i.e., artificial antibodies) for pesticide immunoassays and immunosensors. These reagents have the potential to provide large amounts of reagents for the development of methods and to support their widespread use. Some MIP-based affinity separation methods and biosensors have already been developed for the extraction and determination of pesticides in aqueous samples. Aptamers are artificial nucleic acid ligands that can be generated to detect biomacromolecules, such as proteins, and small molecules, such as amino acids, drugs, and pesticides. Currently, aptamer-based bioanalytical methods are mainly employed for clinical applications. Additional studies of molecular recognition-based MIPs and aptamers could facilitate the development of more cost-effective methods including immunoaffinity separation techniques for pesticides.

Future research may also be directed to new immunoassay formats. The development of microimmunoassays, using compact discs (CDs) as an analytical platform, has recently drawn much attention from researchers. An indirect competitive procedure is conducted on the polycarbonate surface of a CD and a modified CD reader performs as a laser scanner for the detection of microscopic reaction products [93–95]. These test systems hold promise for the simultaneous determination of multiple pesticide residues in environmental samples in a rapid and cost-effective format. New platforms may also be integrated with new labels such as more robust enzymes or highly sensitive visualization techniques, such as laser-induced fluorescence detection (LIF) to produce even lower limits of detection.

Nanotechnology is a rapidly growing discipline of scientific research and is applied to a wide variety of fields. Nanomaterials with dimensions of <100 nm have physical and chemical properties that make them attractive for many applications requiring high strength, conductivity, durability, and reactivity. The application of nanotechniques in immunoassays is also of great interest to researchers [93,96]. New detection strategies based on gold and silver particles have been successfully demonstrated for immunoassay labeling to meet the needs of diverse detection methods. These particles have been used for various techniques such as scanning and transmission electron microscopy, Raman spectroscopy, and sight visualization due to their easily controlled size distribution, and long-term stability and compatibility with biomacromolecules.

Initial studies on nanoparticle-labeled microfluidic immunoassays have shown their unique advantages over conventional immunoassay formats for the detection of small molecules, macromolecules, and microorganisms. Submicron-sized striped metallic rods intrinsically encoded through differences in reflectivity of adjacent metal stripes have been used in autoantibody immunoassays. These bar-coded particles act as supports with antigens attached to the surface providing a permanent tag for the tracking of analyte [97].

Nanomaterials including gold, zirconia (ZrO_2), and carbon nanotubes have been applied as biosensors for monitoring OP pesticides [76,77,98]. An optical sensor based on fumed silica gel functionalized with gold nanoparticles has also been reported for OP pesticides [98]. Nanoparticles possess extraordinary optical properties that may offer alternative strategies for the development of optical sensors. An electrochemical sensor for detection of OP pesticides has been developed using ZrO_2 nanoparticles as selective sorbents, possessing a strong affinity for the phosphoric group. The nitroaromatic OPs strongly bind to the ZrO_2 surface. A square-wave voltammetric analysis was used to monitor the amount of bound OP pesticide. Another sensitive flow-injection amperometric biosensor for OP pesticides and nerve agents was developed using self-assembled acetylcholinesterase (AchE) on a carbon nanotube (CNT)-modified glassy carbon electrode [77]. The CNTs have two main functions for the biosensor; first, as platforms for AchE immobilization by providing a microenvironment that can maintain the bioactivity of AchE, and second, as a transducer for amplifying the electrochemical signal of the product of the enzymatic reaction. The integration of nano- and biomaterials could be extended to other biological molecules for future biosensor or immunoassay research.

Advancements in biosensor technology will continue with expansion of multi-analyte detection and more rapid analytical capability. For example, a chip containing 92,000 electrodes with a 30 μs read is already investigated. With a 30 μs read time, enzymatic kinetic reads could be performed directly on the chip. However, the capability of 92,000 electrodes $\times$ 1000 reads presents storage, data acquisition, and conversion issues. The limiting factor at this time is computer capability. Other technologies such as a 40 s kinetic read of 12,000 electrodes with 4 or 8 electrodes discharged at one time in microsecond intervals are near realization.

Through future research, immunoassays and biosensors for pesticides may find critical applications related to in vitro and in vivo studies in the diverse field of environmental science and human exposure.

REFERENCES

1. Baker, S.R. and Wilkinson, C.F. The effects of pesticides on human health, Vol. XVIII, in *Advances in Modern Environmental Toxicology*, Princeton Scientific Publishing, Princeton, NJ, pp. 438, 1990.
2. Van Emon, J.M., Ed. *Immunoassay and Other Bioanalytical Techniques*, CRC Press, Taylor & Francis Group, Boca Raton, FL, 2007.
3. NRC, National Research Council, *Pesticides in the Diets of Infants and Children*, National Academy Press, Washington, D.C., p. 386, 1993.
4. FQPA, Food Quality Protection Act of 1996. *Public Law*, 104–170, 1996.
5. Hammock, B.D. and Mumma, R.O. Potential of immunochemical technology for pesticide analysis, in *Pesticide Analytical Methodology*, Harvey, J.J., and Zweig, G., Eds., American Chemical Society, Washington, D.C., pp. 321–352, 1980.
6. Van Emon, J.M. and Lopez-Avila, V. Immunochemical methods for environmental analysis. *Anal. Chem.*, 64(2), 79A–88A, 1992.
7. Chuang, J.C., Miller, L.S., Davis, D.B., Peven, C.S., Johnson, J.C., and Van Emon, J.M. Analysis of soil and dust samples for polychlorinated biphenyls by enzyme-linked immunosorbent assay (ELISA). *Anal. Chim. Acta*, 376, 67–75, 1998.
8. Chuang, J.C., Van Emon, J.M., Chou, Y.-L., Junod, N., Finegold, J.K., and Wilson, N.K. Comparison of immunoassay and gas chromatography–mass spectrometry for measurement of polycyclic aromatic hydrocarbons in contaminated soil. *Anal. Chim. Acta*, 486, 31–39, 2003.

9. Khosraviani, M., Pavlov, A.R., Flowers, G.C., and Blake, D.A. Detection of heavy metals by immunoassay: optimization and validation of a rapid, portable assay for ionic cadmium. *Environ. Sci. Technol.*, 32, 137–142, 1998.

10. Nichkova, M., Park, E., Koivunen, M.E., Kamita, S.G., Gee, S.J., Chuang, J.C., Van Emon, J.M., and Hammock, B.D. Immunochemical determination of dioxins in sediment and serum samples. *Talanta*, 63, 1213–1223, 2004.

11. Van Emon, J.M. and Gerlach, C.L. A status report on field-portable immunoassay. *Environ. Sci. Technol.*, 29(7), 312A–317A, 1995.

12. Rodriguez-Mozaz, S., Lopez de Alda, M.J., and Barcelo, D. Fast and simultaneous monitoring of organic pollutants in a drinking water treatment plant by a multi-analyte biosensor followed by LC-MS validation. *Talanta*, 69(2), 377–384, 2006.

13. Marco, M. and Barcelo, D. Environmental applications of analytical biosensors. *Meas. Sci. Technol.*, 7, 1547–1562, 1996.

14. Yang, Y., Guo, M., Yang, M., Wang, Z., Shen, G., and Yu, R. Determination of pesticides in vegetable samples using an acetylcholinesterase biosensor based on nanoparticles ZrO_2/chitosan composite film. *Int. J. Environ. Anal. Chem.*, 85(3), 163–175, 2005.

15. Zhang, J. Tube-immunoassay for rapid detection of carbaryl residues in agricultural products. *J. Environ. Sci. Health, Part B*, 41(5), 693–704, 2006.

16. Garcia-Reyes, J.F., Llorent-Martinez, E.J., Ortega-Barrales, P., and Molina-Diaz, A. Determination of thiabendazole residues in citrus fruits using a multicommuted fluorescence-based optosensor. *Anal. Chim. Acta*, 557(1–2), 95–100, 2006.

17. Zacco, E., Galve, R., Marco, M.P., Alegret, S., and Pividori, M.I. Electrochemical biosensing of pesticide residues based on affinity biocomposite platforms. *Biosens. Bioelectron.*, 22(8), 1707–1715, 2007.

18. Van Emon, J.M. and Gerlach, C.L. Environmental monitoring and human exposure assessment using immunochemical techniques. *J. Microbiol. Methods*, 32, 121–131, 1998.

19. Van Emon, J.M. Immunochemical applications in environmental science. *J. AOAC Int.*, 84(1), 125, 2001.

20. Voller, A., Bidwell, D.E., and Bartlett, A. Microplate enzyme immunoassays for the immunodiagnosis of virus infections, in *Manual of Clinical Immunology*, Rose, N., and Friedman, H., Eds., American Society for Microbiology, Washington, D.C., pp. 506–512, 1976.

21. Gee, S.J., Hammock, B.D., and Van Emon, J.M. *A User's Guide to Environmental Immunochemical Analysis*. EPA/540/R-94/509, March 1994.

22. Jaeger, L.L., Jones, A.D., and Hammock, B.D. Development of an enzyme-linked immunosorbent assay for atrazine mercapturic acid in human urine. *Chem. Res. Toxicol.*, 11, 342–352, 1998.

23. Chuang, J.C., Van Emon, J.M., Durnford, J., and Thomas, K. Development and evaluation for an enzyme-linked immunosorbent assay (ELISA) method for the measurement of 2,4-dichlorophenoxyacetic acid in human urine. *Talanta*, 67, 658–666, 2005.

24. Thurman, E.M., Meyer, M., Pomes, M., Perry C.A., and Schwab, A.P. Enzyme-linked immunosorbent assay compared with gas chromatography/mass spectrometry for the determination of triazine herbicides in water. *Anal. Chem.*, 62, 2043–2048, 1990.

25. Wong J.M., Li, Q.X., Hammock, B.D., and Seiber, J.N. Method for the analysis of 4-nitrophenol and parathion in soil using supercritical fluid extraction and immunoassay. *J. Agric. Food Chem.*, 39, 1802–1807, 1991.

26. Chuang, J.C., Pollard, M.A., Misita, M., and Van Emon, J.M. Evaluation of analytical methods for determining pesticides in baby food. *Anal. Chim. Acta*, 399, 135–142, 1999.

27. Ahn, K.C., Ma, S., and Tsai, H. An immunoassay for a urinary metabolite as a biomarker of human exposure to the pyrethroid insecticide permethrin. *Anal. Bioanal. Chem.*, 384, 713–722, 2006.

28. Koivunen, M.E., Dettmer, K., Vermeulen, R., Bakke, B., Gee, S.J., and Hammock, B.D. Improved methods for urinary atrazine mercapturate analysis—assessment of an enzyme-linked immunosorbent assay (ELISA) and a novel liquid chromatography–mass spectrometry (LC-MS) method utilizing online solid phase extraction (SPE). *Anal. Chim. Acta*, 572, 180–189, 2006.

29. Chuang, J.C., Van Emon, J.M., Finegold, K., Chou, Y.-L., and Rubio, F. Immunoassay method for the determination of pentachlorophenol in soil and sediment. *Bull. Environ. Contam. Toxicol.*, 76(3), 381–388, 2006.

30. Van Emon, J.M., Hammock, B., and Seiber, J.N. Enzyme-linked immunosorbent assay for paraquat and its application to exposure analysis. *Anal. Chem.*, 58, 1866–1873, 1986.

31. Van Emon, J.M., Seiber, J.N., and Hammock, B.D. Application of an enzyme-linked immunosorbent assay to determine paraquat residues in milk, beef, and potatoes. *Bull. Environ. Contam. Toxicol.*, 39, 490–497, 1987.

32. Van Emon, J.M., Seiber, J.N., and Hammock, B.D. Immunoassay techniques for pesticide analysis in analytical methods for pesticides and plant growth regulators, in *Advanced Analytical Techniques, Vol. XVII*, Sherma, J., Ed., Academic Press, New York, pp. 217–263, 1989.

33. Nam, K. and King, J.W. Supercritical fluid extraction and enzyme immunoassay for pesticide detection in meat products. *J. Agric. Food Chem.*, 42, 1469–1474, 1994.

34. Richman, S.J., Karthikeyan, S., Bennett, D.A., Chung, A.C., and Lee, S.M. Low-level immunoassay screen for 2,4-dichlorophenoxyacetic acid in apples, grapes, potatoes, and oranges: circumventing matrix effects. *J. Agric. Food Chem.*, 44, 2924–2929, 1996.

35. Yeung, J.M., Mortimer, R.D., and Collins, P.G. Development and application of a rapid immunoassay for difenzoquat in wheat and barley products. *J. Agric. Food Chem.*, 44, 376–380, 1996.

36. Bashour, I.I., Dagher, S.M., Chammas, G.I., and Kawar, N.S. Comparison of gas chromatography and immunoassay methods for analysis of total DDT in calcareous soils. *J. Environ. Sci. Health, Part B: Pestic. Food Contam. Agric. Wastes*, B38(2), 111–119, 2003.

37. Botchkareva, A.E., Eremin, S.A., Montoya, A., Marcius, J.J., Mickova, B., Rauch, P., Fini, F., and Girotte, S. Development of chemiluminescent ELISAs to DDT and its metabolites in food and environmental samples. *J. Immuno. Methods*, 283(1–2), 45–57, 2003.

38. Chuang, J.C., Van Emon, J.M., Reed, A.W., and Junod, N. Comparison of immunoassay and gas chromatography-mass spectrometry methods for measuring 3,5,6-trichloro-2-pyridinol in multiple sample media. *Anal. Chim. Acta*, 517(1–2), 177–185, 2004.

39. Lee, J.K., Ahn, K.C., Stoutamire, D.W., Gee, S.J., and Hammock, B.D. Development of an enzyme-linked immunosorbent assay for the detection of the organophosphorus insecticide acephate. *J. Agric. Food Chem.*, 51, 3695–3703, 2003.

40. Lee, J.K., Kim,Y.J., Lee, E.Y., Kim, D.K., and Kyung, K.S. Development of an ELISA for the detection of fenazaquin residues in fruits. *Agric. Chem. Biotech.*, 48(1), 16–25, 2005.

41. Kolosova, A.Y., Park, J., Eremin, S.A., Park, S., Kang, S., Shim, W., Lee, H., Lee, Y., and Chung, D. Comparative study of three immunoassays based on monoclonal antibodies for detection of the pesticide parathion-methyl in real samples. *Anal. Chim. Acta*, 511(2), 323–331, 2004.

42. Brun, E.M., Marta, G., Puchades, R., and Maquieira, A. Highly sensitive enzyme-linked immunosorbent assay for chlorpyrifos, application to olive oil analysis. *J. Agric. Food Chem.*, 53(24), 9352–9360, 2005.

43. Morozova, V.S., Levashova, A.I., and Eremin, S.A. Determination of pesticides by enzyme immunoassay. *J. Anal. Chem.*, 60(3), 202–217, 2005.

44. Koivunen, M.E., Gee, S.J., Park, E.-K., Lee, K., Schenker, M.B., and Hammock, B.D. Application of an enzyme-linked immunosorbent assay for the analysis of paraquat in human-exposure samples. *Arch. Environ. Contam. Toxicol.*, 48, 184–190, 2005.

45. Gabaldon, J.A., Maquieria, A., and Puchades, R. Development of a simple extraction procedure for chlorpyrifos determination in food samples by immunoassay. *Talanta*, 71(3), 1001–1010, 2007.

46. Watanabe, E., Miyake, S., Baba, K., Eun, H., and Endo, S. Immunoassay for acetamiprid detection: application to residue analysis and comparison with liquid chromatography. *Anal. Bioanal. Chem.*, 386(5), 1441–1448, 2006.

47. Watanabe, E., Baba, K., Eun, H., Arao, T., Ishii, Y., Ueji, M., and Endo, S. Evaluation of performance of a commercial monoclonal antibody-based fenitrothion immunoassay and application to residual analysis in fruit samples. *J. Food Prot.*, 69(1), 191–198, 2006.

48. Watanabe, E. and Miyake, S. Immunoassay for iprodione: key estimation for residue analysis and method validation with chromatographic technique. *Anal. Chim. Acta*, 583(2), 370–376, 2007.

49. Watanabe, E., Baba, K., Eun, H., and Miyake, S. Application of a commercial immuno-assay to the direct determination of insecticide imidacloprid in fruit juices. *Food Chem.*, 102(3), 745–750, 2007.

50. Garces-Garcia, M., Morais, S., Gonzaliz-Martinez, F.A., Puchades, R., and Maquieira, A. Rapid immunoanalytical method for the determination of atrazine residues in olive oil. *Anal. Bioanal. Chem.*, 378(2), 484–489, 2004.

51. Furzer, G.S., Veldhuis, L., and Hall, J.C. Development and comparison of three diag-nostic immunoassay formats for the detection of azoxystrobin. *J. Agric. Food Chem.*, 54(3), 688–693, 2006.

52. Zhang, C., Shang, Y., and Wang, S. Development of multianalyte flow-through and lateral-flow assays using gold particles and horseradish peroxidase as tracers for the rapid determination of carbaryl and endosulfan in agricultural products. *J. Agric. Food Chem.*, 54, 2502–2507, 2006.

53. Cho, Y.A., Kim, Y.J., Hammock, B.D., Lee, Y.T., and Lee, H. Development of a microtiter plate ELISA and a dipstick ELISA for the determination of the organophos-phorus insecticide fenthion. *J. Agric. Food Chem.*, 57, 7854–7860, 2003.

54. Lee, J.K., Ahn, K.A., Park, O.S., Kang, S.Y., and Hammock, B.D. Development of an ELISA for the detection of the residues of the insecticide imidacloprid in agricultural and environmental samples. *J. Agric. Food Chem.*, 49, 2159–2167, 2001.

55. Lee, W.Y., Lee, E.K., Kim, Y.J., Park, W.C., Chung, T., and Lee, Y.T. Monoclonal antibody-based enzyme-linked immunosorbent assays for the detection of the organo-phosphorus insecticide isofenphos. *Anal. Chim. Acta*, 557(1–2), 169–178, 2006.

56. Skerritt, J.H., Guihot, S.L., Asha, M.B., Rani, B.E.A., and Karanth, N.G.K. Sensitive immunoassays for methyl-parathion and parathion and their application to residues in foodstuffs. *Food Agric. Immunol.*, 15(1), 1–15, 2003.

57. Yang, Z., Kolosova, W.S., and Chung, D. Development of monoclonal antibodies against pirimiphos-methyl and their application to IC-ELISA. *J. Agric. Food Chem.*, 54, 4551–4556, 2006.

58. Irwin, J.A., Tolhurst, R., Jackson, P., and Gale, K.R. Development of an enzyme-linked immunosorbent assay for the detection and quantification of the insecticide tebufenozide in wine. *Food Agric. Immunol.*, 15(2), 93–104, 2003.

59. Shivaramaiah, H.M., Odeh, I.O.A., Kennedy, I.R., and Skerritt, J.H. Mapping the distribution of DDT residues as DDE in the soils of the irrigated regions of northern new South Wales, Australia using ELISA and GIS. *J. Agric. Food Chem.*, 50(19), 5360–5367, 2002.

60. Biagini, R.E., Smith, J.P., Sammons, D.L., MacKenzie, B.A., Striley, C.A.F., Robertson, S.K., and Snawder, J.E. Development of a sensitivity enhanced multiplexed fluorescence covalent microbead immunosorbent assay (FCMIA) for the measurement of glyphosate, atrazine and metolachlor mercapturate in water and urine. *Anal. Bioanal. Chem.*, 379(3), 368–374, 2004.

61. Shan, G., Huang, H., Stoutamire, W., Gee, J., Leng, G., and Hammock, B.D. A sensitive class specific immunoassay for the detection of pyrethroid metabolites in human urine. *Chem. Res. Toxicol.*, 17, 218–225, 2004.

62. CDC. Third National Report on Human Exposure to Environmental Chemicals. Census for Diseases Control and Prevention, Atlanta, GC 30341, 2005. http://www.cdc.gov/exposurereport/

63. Van Emon, J.M., Gerlach, C.L., and Bowman, K. Bioseparation and bioanalytical techniques in environmental monitoring. *J. Chromatogr. B*, 715, 211–228, 1998.

64. Chuang, J.C., Van Emon, J.M., Jones, R., Durnford, J., and Lordo, R. Development and application of immunoaffinity column chromatography for atrazine in complex sample media. *Anal. Chim. Acta*, 583, 32–39, 2007.

65. *Molecular Devices SOFTmax PRO User Manual*. Molecular Devices, Sunnyvale, CA, 1998.

66. Sendecor, G.W. and Cochran, W.G. *Statistical Methods*, 8th edn, Iowa State University Press, Ames, IA, 1989.

67. Hollander, M. and Wolfe, D.A. *Nonparametric Statistical Methods*, John Wiley & Sons, New York, 1973.

68. Rosner, B. *Fundamentals of Biostatistics*, 5th edn, Duxbury Press, North Scituate, MA, 2000.

69. Zacco, E., Pividori, M.I., Alegret, S., Galve, R., and Marco, M.P. Electrochemical magnetoimmunosensing strategy for the detection of pesticides residues. *Anal. Chem.*, 78(6), 1780–1788, 2006.

70. Zhang, Y., Muench, S.B., Schulze, H., Perz, R., Yang, B., Schmid, R.D., and Bachmann, T.T. Disposable biosensor test for organophosphate and carbamate insecticides in milk. *J. Agric. Food Chem.*, 53(13), 5110–5115, 2005.

71. Nikolelis, D.P., Simantiraki, M.G., Siontorou, C.G., and Toth, K. Flow injection analysis of carbofuran in foods using air stable lipid film based acetylcholinesterase biosensor. *Anal. Chim. Acta*, 537(1–2), 169–177, 2005.

72. Zheng, Y., Hua, T., Sun, D., Xiao, J., Xu, F., and Wang, F. Detection of dichlorvos residue by flow injection calorimetric biosensor on immobilized chicken liver esterase. *J. Food Eng.*, 74(1), 24–29, 2006.

73. Longobardi, F., Solfrizzo, M., Compagnone, D., Del Carlo, M., and Visconti, A. Use of electrochemical biosensor and gas chromatography for determination of dichlorvos in wheat. *J. Agric. Food Chem.*, 53(24), 9389–9394, 2005.

74. Cho, Y.A., Cha, G.S., Lee, Y.T., and Lee, H. A dipstick-type electrochemical immuno-sensor for the detection of the organophosphorus insecticide fenthion. *Food Sci. Biotech.*, 14(6), 743–746, 2005.

75. Viveros, L., Paliwal, S., McCrae, D., Wild, J., and Simonian, A. A fluorescence-based biosensor for the detection of organophosphate pesticides and chemical warfare agents. *Sens. Actuators, B: Chemical*, B115(1), 150–157, 2006.

76. Liu, G. and Lin, Y. Electrochemical sensor for organophosphate pesticides and nerve agents using zirconia nanoparticles as selective sorbents. *Anal. Chem.*, 77(18), 5894–5901, 2005.

77. Liu, G. and Lin, Y. Biosensor based on self-assembling acetylcholinesterase on carbon nanotubes for flow injection/amperometric detection of organophosphate pesticides and nerve agents. *Anal. Chem.*, 78, 835–843, 2006.

78. Minunni, M. and Mascini, M. Detection of pesticide in drinking water using real-time biospecific interaction analysis (BIA). *Anal. Lett.*, 26, 1441–1460, 1993.

79. Fu, Y., Yuan, R., Chai, Y., Zhou, L., and Zhang, Y. Coupling of a reagentless electrochemical DNA biosensor with conducting polymer film and nanocomposite as matrices for the detection of the HIV DNA sequences. *Anal. Lett.*, 39(3), 1532–1236, 2006.

80. Dill, K., Grodzinsky, P., and Liu, R., Eds. *Recent Advances in Microarray Technology*, Springer-Verlag, Heidelberg, 2007.

81. Dill, K. Sensitive analyte detection and quantitation using the threshold immunoassay system, in *Environmental Immunochemical Methods*, Van Emon, J.M., Gerlach, C.L., and Johnson, J.C., Eds., ACS symposium series 646, American Chemical Society, Washington, D.C., 1996, chap. 9.

82. Mulchandani, A., Chen, W., Mulchandani, P., Wang, J., and Rogers, K.R. Biosensors for direct determination of organophosphate pesticides. *Biosens. Bioelectron.*, 16, 225–230, 2001.

83. Makower, A., Hlamek, J., Skladal, P., Kerchen, F., and Scheller, F.W. New principle of direct real-time monitoring of the interaction of cholinesterase and its inhibitors by piezoelectric biosensor. *Biosens. Bioelectron.*, 18, 1329–1337, 2003.

84. Nakamura, C., Hasegawa, M., Nakamura, N., and Miyake, J. Rapid and specific detection of herbicides using a self-assembled photosynthetic reaction center from purple bacterium on a SPR chip. *Biosens. Bioelectron.*, 18, 599–603, 2003.

85. Strachan, G., Grant, S.D., Learmonth, D., Longstaff, M., Porter, A.J., and Harris, W.J. Binding characteristics of anti-atrazine monoclonal antibodies and their fragments synthesized in bacteria and plants. *Biosens. Bioelectron.*, 13, 665–673, 1998.

86. Liu, X., Neoh, K.G., and Kang, E.T. Enzymatic activity of glucose oxidase covalently wired via viologen to electrically conductive polypyrrole films. *Biosens. Bioelectron.*, 19, 823–834, 2004.

87. Retama, J.R., Cabarcos, E.L., Mecerreyes, D., and Lopez-Ruiz, B. Design of an amperometric biosensor using polypyrrole-microgel composites containing glucose oxidase. *Biosens. Bioelectron.*, 20, 1111–1117, 2004.

88. Chen, C., Jiang, Y., and Kan, J. A noninterference polypyrrole glucose biosensor. *Biosens. Bioelectron.*, 22, 639–643, 2006.

89. Vilkanauskyte, A., Erichsen, T., Marcinkeviciene, L., Laurinavicius, V., and Schuhmann, W. Reagentless biosensors based on co-entrapment of a soluble redox polymer and an enzyme within an electrochemically deposited polymer film. *Biosens. Bioelectron.*, 17, 1025–1031, 2002.

90. Lopez, M.A., Ortega, F., Dominguez, E., and Katakis, I. Electrochemical immunosensor for the detection of atrazine. *J. Mol. Recognit.*, 11, 178–181, 1998.

91. Gross, P. and Comtat, M. A bioelectrochemical polypyrrole-containing $Fe(CN)_6^{3+}$-interface for the design of a NAD-dependent reagentless biosensor. *Biosens. Bioelectron.*, 20, 204–210, 2004.

92. Maly, J., Masojidek, J., Masci, A., Ilie, M., Cianci, E., Foglietti, V., Vastarella, W., and Pilloton, R. Direct mediatorless electron transport between monolayer of photosystem II and poly(mercapto-*p*-benzoquinone) modified gold electrode—new design of biosensor for herbicide detection. *Biosens. Bioelectron.*, 21, 923–932, 2005.

93. Morais, S., Puchades, R., and Maquieira, A. Compact discs as analytical platform for multi-residue immunosensing, Abstracts of Papers, 232nd ACS National Meeting, San Francisco, CA, United States, September 10–14, 2006.

94. Lai, S., Wang, S., Luo, J., Lee, J., Yang, S., and Madou, M. Design of a compact disk-like microfluidic platform for enzyme-linked immunosorbent assay. *Anal. Chem.*, 76(7), 1832–1837, 2004.

95. Nolte, D., Varma, M., Peng, L., Inerowicz, H., Halina, D., and Regnier, F. Spinning-disk laser interferometers for immuno-assays and proteomics: the BioCD. *Proc. SPIE Int. Soc. Opt. Eng.* (*220*), 5328, 41–48, 2004.

96. Lin, F., Sabri, M., Alirezaie, J., Li, D., and Sherman, P. Development of a nanoparticle-labeled microfluidic immunoassay for detection of pathogenic microorganisms. *Clin. Diag. Lab. Immunol.*, 12(3), 418–425, 2005.

97. Gonzalez-Buitrago, J.J. Multiplexed testing in the autoimmunity laboratory. *Clin. Chem. Lab. Med.*, 44(10), 1169–1174, 2006.

98. Newman, J.D.S., Roberts, J.M., and Blanchard, G.J. Optical organophosphate sensor based upon gold nanoparticle functionalized fumed silica gel. *Anal. Chem.*, 79(9), 3448–3454, 2007.

5 Quality Assurance

Árpád Ambrus

CONTENTS

5.1 INTRODUCTION

The results of measurements should provide reliable information and the laboratory should be able to prove the correctness of measurements with documented evidence. Analysts carry serious responsibilities to produce correct and timely analytical results, and are fully accountable for the quality of their work. The expanding national and international trade, the responsibility of national registration authorities permitting the use of various chemicals required long ago reliable test methods, which were acceptable by all parties concerned. The accuracy and precision of the analytical results may be assured by proficient analysts applying properly validated methods, which are fit for the purpose, in a laboratory accredited according to the relevant standards or guidelines.[1,2] Several documents and guidelines had been, and are developed to assist the analysts to apply the relevant analytical quality control (AQC)[3] quality assurance (QA) principles in their diverse daily work, and to provide guidance for accreditation purposes. The Codex Committee on Pesticide Residues (CCPR) continuously updates the *Guidelines on Good Laboratory Practice*,[4] which also includes detailed information on the minimum criteria for validation of methods. The EURACEM/CITAC* published additional guidelines on application of quality assurance in nonroutine laboratories,[5] interpretation of proficiency test results,[6] and traceability of measurements.[7] These documents and GLs are complimentary to the requirements of the ISO[†]/IEC[‡] 17025 and OECD[§] GLP Principles, and can be freely downloaded from the Internet.

5.1.1 QUALITY SYSTEMS

The Good Laboratory Practice (GLP) is a quality system concerned with the organizational processes and the conditions under which nonclinical health and environmental safety studies are planned, performed, monitored, archived, and reported. The ISO/IEC 17025:2005 Standard, replacing the previous standards (ISO/IEC Guide25 and EN 45001), contains all the general requirements for the technical competence to carry out tests, including sampling, that laboratories have to meet if they wish to demonstrate that they operate a quality system, and are able to generate technically valid results. It covers analytical tasks performed using standard methods, nonstandard methods, and laboratory-developed methods, and incorporates all those requirements of ISO 9001and ISO 9002 that are relevant to the scope of the services that are covered by the laboratory's quality system. The OECD GLP GLs and the ISO/IEC Standard focus on different fields of activities, but they have been developed simultaneously, and they are specifying basically the same requirements in terms of AQC.

The *quality assurance (QA) program* aimed at achieving the required standard of analysis. It means a defined system, including personnel, which is independent of the study conduct and designed to assure test facility management that the analyses of samples or conduction of the studies comply with the established procedures.

Measurements of any type contain a certain amount of error. This error component may be introduced when samples are collected, transported, stored, and analyzed or when data are evaluated, reported, stored, or transferred electronically. It is the responsibility of quality assurance programs to provide a framework for determining and minimizing these errors through each step of the sample collection, analysis, and data management processes. The process must ensure that we do the right experiment as well as doing the experiment right.[8] Systems alone cannot deliver quality. Staff must be trained, involved with the tasks in such a way that they can contribute their skills and ideas and must be provided with the necessary resources. Accreditation of the laboratory by the appropriate national accreditation scheme, which itself should conform to accepted standards, indicates that the laboratory is applying sound quality assurance principles.

The *internal quality control (QC)* and proficiency testing are important parts of the quality assurance program, which must also include the staff training,

---

* Co-operation on International Traceability in Analytical Chemistry.
† International Standard Organisation.
‡ International Electrotechnical Commission.
§ Organization for Economic Co-operation and Development.

administrative procedures, management structure, auditing, and so on. The laboratory shall document its policies, systems, programs, procedures, and instructions to the extent necessary to assure the quality of the results. The system's documentation shall be communicated to, understood by, available to, and implemented by the appropriate personnel.

The laboratory shall have *quality control procedures** for monitoring the batch to batch validity, accuracy, and precision of the analyses undertaken. Measurement and recording requirements intended to demonstrate the performance of the analytical method in routine practice. The resulting data shall be recorded in such a way that trends are detectable and, where practicable, statistical techniques shall be applied for evaluating the results. This monitoring shall be planned and reviewed and may include, but not be limited to, the regular use of certified reference materials and/or internal quality control using secondary reference materials; participating in interlaboratory comparison or proficiency-testing programs; performing replicate tests using the same or different methods; and retesting of retained items.[1]

The analytical methods must be thoroughly validated before use according to recognized protocol. These methods must be carefully and fully documented, staff adequately trained in their use, and control charts should be established to ensure the procedures are under proper statistical control. Successful participation in proficiency test programs does not replace the establishment of within laboratory performance of the method. The performance of the method should be fit for the purpose and fulfill the quality requirements in terms of accuracy, precision, sensitivity, and specificity. Where possible, all reported data should be traceable to international standards by applying calibrated equipment and analytical standards with known purity certified by ISO accredited supplier.

Presently, it is definitely more economical to contract out a few samples requiring tests with special methodology and expertise to well-established and experienced (preferably accredited) laboratories, than to invest a lot of time, instruments, and so on to set up and maintain a validated method (and experience to apply it) for incidental samples in a laboratory.

As an *external quality control*, participating in proficiency-testing schemes, provides laboratories with an objective means to demonstrate their capability of producing reliable results.

5.1.2 CHARACTERIZATION OF THE UNCERTAINTY AND BIAS OF THE METHODS

The interpretation of the results and making correct decisions require information on the accuracy and precision of the measurements. The measurement process is subjected to a number of influencing factors which may contribute to random, systematic, and gross errors.[9,10] The quality control of the process aims to monitor the uncertainty (repeatability, reproducibility) and trueness of the measurement results.

---

* Synonymous with the term analytical quality control (AQC) and performance verification.

5.1.2.1   Uncertainty of the Measurement Results

The uncertainty of the measurements is mainly due to some random effects. The uncertainty "estimate" describes the range around a reported or experimental result[11] within which the true value can be expected to lie within a defined level of probability. This is a different concept to measurement error (or accuracy of the result) which can be defined as the difference between an individual result and the true value. It is worth noting that, while the overall random error cannot be smaller than any of its contributing sources, the resultant systematic error can be zero even if each step of the determination of the residues provides biased results. Another important difference between the random and systematic errors is that once the systematic error is quantified the results measured can be corrected for the bias of the measurement, while the random error of a measurement cannot be compensated for, but its effects can be reduced by increasing the number of observations.

The combined uncertainty is calculated as[10]

$$u(y(x_{i,\,j,\,...})) = \sqrt{\sum_{i=1}^{n} c_i^2 u_i^2 + \sum_{i,k=1}^{n} c_i c_k u(x_i, y_k)}, \qquad (5.1)$$

where

u_i is the standard uncertainty of the ith component
c_i and c_k are the sensitivity coefficients
$u(x_i, y_k)$ is the covariance between x_i and y_k $(i \neq k)$

The covariance can be calculated with the regression correlation coefficient $r_{i,k}$: $u(x_i, x_k) = u(x_i) \times u(x_k) \times r_{ik}$.

The uncertainty components of a residue analytical result may be grouped according to the major phases of the determination[12] (external operations: sampling (S_S), packing, shipping, and storage of samples; preparation of test portion: sample preparation and sample processing (S_{Sp}); analysis (S_A): extraction, cleanup, evaporation, derivatization, instrumental determination). The major sources of the random and systematic errors[13] are summarized in Table 5.1. Their nature and contribution to the combined uncertainty of the results will be discussed in the following sections. The general equation can be simplified for expression of the combined relative standard uncertainty (CV_{Res}) of the results of pesticide residue analysis.

$$CV_{Res} = \sqrt{CV_S^2 + CV_L^2} \quad \text{and} \quad CV_L = \sqrt{CV_{Sp}^2 + CV_A^2}, \qquad (5.2)$$

where CV_S is the uncertainty of sampling and CV_L is the combined uncertainty of the laboratory phase including sample processing (Sp) and analysis (A). The preparation of portion of sample to be analyzed[14] as part of the sample preparation step (such as gentle rinsing or brushing to remove adhering soil, or taking outer withered loose leaves from cabbages) cannot be usually validated and its contribution to the uncertainty of the results cannot be estimated. If the combined uncertainty is calculated from

TABLE 5.1

Major Sources of Random and Systematic Errors in Pesticide Residue Analysis[a]

	Sources of Systematic Error	Sources of Random Error
Sampling	Wrong sampling design or operation Degradation, evaporation of analytes during preparation, transport and storage	Inhomogeneity of analyte in sampled object Varying ambient (sample material) temperature during transport and storage Varying sample size
Sample preparation	The portion of sample to be analyzed (analytical sample) may be incorrectly selected	The analytical sample is in contact and contaminated by other portions of the sample Rinsing, brushing is performed to various extent; stalks and stones may be differentially removed Nonhomogeneity of the analyte in single units of the analytical sample
Sample processing	Decomposition of analyte during sample processing, cross-contamination of the samples	Nonhomogeneity of the analyte in the ground/chopped analytical sample Variation of temperature during the homogenization process Texture (maturity) of plant materials affecting the efficiency of homogenization process Varying chopping time, particle size distribution
Extraction/cleanup	Incomplete recovery of analyte Interference of coextracted materials (load of the adsorbent) Interference of coextracted compounds	Variation in the composition (e.g., water, fat, and sugar content) of sample materials taken from a commodity Temperature and composition of sample/solvent matrix
Quantitative determination	Incorrectly stated purity of analytical standard Biased weight/volume measurements Determination of substance which does not originate from the sample (e.g., contamination from the packing material) Determination of substance differing from the residue definition Biased calibration	Variation of nominal volume of devices within the permitted tolerance intervals Precision and linearity of balances Variable derivatization reactions Varying injection, chromatographic and detection conditions (matrix effect, system inertness, detector response, signal-to-noise variation, etc.) Operator effects (lack of attention) Calibration

[a] Some processes and actions may cause both systematic and random error. They are listed where the contribution is larger.

the linear combination of the variances of its components, according to the Welch–Satterthwaite formula the degree of freedom of the estimated uncertainty is

$$\nu_{\text{eff}} = \frac{S_{c(y)}^4}{\sum_{i=1}^{N} \frac{S_{i(y)}^4}{\nu_i}}, \tag{5.3}$$

with $\nu_{\text{eff}} \leq \sum_{i=1}^{N} \nu_i$. The $S_{c(y)} = u_{c(y)}$ values may be replaced with $S_{c(y)}/y$ (CV) values where the combined uncertainty is calculated from the relative standard deviations.[11]

The CV_L can be calculated from CV_{Sp} and CV_A obtained during the method validation, or from the results of reanalysis of replicate test portions of samples containing field-incurred residues, as part of the internal quality control. Reference materials are not suitable for this purpose as they are thoroughly homogenized. If the relative difference of the residues measured in replicate portions is $R_{\Delta i} = 2(R_{i1} - R_{i2})/(R_{i1} + R_{i2})$, then CV_L is

$$CV_L = \sqrt{\frac{\sum_{i=1}^{n} R_{\Delta i}^2}{2n}}, \tag{5.4}$$

where n is the number of measurement pairs, and the degree of freedom of the corresponding standard deviation is equal to n.

The analytical phase may include, for instance, the extraction, cleanup, evaporation, derivatization, and quantitative determination. Their contribution to the uncertainty of the analysis phase (CV_A) can only be conveniently determined by applying ^{14}C-labeled compounds,[15,16] but it is usually sufficient to estimate their combined effects by the recovery studies. The repeatability of instrumental determination, which does not take into account the effect of preparation of calibration from different sets of standard solutions, can be easily quantified. However, the determination of the total uncertainty of the predicted concentration based on the approximations described, for instance, by J.N. Miller and J.C. Miller,[9] or Meier and Zünd[17] require special software to avoid tedious manual calculations.

5.1.2.2 Systematic Error—Bias of the Measurements

The systematic errors can occur in all phases of the measurement process. However, it practically cannot be quantified during the external, field phase of the process. Once the sample is taken, the most accurate and precise determination of the systematic error including that caused by the efficiency of extraction and dispersion of residues in the treated material can be carried out with radiolabeled compounds. Unfortunately, routine pesticide residue laboratories very rarely have access to facilities suitable for working with radioisotopes. Nevertheless, very useful information on stability of residues during storage, efficiency of extraction, and distribution of residues can be found in the FAO/WHO series of *Pesticide Residues—Evaluations*, which are published annually by FAO, and can be freely downloaded from the

Web site of the Pesticide Management Group.[18] Another source of information is the data submitted to support the claim for registration of the pesticides. Though the whole package is confidential, that part relating to the analysis of residues could be made accessible for laboratories analyzing pesticide residues.

Alternately, laboratories may test the bias of their measurement results with performing recovery studies usually spiking the test portion of the homogenized sample with a known amount of the analyte (R_0) before the extraction. It should be born in mind that the recovery tests can provide information on the systematic error and precision of the procedure only from the point of spiking. Thus, following the usual procedure it will not indicate the loss of residues during storage and sample processing. The recovery studies are normally performed with untreated samples. Where untreated samples are not available or the final extract of blank sample gives detectable response, the analyte equivalent of the average instrument signal obtained from the unspiked sample shall be taken into account. When the average recovery is statistically significantly different from 100%, based on t-test, the results should generally be corrected for the average recovery.[10,19] It should be noted, however, that currently some regulatory authorities require results which are not adjusted for the recovery. It may lead to a dispute situation when parties testing the same lot applying methods producing different recoveries. For instance, the shipment may be simply rejected due to the lower recoveries of analytical method used in the exporting country. Another area, where reporting the most accurate result is necessary, is providing data for the estimation of exposure to pesticide residues. In this case the residues measured should be corrected for the mean recovery, if that is significantly different from 100%. In order to avoid any ambiguity in reporting results, when a correction is necessary, the analyst should give the uncorrected as well as the corrected value, and the reason for and the method of the correction.[20]

The uncertainty of the mean recovery, $CV_{\bar{Q}} = \dfrac{CV_A}{\sqrt{n}}$, affects the uncertainty of the corrected results $\left(CV_{Acor} = \sqrt{CV_A^2 + CV_{\bar{Q}}^2} \right)$. On the one hand, the increase of the uncertainty of the residue values adjusted for the recovery can be practically eliminated if the mean recovery is determined from ≥ 15 measurements ($CV_{Acor} \leq 1.03\, CV_A$). On the other hand, if corrections would be made with a single procedural recovery, the uncertainty of the corrected result would be $1.41\, CV_A$. Therefore, such correction should be avoided as far as practical.

The recovery values obtained from performance verification usually symmetrically fluctuate around their mean, which indicates that the measured values are subjected to random variation. If the procedural recovery performed with an analytical batch is within the expected range, based on the mean recovery and within-laboratory reproducibility of the method, the analyst demonstrated that the method was applied with expected performance. Therefore, the correct approach is to use the typical recovery established from the method validation and the long-term performance verification (within laboratory reproducibility studies) for correction of the measured residue values, if necessary.

Under certain circumstances, such as extraction of soil samples, the extraction conditions cannot be fully reproduced from one batch of samples to the next, leading occasionally to much higher within laboratory reproducibility than repeatability

$(3S_r < S_R)$. In this case, the use of concurrent recovery for adjusting the measured residues may provide more accurate results. Where correlation between the residue values observed, the uncertainty of the residue value adjusted for the recovery should be calculated according to Equation 5.1. Where correlation between the results is quantifiable, it may be necessary to perform at least two recovery tests in one analytical batch covering the expected residue range, and use their average value for correction to reduce the uncertainty and improve the accuracy of the results.

5.2 SAMPLING

The analytical results cannot be better than the sample which is analyzed. Even though the importance of reliable sampling has long been recognized, the majority of regulatory laboratories concentrated only on the validation and establishing performance characteristics of the methods. Very little attention was paid to the quality of the sample as the results of measurements were related only to the sample "as received" and not to the sampled commodity. The ISO/IEC Standard 17025 has changed the situation requiring the incorporation of sampling uncertainty in the combined uncertainty when relevant.

Methods of sampling for the analysis of pesticide residues cannot be validated. Obtaining representative sample which reflects the residue content of the sampled commodity or object can only be assured by careful planning of the sampling program, providing clear instructions for the actual sampling operation including packing and shipping of samples.

The sampling method depends on the objectives of the analysis, and hence the sampling plan and protocol should be prepared jointly by the managers making decision based on the results, the analysts, and the sampling officers responsible for taking the samples. The objectives of the investigation and the corresponding acceptable uncertainty of the measurement results (CV_{Res}), expressed with Equation 5.2, will determine the size, frequency (time or distance), spacing, mixing, dividing of samples, and consequently the time required for sampling and the cost of sampling, shipping, and analysis of samples. Careful balancing of cost and benefit is a key component of designing sampling plans.

The information on the uncertainty of sampling, subsampling, and sample processing is equally important as the information on the uncertainty of analyses.

5.2.1 QUALITY OF SAMPLES

The purpose of sampling is to provide for a specific aim (determine one or some of the characteristic properties) a part of the object that is representative and suitable for analysis. The part of the object taken for further examination is the sample which is usually a very small portion (10^{-5} to 10^{-6}) of the sampled object (e.g., 1–2 kg of apples taken from an orchard of 2 ha yielding 50,000–60,000 kg fruits, or taking 20 soil cores from 5 ha field). The sample may be a single unit or an increment, or it may contain a number of primary samples* defined by the sample size in case of a

---

* One or more units taken from one position in a lot.

composite bulk sample, from which the laboratory sample may be prepared. The test portion (usually 2–50 g) is a representative part of the laboratory sample, which is extracted.

To prepare such a small fraction of the sampled object providing unbiased information with quantifiable uncertainty requires well-defined procedures performed by very responsible and technically highly qualified staff. The samples and the test portions analyzed should satisfy some basic quality requirements:

- Represent the properties of the object under investigation (composition, particle size distribution)
- Be of the size that can be handled by the sampler and the analyst; keep the properties the object had at the time of sampling; be suitable to give the required information (e.g., mean composition, composition as a function of time or place); and keep its identity throughout the whole procedure of transport and analysis[21]

To develop a quality sampling plan, the following actions should be taken and points may be considered:

- Purpose of the study (different sampling procedure would be required if we want to obtain information on the average residue in a commodity or the distribution of residues in crop units, within one field (or lot) or between fields)
- Clear definition of the object, which can usually be properly defined by the lot/batch number, the space coordinates and the time
- Collection of information of the properties of the objects before sampling (it may be necessary to inspect the site to determine the conditions and equipment required)
- Selection of suitable sampling method and tools; testing the suitability of containers to be used to collect, pack, and ship the samples, taking also into account the health, safety, and security precautions
- Determination of the time required for reaching the sampling site and handling the samples
- Provisions for prevention of contamination and deterioration of samples at all stages, including size reduction of bulk sample
- Arrangement for sealing, labeling, delivering the samples and the sampling record to the laboratory in unchanged conditions, and assuring integrity of the whole operation
- Preparation of preprinted sampling record sheet which guides the operator to collect and record all essential information including deviations from the sampling protocol
- Training of sampling personnel to assure that they are aware of the purpose of the operation and the provisions to be taken for obtaining reliable samples (e.g., permitted flexibility to adapt the sampling method for the particular conditions, recording requirements, legal actions, etc.)

5.2.2 SAMPLING OF COMMODITIES OF PLANT AND ANIMAL ORIGIN

For testing compliance with maximum residue limits (MRL), the CCPR elaborated a procedure which became widely accepted and used in many countries.[22] A Codex MRL for a plant, egg, or dairy product refers to the maximum level permitted to occur in a composite bulk sample,* which has been derived from multiple units of the treated product, whereas the MRLs for meat and poultry refers to the maximum residue concentration in the tissues of individual treated animals or birds. Each identifiable lot† to be checked for compliance must be sampled separately. The minimum number of primary samples to be taken depends on the size of the lot. Each primary sample should be taken from a randomly chosen position as far as practicable. The primary samples must consist of sufficient materials to provide the laboratory sample(s) required. The primary samples should be combined and mixed well, if practicable, to form the bulk sample. Where the bulk sample is larger than is required for a laboratory sample,‡ it should be divided to provide a representative portion. A sample divider, quartering, or other appropriate size reduction process may be used but units of fresh plant products or whole eggs should not be cut or broken. Where units may be damaged (and thus residues may be affected) by the processes of mixing or subdivision of the bulk sample, or where large units cannot be mixed to produce a more uniform residue distribution, replicate laboratory samples should be withdrawn or the units should be allocated randomly to replicate laboratory samples at the time of taking the primary samples. In this case, the result to be used should be the mean of valid results obtained from the laboratory samples analyzed.

Further details for the minimum mass and the number of primary samples to be taken depending on the size of the sampled lot or the targeted (acceptable) violation rate are given in the guidelines.

Samples taken for residue analysis in supervised trials are usually larger than specified in the Codex GLs, as the main objective is to obtain the best estimate for the average residues. Sample may be taken from the experimental site randomly, or following some stratified random sampling design. It was shown that, where samples should be taken at different time intervals after the application of the pesticide for establishing decline curves, the least variation can be obtained if the primary sampling positions are selected randomly and marked before the first sampling, and the primary samples are collected from the close vicinity of the marked positions at the various sampling times.[23]

---

* For products other than meat and poultry, the combined and well-mixed aggregate of the primary samples taken from a lot. For meat and poultry, the primary sample is considered to be equivalent to the bulk sample.

† A quantity of a food material delivered at one time and known, or presumed, by the sampling officer to have uniform characteristics such as origin, producer, variety, packer, type of packing, markings, consignor, and so on.

‡ The sample sent to, or received by, the laboratory. A representative quantity of material removed from the bulk sample.

5.2.3 ESTIMATION OF UNCERTAINTY OF SAMPLING

As it was shown, the average residues and CV of residues in individual crop units in samples of size 100–120 (that is each sample consists of 100–120 individual crop units, e.g., oranges) taken repeatedly from the same parent population (e.g., from a field or a lot) may vary significantly. The best estimate of the uncertainty of sampling is provided by the average of CV values.[24]

The sampling uncertainty depends on the size of composite samples and the distribution of residues in the sampled commodity. Based on 174 residue data sets consisted of 22,665 valid residue data derived from specifically designed supervised trials[25] and sampling lots from the market,[26] the estimated typical sampling uncertainty for different kinds of plant commodities and sample sizes specified by the Codex Standard for sampling[22] are summarized in Table 5.2.

There were no data for estimation of the uncertainty of sampling cereal grains, eggs, and processed products. The variation of residues in composite samples taken from different fields is much larger. The typical CV values of between fields variation of residues in composite samples ranged between 80% and 120%. The data evaluation revealed that the coefficients of variation of residues within field and between fields are practically independent of the pesticide, the preharvest interval, and dosage rate.[27]

TABLE 5.2

Typical Sampling Uncertainty for Various Fresh Plant Commodities with Lower (LC) and Upper (UC) Confidence Intervals

Commodity Groups	No. of Sample Sets[a]	SD of CV_S Values	Sample Size	Confidence Limits of CV_{Styp}				
				$LC_{0.99}$	$LC_{0.95}$	CV_{Styp}	$UC_{0.95}$	$UC_{0.99}$
Small commodities (unit mass ≤25 g)	18	0.31	1	0.57	0.62	0.78	0.93	0.99
		0.10	10^b	0.18	0.20	0.25	0.29	0.31
Medium-size commodities (unit mass 25–250 g)	76	0.25	1	0.73	0.75	0.81	0.86	0.88
		0.08	10^b	0.23	0.24	0.25	0.27	0.28
Large commodities (unit mass >250 g)	64	0.27	1	0.58	0.60	0.67	0.73	0.76
		0.12	5^b	0.26	0.27	0.30	0.33	0.34
		0.09	10	0.18	0.19	0.21	0.23	0.24
Brassica leafy vegetables	17	0.16	1	0.33	0.37	0.45	0.53	0.56
		0.07	5^b	0.15	0.16	0.20	0.24	0.25
		0.05	10	0.11	0.12	0.14	0.17	0.18

[a] Primary samples making up the sample sets were taken from different fields/lots.

[b] Minimum number of primary samples to be taken for a composite sample for testing compliance with Codex MRLs.[25]

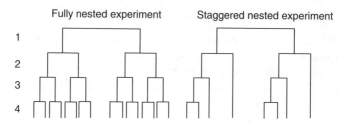

FIGURE 5.1 Experimental designs for estimation of the uncertainty of sampling steps. 1: Lots; 2: sampling of lots; 3: subsampling; and 4: measurements.

The ISO Standard 11648-1 for sampling bulk materials[28] recommends to apply fully nested or staggered nested experimental design to obtain information on the uncertainty of withdrawing the bulk samples from different lots, reducing the sample size with subsampling (sample preparation) and analysis. The procedures are illustrated in Figure 5.1. The standard recommends that for obtaining sufficient information about the variability of the analyte, ~20 lots should be sampled, preferably several pairs of samples taken from each lot.

Sampling of the same residue data population by withdrawing random composite samples with replacement[24] or applying the fully nested experimental design gave very similar results for the average residue and the average CV of the residues. For instance, even if 30 pairs of random composite samples of size 10 were withdrawn 100 times from a data population having a CV of 0.28, the minimum and maximum CV values observed were 0.205 and 0.365, respectively, which is in agreement with the confidence limits shown in Table 5.2.

Concerning the sampling uncertainty, one should always remember that the MRLs refer to the residues in the bulk sample. Hence, for testing compliance with an MRL any amount of material satisfying the minimum sample size is sufficient and the sampling uncertainty need not be taken into account. On the other hand, where the compliance of a lot before shipment has to be verified, then the sampling uncertainty must be included in the combined expanded uncertainty of the measured residue value.

5.3   SAMPLE PREPARATION AND PROCESSING

For food commodities, the Codex MRLs refer to the specified portion of the commodity which is analyzed.[14] The preparation of the analytical sample* may require removal of foreign materials and certain parts of the sampled material (such as shell of nuts, stone of mango or peach, adhering soil, outer withered loose leaves in case of plant materials, and peddles and remains of plants from soil, etc.). These procedures may significantly affect the residue level. As they cannot be

---

* The material prepared for analysis from the laboratory sample, by separation of the portion of the product to be analyzed.

validated and their contribution to the uncertainty of the results cannot be estimated, the sample preparation procedure should be clearly written and consistently followed without any deviation to obtain comparable results.

The residues in individual crop units are not uniformly distributed. Therefore the whole laboratory sample must be prepared and the entire analytical sample should be chopped, ground, or mixed to obtain a well-mixed material from which the representative test portions can be withdrawn for extraction. The large crops making up the laboratory sample (e.g., five watermelons) may not be processed together due to the limited capacity of the equipment. In these cases, representative portions should be cut from the individual units in such a way that the ratio of the surface and inner part remains the same.

The efficiency of the comminuting procedure depends on the equipment, maturity, and variety of the crops, but it is independent of the concentration and nature of the analyte, and the extraction method. The efficiency of processing is characterized with CV_{Sp} (Equation 5.2). It is more difficult to obtain a well-mixed matrix from plant materials with hard peal and soft pulp (tomato) than from a soft fruit (orange). The homogeneity (well-mixed status) of the processed analytical sample cannot be verified with the usual recovery studies. It should either be tested with samples treated with pesticides according to the normal practice,[29] or a small part of the surface of the crops should be treated with suitable test compounds.[30] A third alternative is to treat a small portion of the sample matrix with the test compound and then mix it with the rest of the sample.[16]

The uncertainty of sample processing can be quantified as part of the method validation by applying fully nested or staggered nested[31] experimental design and evaluating the results with ANOVA. The scheme of the process is very similar to that shown in Figure 5.1. In this case, uncertainty information can be obtained only for the size of the test portion. If the expectable uncertainty should be determined for a given range of test portion size to optimize the analytical procedure, the concept of sampling constant[29] can be used. The sampling constant, K_s, is defined as[32]

$$K_s = mCV^2, \qquad (5.5)$$

where m is the mass of a single increment and CV is the relative standard deviation of the concentration of the analyte in the test portions of size m.

If the analytical sample is well-mixed, the sampling constant should be the same for small (Sm) and large (Lg) portions, and Equation 5.5 can be written as

$$m_{Sm}S_{Sm}^2 = m_{Lg}S_{Lg}^2. \qquad (5.6)$$

If the ratio of $S_{Sm}^2 m_{Sm}/S_{Lg}^2 m_{Lg}$ is smaller than the critical F value, the processed sample can be considered well-mixed, and the expected sample processing uncertainty can be calculated for any test portion size $\geq m_{Sm}$.

The acceptable variability of sample processing depends on the variability of the other steps of the determination. When the combined uncertainty of the measurement results should include the sampling uncertainty, then the CV_{Sp} should be $\leq 8\%$–10% depending on the crop analyzed. Where only the CV_L is taken into account, the CV_{Sp} should be less than $0.25–0.3 \times CV_A$. Under these conditions, the sample processing will not significantly contribute to the combined uncertainty of the measurement.

The efficiency of sample processing depends on, among others, the type and variety of the sample and the implementation of the process. As it can be a significant contributor to the combined uncertainty, it should be tested regularly as part of the performance verification of the laboratory phase of the determination of pesticide residues. It can be most economically carried out with the reanalysis of the replicate test portions of samples containing the analyte in a different analytical batch. The within laboratory reproducibility of the procedure (CV_L) can be calculated with Equation 5.4. As it was shown in Section 5.2.3, a minimum of 20 measurement pairs are needed to obtain a reliable estimate.

5.4 STABILITY OF RESIDUES

The pesticide residues may be subjected to different chemical reactions or evaporated after the samples are taken. The change of concentration of the residues should be avoided as far as possible to assure the representativeness of the samples and the results.

5.4.1 Stability during Storage

The supervised trial samples are usually deep-frozen shortly after the sampling and shipped deep-frozen to the laboratory within the shortest possible time, where they are kept deep-frozen until analysis. During this storage period, the concentration of residues of the pesticides and their metabolites may decline due to processes such as volatilization or reaction with enzymes. Storage stability tests are carried out with representative commodities to demonstrate the stability of residues during frozen storage before analysis. The storage stability studies are part of the data package submitted to support registration of a compound. The FAO/WHO *Pesticide Residues—Evaluations*[18] also include information on the stability of residues during storage.

Where it is foreseen that the samples have to be stored in the laboratory over 1 month, and appropriate information on the stability of residues is not available on representative sample matrices under similar conditions as the samples will be stored, storage stability test should be carried out. The basic principles[20,33] to be considered for planning storage stability studies are briefly summarized later.

Stability data obtained on one commodity from a commodity group (see Section 5.5) can be extrapolated within the same group, provided that the storage conditions are comparable. The study can be performed with sample containing field-incurred residues, if the suitable homogeneity of the material had been verified before ($CV_{Sp} < 0.25–0.3 \times CV_A$). Alternately, the test portions withdrawn from the homogenized untreated sample matrix should be spiked individually. Untreated test material should be prepared and stored under the same conditions. The treated and blank test material should be sufficient for a minimum of 8×4 treated as well as untreated test portions for analyses with some extra material as reserve. The total number of test portions should be larger, if the extension of the study period may be necessary. The active substance and its metabolites or degradation products included in the residue definition should be tested separately if spiked test portions are used. The initial residue concentration should be sufficiently high to enable the accurate determination of the residues if their concentration decreases during storage. Normally, analyses at five time points are sufficient. The first test should be performed at day 0 to verify the initial concentration, and the others selected according to approximate geometrical progression (e.g., 0, 1, 3, 6, and 12 months or 0, 2, 4, 8, and 16 weeks if decline of residues are suspected). At each time, two treated test portions and at least one freshly spiked untreated sample should be analyzed.

The results should be reported in the form of individual residue concentration (milligram/kilogram) measured in the treated stored samples (survived residues), the concurrent recoveries expressed in percentage of the spiked amount, and the standard uncertainty of the measurement determined independently as part of the validation of the analytical method. The individual recoveries obtained should preferably be within the warning limits of the established method. If that is not the case, the analysis of residues in additional test portions of the stored samples should be repeated together with additional recovery studies.

Where the storage stability study carried out with samples belonging to the representative commodity groups indicates that the residue is stable, then it can be assumed that the residues would be stable in other matrices stored under similar conditions.

5.4.2 STABILITY OF RESIDUES DURING SAMPLE PROCESSING

The laboratory sample processing received unproportionally little attention in the past, though its contribution to the uncertainty and the bias of the results can be quite large. In general, supervised trial samples are transported, stored, and processed under deep-frozen conditions, whereas monitoring laboratories usually receive and process the samples at ambient temperature. The analysts were aware of the rapid decomposition of dithiocarbamates or daminozide if they were in contact with the macerated samples and eliminated the homogenization step from the method, but did not test the stability of other residues or associate the loss of residues with their potential decomposition, until some publications indicated the substantial decomposition (50%–90%) of certain compounds (chlorothalonil, phthalimides, thiabendazole, dichlofluanid).[15,34] Further studies revealed that processing in the presence of dry ice (cryogenic milling) at or below $-20°C$ reduced or practically eliminated the loss of all pesticides which decomposed at ambient temperature.[35] Furthermore, cryogenic processing may provide more homogeneous sample matrix and reduce the uncertainty of sample processing. Notable that the decomposition of pesticides in test portions spiked after the homogenization of the sample is much smaller and does not affect the recovery of most of the compounds substantially. It may probably be attributed to the inactivation of the enzymes by the extracting solvent and the different concentration of the chemicals in the diluted extract.

The decomposition of the residues depends on the composition of the sample material and the homogenization process. When intensive and extended comminution in high-speed blender is carried out to reduce the sample processing uncertainty, a significant bias can be introduced due to the decomposition of the residues.

Because the rate of decomposition may depend on the laboratory equipment, the variety and maturity of the processed crop, and many other factors, currently there is no sufficient knowledge to extrapolate findings from one laboratory to another. Consequently the laboratories, analyzing wide range of pesticides in large number of various commodities, should apply cryogenic processing as standard procedure to reduce the chance of producing biased results. Furthermore, the laboratories should verify the suitability of their procedures as part of the method validation with testing the stability of those compounds, which are known to rapidly decompose under unfavorable conditions.

The cryogenic processing applied successfully in some laboratories[35] includes: preparation of the portion of sample to which the MRL applies (analytical sample) on receiving the fresh sample in the laboratory; placing the analytical sample into the deep-freezer within the shortest possible time; chopping, grinding the sample in the presence of sufficient amount of dry ice (about 1:1 sample/dry ice ratio) to keep the temperature below $-20°C$ (this process requires robust choppers with stainless steel bowl and lid); withdrawing the test portions needed for various extractions and confirmation of residues into appropriate unsealed containers and placing them in deep-freezer for a minimum of 16 h to allow the carbon dioxide to evaporate; weighing the mass of the test portion, adding extraction solvent and warming the test portion up to room temperature before proceeding with the extraction.

The stability of residues can be tested with a mixture of pesticides, which contains a reference compound (R) known to be stable (e.g., chlorpyrifos), at least one compound decomposing rapidly (chlorothalonil, dichlofluanid, captan) and the other compounds to be tested. The test mixture should be carefully applied on the surface of the plant material avoiding runoff. The treated sample should be kept in fume cupboard until the solvent completely evaporates. The processing under ambient temperature can now be started, while the treated sample should be placed in deep-freezer before cryogenic milling for a minimum of 16 h.

A minimum of three test portions should be withdrawn from the comminuted material, and the extract should be analyzed in duplicate. The number of test portions and the replicate analyses depend on the CV_A of the analytical method and the percentage decomposition, which should be quantified with a selected probability. The result of the analysis is evaluated by comparing the measured residues to the expected ones.

The significance of the difference between the expected and survived residues $(A_{11}-A'_{11}, A_{12}-A'_{12}, \text{etc.})$ can be calculated with the one-tail Student's t-test for differences.

$$t_{calc} = \frac{D}{S_d/\sqrt{n}} \quad \text{where} \quad S_d = \sqrt{\frac{\sum (d_i - D)^2}{n - 1}}, \tag{5.7}$$

where D is the average of the differences and n is the number of measurement pairs.

Before the start of the study, one can also calculate with Equation 5.6 the minimum difference between the expected and survived residues (Table 5.3), which can be considered statistically significant, and decide on the number of test portions and replicate analysis to be performed in the study depending on the

TABLE 5.3

Examples for the Quantifiable Differences between the Expected and Survived Residues

Residue[a]	n[b]	CV_A[c]	CV_{qR}[d]	CV_d[e]	SD[f]	x_d[g]	Rel. dif. %[h]
0.5	6	0.1	0.1	0.141	0.071	0.058	11.6
0.5	10	0.1	0.1	0.141	0.071	0.041	8.2
0.5	15	0.1	0.1	0.141	0.071	0.032	6.4
0.5	6	0.15	0.1	0.180	0.090	0.074	14.8
0.5	10	0.15	0.15	0.212	0.106	0.061	12.3
0.5	6	0.08	0.08	0.113	0.057	0.047	9.3

[a] Expected residue concentration.
[b] Number of valid residue values determined in analytical portions.
[c] Repeatability of the recovery of the analyte (A, B, C, . . . , X).
[d] Repeatability of the recovery of reference compound.
[e] Relative uncertainty of the calculated difference.
[f] Standard deviation of the difference.
[g] Quantifiable significant difference between the expected and survived residues is $>x_d$.
[h] Quantifiable relative difference.

percentage of the decomposition which should be quantified. The acceptable decrease of residues during sample processing has not been officially specified. As a guidance value 5%–10% may be used, as it is considered acceptable difference between two standard solutions.[3]

5.5 METHOD VALIDATION

The concepts of method validation have been developed simultaneously by AOAC International, EURACHEM, IUPAC Working Party, and several national organizations. The general criteria set by the different guidelines are similar and provide the basis for assuring reliability of the methods validated for one or a few analyte–sample matrix combinations. However, these general guidelines are not directly applicable to the methods used in pesticide residue analysis as they cannot address the specific requirements and limitations. To provide guidance on in-house method validation to analysts, national authorities, and accreditation bodies, a *Guideline for Single-laboratory Method Validation* was developed and discussed at an International Workshop.[36] The Guidelines were included in the GLP GLs of CCPR.[4] The Guidelines also provide specific information for extension of the method to a new analyte and/or new sample matrix, and adaptation of a fully validated method in another laboratory.

According to the Guidelines the method validation is not a one-time, but continuous operation including the performance verification during the use of the method. Information essential for the characterization of a method may be gathered during the development or adaptation of an analytical procedure; establishment of acceptable performance; regular performance verification of methods applied in the laboratory, demonstration of acceptable performance in second or third laboratory

(AOAC Peer-Verified Method), and participation in proficiency test or interlaboratory collaborative study.

Before validation of a method commences, the method must be optimized, standard operation procedure (SOP) describing the method in sufficient detail should be prepared, and the staff performing the validation should be experienced with the method. Parameters to be studied are: stability of residues during sample storage, sample processing, and in analytical standards; efficiency of extraction; homogeneity of analyte in processed samples; selectivity of separation; specificity of analyte detection; calibration function; matrix effect; analytical range, limit of detection, limit of quantitation (LOQ), and ruggedness of the method.

The validation should be performed in case of *individual methods* with the specified analyte(s) and sample materials, or using sample matrices representative of those to be tested by the laboratory; *group specific methods* with representative commodity(ies)* (Table 5.4) and a minimum of two representative analytes[†] selected from the group; MRMs with representative commodities and a minimum of 10 representative analytes. For method validation purposes, commodities should be differentiated sufficiently but not unnecessarily. The concentration of the analytes used to characterize a method should be selected to cover the analytical ranges of all analytes. Full method validation shall be performed in all matrices and for all compounds specified, if required by relevant legislation.

The method is considered applicable for an analyte if its performance satisfies the basic requirements summarized in Table 5.5. The repeatability and reproducibility criteria given in the table are based on the Horwitz equation: $RSD = 2C^{(-0.1505)}$. In the equation, the concentration C is expressed in dimensionless mass ratios (e.g., $1 \text{ mg/kg} \equiv 10^{-6}$). Recent studies indicated that the Horwitz equation would probably overestimate the variability of the results at low concentrations ($<0.1 \text{ mg/kg}$).[37] Therefore, the tabulated data should be considered as the upper limit of the acceptable reproducibility.

5.5.1 INTERNAL QUALITY CONTROL

Based on the validation and optimization data generated, a QC scheme should be designed.

The performance of the method shall be regularly verified during its use as part of the internal quality control program of the laboratory.

The internal quality control/performance verification is carried out to: monitor the performance of the method under the actual conditions prevailing during its use, and take into account the effect of inevitable variations caused by, for instance, the composition of samples, performance of instruments, quality of chemicals, varying performance of analysts, and laboratory environmental conditions; demonstrate that

---

* Single food or feed used to represent a commodity group for method validation purposes. A commodity may be considered representative on the basis of proximate sample composition, such as water, fat/oil, acid, sugar and chlorophyll contents, or biological similarities of tissues, and so on.

[†] Analyte chosen to represent a group of analytes which are likely to be similar in their behavior through a multiresidue analytical method, as judged by their physicochemical properties, for example, structure, water solubility, K_{ow}, polarity, volatility, hydrolytic stability, pK_a, and so on.

TABLE 5.4
Representative Commodities for Multiresidue Methods[a] and Storage Stability Tests

Group	Common Properties	Commodity Group	Representative Species
Plant products			
I	High water and chlorophyll content	Leafy vegetables	Spinach or lettuce
		Brassica leafy vegetables	Broccoli, cabbage, kale
		Legume vegetables	Green beans, green peas
		Fodder crops	
II	High water and low or no chlorophyll content	Pome fruits	Apple, pear, peach, cherry
		Stone fruits	Strawberry
		Berries, small fruits	Grape
		Fruiting vegetables	Tomato, bell pepper, melon
		Root vegetables	Potato, carrot, parsley
		Fungi	Mushroom
III	High acid content	Citrus fruits	Orange, lemon
		Berries, pineapple	Blueberry, current
IV	High sugar content		Raisins, dates
V	High oil or fat	Oil seeds	Avocado, sunflower seed
		Nuts	Walnut, nuts, pistachios, peanut
VI	Dry materials	Cereals	Wheat, rice, or maize grains
		Cereal products	Wheat bran, wheat flour
	Commodities requiring individual test		e.g., Garlic, hops, tea, spices, cranberry
Commodities of animal origin			
		Mammalian meat (muscle)	Any of the major species
		Poultry meat, edible offals, fat	
		Eggs	
		Milk	

[a] For storage stability tests groups I and II may be combined, and crops of high protein or starch content should be considered separately.

the performance characteristics of the method are similar to those obtained during method validation, the application of the method is under "statistical control," and the accuracy and uncertainty of the results are comparable to the performance characteristics established during method validation.

The results of internal quality control provide essential information for the confirmation and refinement of performance characteristics established during the initial validation, and extension of the scope of the method. Some key components of the QC scheme are summarized later.

The correct preparation of analytical standards should be verified by comparing its analyte content to the old standard, or preparing the new standard in duplicate at

TABLE 5.5

Acceptable within Laboratory Performance Characteristics of a Method[a]

Concentration mg/kg	Repeatability		Reproducibility		Trueness[b]
	$CV_A\%^c$	$CV_L\%^d$	$CV_A\%^c$	$CV_L\%^d$	Range of Mean Recovery, %
≤0.001	35	36	53	54	50–120
>0.001 to ≤0.01	30	32	45	46	60–120
>0.01 to ≤0.1	20	22	32	34	70–120
>0.1 to ≤1	15	18	23	25	70–110
>1	10	14	16	19	70–110

[a] With multiresidue methods, there may be certain analytes where these quantitative performance criteria cannot be strictly met. The acceptability of data produced under these conditions will depend on the purpose of the analyses, for example, when checking for MRL compliance the indicated criteria should be fulfilled as far as technically possible, while any data well below the MRL may be acceptable with the higher uncertainty.

[b] These recovery ranges are appropriate for multiresidue methods, but strict criteria may be necessary for some purposes, for example, methods for single analyte.

[c] CV_A: Coefficient of variation of analysis excluding sample processing.

[d] CV_L: Overall coefficient of variation of a laboratory result, allowing up to 10% variability of sample processing.

the first time. A balance with 0.01 mg sensitivity should not be used to weigh <10 mg standard material. The dilutions of standard solutions should be made independently based on weight measurement except the last step for which an A-grade volumetric flask should be used.[12]

Weighted regression (WLR) should be applied for evaluation of the linear calibration function for GLC and HPLC measurements especially at the lower third of the calibrated concentration range. The confidence limits at the middle and upper calibrated range are about the same with the weighted and ordinary (OLR) regression calculation (Figure 5.2).

The goodness of the calibration should be characterized with the standard deviation of the relative residuals (S_{rr}), as it is much more sensitive indicator than the regression coefficient (see Figure 5.2 and Table 5.6). The relative residuals' (residuals/predicted response $\Delta y_i = y_i - \hat{y}$; $\bar{Y}_{rel,i} = \Delta y_i/\hat{y}$) standard deviation ($S_{rr}$) is calculated as

$$S_{rr} = \sqrt{\frac{\sum(\bar{Y}_{rel,i} - \bar{Y}_{rel})^2}{n-2}}, \tag{5.8}$$

where y_i is the response of standard x_i and $\hat{y}$ is the corresponding response on the regression line. Where the calibration points are spread over the analytical sequence, an S_{rr} of 0.1 and 0.6 may be considered acceptable for GC and HPLC methods, respectively.

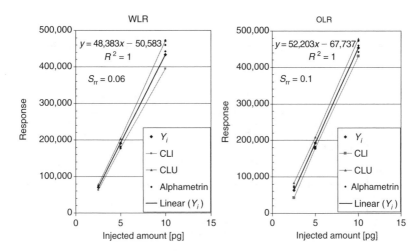

FIGURE 5.2 Evaluation of calibration with weighted (WLR) and ordinary (OLR) linear regression.

For the most effective internal quality control, analyze samples concurrently with quality control check samples. For checking acceptability of individual recovery results, the initial control charts is constructed with the average recovery (Q) of representative analytes in representative matrices and the typical within laboratory reproducibility coefficient of variation (CV_{typ}) of analysis. The warning and action limits are $Q \pm 2 \times CV_{Atyp} \times Q$ and $Q \pm 3 \times CV_{Atyp} \times Q$, respectively. At the time of the use of the method, the recoveries obtained for individual analyte/sample matrices are plotted in the chart.

Based on the results of internal quality control tests, refine the control charts at regular intervals if necessary. If the analyte content measured in the quality control check samples is outside the action limits, the analytical batch (or at least the analysis of critical samples in which residues found are ≥ 0.7 MRL and 0.5 MRL for regularly and occasionally detected analytes, respectively) may have to be repeated. When the

TABLE 5.6

Comparison of the Standard Deviation of the Relative Residuals and the Regression Coefficient

S_{rr}	R^2
0.042	0.9937
0.061	0.9976
0.062	0.9882
0.085	0.9988

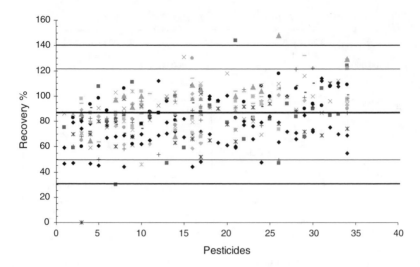

FIGURE 5.3 Illustration of long-term reproducibility of a MRM with different pesticide sample combinations.

results of quality control check samples fall repeatedly outside the warning limits (1 in 20 measurements outside the limit is acceptable), the application conditions of the method have to be checked, the sources of error(s) have to be identified, and the necessary corrective actions have to be taken before the use of the method is continued.

The differences of the replicate measurements of test portions of positive samples can be used to calculate the overall within laboratory reproducibility of the method (CV_{Ltyp} calculated with Equation 5.4). The CV_{Ltyp} will also include the uncertainty of sample processing, but will not indicate if the analyte is lost during the process.

The long-term reproducibility of the MRM can be demonstrated by plotting on the control chart all recovery values of compounds, that can be characterized with the same typical average recovery and CV_A, obtained during the use of the method. Figure 5.3 shows the quality control chart including 394 recoveries of 35 GC amenable residues in 21 commodities at spiking levels of 0.01–1 mg/kg over 1 year (F. Zakar, personal communication, 2000).

The applicability of the method for the additional analytes and commodities shall be verified as part of the internal quality control program. All reported data for a specific pesticide matrix combination should be supported with either validation or performance verification performed on that particular combination.

5.6 INTERLABORATORY STUDIES

Regular participation in interlaboratory studies and proficiency tests is an important part of the quality assurance programs and a basic requirement for accreditation, as it provides the opportunity for the laboratories to prove the suitability of their methods and proficiency of the staff in their application. Good results obtained in these studies are only showing the capability of the laboratory, but they do not verify similar performance during the daily work, which must be shown with the results of internal quality control.

The harmonized criteria for testing the proficiency of laboratories had been jointly elaborated by ISO, AOAC, and IUPAC.[38,39] The current proficiency test programs[40,41] organized by several organizations are based on the revised version of the harmonized criteria.[42]

Before the samples issued the participants are normally informed about the list of pesticides that may be present in the sample. The carefully homogenized and tested material[43] may contain field-incurred residues or spiked with mixtures of analytes, which are known to be stable during the expected duration of the study. The reported results are first screened for obvious erroneous data, then statistically evaluated for analytical outliers with Cochran test and robust statistics.[44] The assigned value for the mean $(\hat{x})$ is established by calculation of the robust mean, the median, or the mode depending on the distribution of the results after removal of spurious and outlier values. The target standard deviation, σ, should reflect the best practice or "fit for purpose" for the given analyte, which may be derived from the results of collaborative studies[40] or predicted based on Horwitz equation or those suggested by Thompson.[37] The z-score is calculated from the assigned and reported value $z = (x_i - \hat{x})/\sigma$. The interpretation of the z-score is based on the normal statistics. Laboratories performing as expected should produce results within the $2 \times z$ in most of the cases, but 1 result out of 20 may be between $2–3 \times z$. Results above $3 \times z$ should occur very rarely (the probability is 0.3%) and "requires action." In order to evaluate each laboratory's overall performance, and taking into account all pesticides analyzed, the EU proficiency test program[41] used three methods to combine the z-scores: RSZ $= (\Sigma \mid z_i \mid)/n$; SSZ $= \Sigma z^2$, and the "weighted sum of z-scores $= (\Sigma \mid z_i \mid \omega_{zi})/n$ (the ω_{zi} is assigned as $\omega_{zi} = 1$ if $z \leq 2$; $\omega_{zi} = 3$ if $2 < z \leq 3$; and $\omega_{zi} = 5$ if $z > 3$).

Proficiency scheme providers, participants, and end users should avoid classification and ranking of laboratories on the basis of their z-scores.

REFERENCES

1. International Standard Organization, ISO/IEC 17025:2005 General requirements for the competence of testing and calibration laboratories, 2005.
2. OECD Series on Principles of good laboratory practice and compliance monitoring No. 1-11, OECD Environmental Directorate, Paris, 1998, http://www.oecd.org/document/63/0,2340,en_2649_34365_2346175_1_1_1_1,00.html
3. European Commission, Quality control procedures for pesticide residues analysis, Document No. SANCO/10232/2006, 2006.
4. Codex Committee on Pesticide Residues (CCPR), Guidelines on good laboratory practice in residue analysis, CAC/GL 40-1993, Rev. 1-2003, 2003. www.codexalimentarius.net/download/report/655/al29_24e.pdf
5. EURACHEM/CITAC, Quality assurance for research and development and non-routine analysis, 1998.
6. EURACHEM/CITAC, Selection, use and interpretation of proficiency testing (PT) schemes by laboratories—2000, 2000.

7. EURACHEM/CITAC, *Traceability in Chemical Measurement*. A guide to achieving comparable results in chemical measurement, 2003.
8. King, B., Quality in the analytical laboratory, in *6th International Symposium on Quality Assurance and TQM for Analytical Laboratories*, ed. Parkany, M., Royal Society of Chemistry, Cambridge, UK, 1995, pp. 8–18.
9. Miller, J.N. and Miller, J.C., *Statistics and Chemometrics for Analytical Chemistry*, 4th edn, Pearson Education Limited, Harrow, UK, 2000, chap. 1.
10. EURACHEM, *EURACHEM/CITAC Guide Quantifying Uncertainty in Analytical Measurements*, 2nd edn, 2000, http://www.measurementuncertainty.org
11. International Standard Organisation, *Guide to the Expression of Uncertainty of Measurement*, 1995.
12. Ambrus, A., Reliability of measurement of pesticide residues in food, *Accred. Qual. Assur.*, 9, 288, 2004.
13. Codex Committee on Pesticide Residues (CCPR), Proposed draft guidelines on estimation of uncertainty of results, in Report of the 38th Session of the Codex Committee on Pesticide Residues, 2006, http://www.codexalimentarius.net/web/archives.jsp?year = 06
14. Food and Agriculture Organisation, Portion of commodities to which codex maximum residue limits apply and which is analyzed, in *Joint FAO/WHO Food Standards Programme Codex Alimentarius V. 2, Pesticide Residues in Food*, 2nd edn, FAO, Rome, 1993, p. 391.
15. El-Bidaoui, M., et al., Testing the effect of sample processing and storage on stability of residues, in *Principles and Practices of Method Validation*, eds. Fajgelj, A. and Ambrus, A., Royal Society of Chemistry, Cambridge, UK, 2000, pp. 75.
16. Suszter, G., et al., Estimation of efficiency of processing soil samples, *J. Environ. Sci. Health Part B*, 41, 1, 2006.
17. Meier, P.C. and Zünd, R.E., *Statistical Methods in Analytical Chemistry*, John Wiley & Sons, New York, 1993, chap. 2.
18. FA/WHO Pesticide Residues—Evaluations, http://www.fao.org/AG/AGP/AGPP/Pesticid/
19. Thompson, M., et al., Harmonised guidelines for the use of recovery information in analytical measurement, *Pure Appl. Chem.*, 71, 337, 1999.
20. Food and Agriculture Organisation, FAO, Manual on submission and evaluation of pesticide residue data for the estimation of maximum residue levels in food and feed, FAO Plant Production and Protection Paper 170, 2002.
21. Kateman, G. and Buydens, L., *Quality Control in Analytical Chemistry*, 2nd edn, John Wiley & Sons, New York, 1993, chap. 2.
22. Codex Secretariat, Recommended method of sampling for the determination of pesticide residues for compliance with MRLs, CAC GL 33-1999. www.codexalimentarius.net/download/standards/361/CXG_033e.pdf
23. Ambrus, A. and Lantos, J., Evaluation of the studies on decline of pesticide residues, *J. Agric. Food Chem.*, 50, 4846, 2002.
24. Ambrus, A. and Soboleva, E., Contribution of sampling to the variability of pesticide residue data, *JAOAC Int.*, 87, 1368, 2004.
25. Ambrus, Á., Variability of pesticide residues in crop units, *Pest Manag. Sci.*, 62, 693–714, 2006.
26. Hill, A.R.C. and Reynolds, S.L., Unit-to-unit variability of pesticides residues in fruit and vegetables, *Food Addit. Contam.*, 19, 733, 2002.
27. Ambrus, A., Within and between field variability of residue data and sampling implications, *Food Addit. Contam.*, 17(7), 519, 2000.

28. International Standard Organization, ISO 11648-1 Statistical aspects of sampling from bulk materials—Part 1: General principles, 2003.

29. Ambrus, Á., Solymosné, M.E., and Korsós, I., Estimation of uncertainty of sample preparation for the analysis of pesticide residues, *J. Environ. Sci. Health Part B*, 31, 443, 1996.

30. Maestroni, B., et al., Testing the efficiency and uncertainty of sample processing using ^{14}C labelled chlorpyrifos—Part I, in *Principles of Method Validation*, eds. Fajgelj, A. and Ambrus, A., Royal Society of Chemistry, Cambridge, UK, 2000, p. 49.

31. Lyn, J.A., et al., Measurement uncertainty from physical sample preparation: estimation including systematic error, *Analyst*, 1391, 2003.

32. Wallace, D. and Kratochvil, B., Visman equations in the design of sampling plans for chemical analysis of segregated bulk materials, *Anal. Chem.*, 59, 226, 1987.

33. Food and Agriculture Organization, Guidelines on pesticide residue trials to provide data for the registration of pesticides and the establishment of maximum residue limits—Part 1—Crops and Crop Products, 1986.

34. Hill, A.R.C., Harris, C.A., and Warburton, A.G., Effects of sample processing on pesticide residues in fruits and vegetables, in *Principles and Practices of Method Validation*, eds. Fajgelj, A. and Ambrus, A., Royal Society of Chemistry, Cambridge, UK, 2000, p. 41.

35. Fussell, R.J., et al., Assessment of the stability of pesticides during cryogenic sample processing. 1. Apples, *J. Agric. Food Chem.*, 441, 2002.

36. Fajgelj, A. and Ambrus, A., eds., *Principles and Practices of Method Validation*, Royal Society of Chemistry, Cambridge, UK, 2000.

37. Thompson, M., Recent trends in inter-laboratory precision at ppb and sub-ppb concentrations in relations to fitness for purpose criteria in proficiency testing, *Analyst*, 385, 2000.

38. Thompson, M. and Wood, R., The international harmonized protocol for the proficiency testing of (chemical) analytical laboratories, *JAOAC Int.*, 926, 1993.

39. Thompson, M. and Wood, R., Harmonised guidelines for internal quality control in analytical chemistry laboratories, *Pure Appl. Chem.*, 649, 1995.

40. FAPAS, Protocol for the organisation and analysis of data, 2002, http://www.fapas.com/pdfpub/FAPASProtocol.pdf

41. Fernandez-Alba, A.R., CRL Proficiency Test 8, Pesticide residues in augberine homogenate, Final Report, 2006.

42. Thompson, M., Ellison, S.L.R., and Wood, R., International harmonised protocol for the proficiency testing of analytical laboratories, IUPAC Technical Report, *Pure Appl. Chem.*, 145, 2006.

43. Fearn, T. and Thompson, M., A new test for 'sufficient homogeneity', *Analyst*, 1414, 2001.

44. Analytical Methods Committee, Robust statistics: a method of coping with outliers, Technical Brief No. 6, 2001, http://www.rsc.org/pdf/amc/brief6.pdf

6 Determination of Pesticides in Food of Vegetal Origin

Frank J. Schenck and Jon W. Wong

CONTENTS

6.1   INTRODUCTION

According to the European Commission, currently >1100 substances are registered as pesticides [1]. Pesticides are defined by the United States Federal Insecticide Fungicide Rodenticide Act (FIFRA) as "any substance or mixture of substances intended for preventing, destroying or repelling any pest," where pests are defined as organisms that may be deleterious to human or the environment, including vertebrates other than human, invertebrates, plants, fungi, and microorganisms [2]. This gives rise to specific terms such as insecticides, acaricides, herbicides, fungicides, nematicides, and rodenticides.

Pesticides have been traditionally classified based on functional groups (e.g., *N*-methylcarbamates, pyrethroids) or based on specific elements in their molecular structure. The latter classification is a result of the fact that gas chromatography (GC) with element-selective detectors has been used for many years. The U.S. Food and Drug Administration (FDA) still classifies pesticides as organohalogen (OH) or organochlorine (OC), organophosphorus (OP), and organonitrogen (ON), despite the fact that the majority of analytes are screened primarily by gas chromatography with mass spectrometry (GC-MS) [3–5].

Pesticide residues most commonly found in food samples of vegetal origin are pesticides that are intentionally applied to the plants to attack invertebrate pests (insecticides, acaricides, etc.) and plant diseases (fungicides). The FDA *Pesticide Analytical Manual* (PAM) states, "whenever a sample of unknown pesticide treatment history is analyzed, and no residue(s) is targeted, a multiclass multiresidue method (multiclass MRM) should be used to provide the broadest coverage of residues" [3]. Thus, in theory, the sample should be assayed for all chemicals that are currently used as pesticides. Many multiclass MRMs have been developed for foods of vegetal origin, and their various extraction, cleanup, and detection procedures will be the discussed in this chapter.

Three important molecular properties that determine if the pesticide will or will not be recovered and detected through any of the various extraction, cleanup, and detection steps used in the analysis of pesticides are polarity, volatility, and thermal lability. Generally, polarity for nonionic pesticides ranges from the lipophilic OCs (e.g., *p,p'*-DDE) and synthetic pyrethroids (e.g., deltamethrin) to the very polar, water-soluble OPs, methamidophos, and acephate. Thus, a measure of the usage of a comprehensive multiclass MRM is whether it can recover both the nonpolar and polar pesticides. Volatility and thermal lability are important because they determine whether the pesticide can be determined by GC or not. Many pesticides are thermally labile, and will degrade in a GC due to the heated conditions of the injector and the increasing temperature gradients applied to the column. Other separation methods, mainly high-performance liquid chromatography (HPLC), must then be used.

6.2 SAMPLE EXTRACTION

The first step in the sample preparation process is to obtain a homogeneous composite. The FDA PAM [6] states that fresh fruits and vegetables should be comminuted for at least 5 min in a vertical cutter mixer. Studies evaluating the homogeneity of produce samples composited in this fashion have shown that reproducible results can be obtained with sample sizes as small as 10–25 g [7,8]. After a sample is composited, the sequence of steps in pesticide residue analysis usually involves: (1) extraction of the pesticide residues from the sample matrix, (2) removal of coextracted water from the extract, (3) cleanup of the extract, and (4) analytical determination. Generally, for purposes of extraction, food samples of vegetal origin can be broken up into three broad groups: samples of medium and high water content (fresh fruits and vegetables; produce), dry samples (containing <10% water), and fatty samples (containing >2% fat) [3].

6.2.1 ORGANIC SOLVENT EXTRACTION

Extraction with an organic solvent using a blender or a homogenizer is still the most widely used approach for the separation of nonionic pesticides from the plant matrix. The most commonly used extraction solvents are acetonitrile, acetone, ethyl acetate, and methanol. While water-miscible solvents, such as acetonitrile and acetone, will effectively extract pesticide residues from high moisture fresh fruits and vegetables, they will not adequately extract pesticide residues from dry samples, such as grains or feeds. Bertuzzi et al. [9] first demonstrated that acetonitrile–water (65/35, v/v) will effectively extract pesticides from dry products. Later, Luke and Doose [10] likewise showed that acetone–water mixtures (65/35, v/v) could also be used to extract pesticides from dry products.

Traditionally, it has been accepted that pesticide extraction from solid foods must be accomplished using some type of mechanical homogenization. Studies at the Florida Department of Agriculture and Consumer Services Laboratory showed that pesticides could be extracted from well-comminuted produce samples by shaking with acetonitrile [11]. The advantages of shaking over mechanical homogenization are mainly that it is faster and easier, less equipment is needed, and there is less chance of carryover from one sample to the next. A number of methods have been presented that employ organic solvent extraction by shaking instead of using mechanical blenders or homogenizers [8,11–20]. Shaking may not work as well as homogenizing for some of the more nonpolar OCs. Okihashi et al. [21] reported that recoveries of incurred residues of nonpolar OCs like *o,p'*-DDT, dicofol, and endrin were lower when shaking instead of homogenizing with acetonitrile.

6.2.1.1 Acetonitrile Extraction and Liquid–Liquid Partitioning

Acetonitrile was the extraction solvent used for one of the earliest multiclass MRMs, the "Mills method," developed in the 1960s [3,22,23]. Even though the Mills method was developed when pesticide methods were primarily concerned with the recovery of the nonpolar OCs, a water-miscible solvent, acetonitrile, was used. Polar solvents are needed for the extraction of nonpolar OC pesticides from the plant matrix.

Mumma et al. [24] hypothesized that this is caused by the pesticides interacting with surfactant phospholipids, sulfolipids, and glycolipids from the plant. Using the Mills method, water and NaCl are added to the sample extract, and pesticide residues are partitioned from the acetonitrile–water mixture into a very nonpolar solvent, petroleum ether.

While the Mills method worked very well with the nonpolar OC pesticides used in the 1960s, some of the more polar OPs that were developed in the 1970s were not easily recovered. An advantage of using acetonitrile as an extraction solvent is that while it is completely miscible with water, it can be readily separated from water not only by liquid–liquid partition with nonpolar solvents (Mills method) but also by the addition of salts (salting out). Acetonitrile extractions of produce samples followed by salting out were used at the California Department of Food and Agriculture in the early 1990s [25,26]. While the resulting extracts were not as clean as those obtained by the Mills method, both polar and nonpolar pesticides could be recovered. This approach of using acetonitrile extraction followed by salting out has been adopted by regulatory agencies in Florida, Canada, and New York [11,27,28].

In 2003, a new approach to the extraction of pesticides from fresh fruits and vegetables with acetonitrile, called *quick, easy, cheap, effective, rugged*, and *safe* (QuEChERS) was reported [12]. This method entailed shaking the sample with acetonitrile, followed by shaking with sodium chloride (NaCl) and $MgSO_4$ to remove the water. The salts create an exothermic reaction with water, induce phase separation between water and acetonitrile, and bind water to drive the pesticide analytes into the acetonitrile phase, resulting in high recoveries, including the polar and water-soluble pesticide, methamidophos. A modified QuEChERS extraction, using 1% acetic acid–acetonitrile extraction solvent and sodium acetate rather than NaCl, was developed to facilitate the recovery of base-sensitive fungicides like chlorothalonil and captan [16].

Vegetal matrices containing a high lipid content present a challenge for cleanup because the fats and waxes would have an adverse effect on the GC columns and could interfere with the analysis and detection of the pesticides. In 1952, Jones and Riddick [29] found that even the most lipophilic, nonpolar pesticides like *p,p'*-DDT could be separated from fats and waxes by liquid–liquid partition between hexane and acetonitrile. Pesticides have been extracted from olive oil by dissolving the oil in hexane, and then shaking with acetonitrile [30,31]. Similarly soya oil [32] and olive oil [33] have been dissolved in hexane and loaded onto diatomaceous earth columns, with subsequent elution with acetonitrile.

6.2.1.2 Acetone Extraction and Liquid–Liquid Partitioning

The introduction of new water-soluble, very polar, OP insecticides, such as methamidophos and acephate in the 1970s, resulted in the development of the "Luke method," at the FDA [34,35]. Produce samples were extracted with acetone, and water was removed from the extract by a series of liquid–liquid partition steps, first with petroleum ether–dichloromethane, followed by dichloromethane–NaCl. Both polar and nonpolar pesticides could be recovered. Hopper [36] substituted diatomaceous earth columns for separatory funnels used in the Luke method. Acetone–water extracts are adsorbed onto the diatomaceous earth, and the pesticides are eluted from the column with dichloromethane. The dichloromethane used in the Luke method came to be recognized as an environmental hazard, so in Europe, a combination of ethyl acetate and cyclohexane was used instead of dichloromethane [37]. Recently, Luke et al. [38] demonstrated that salting out with a combination of fructose, anhydrous $MgSO_4$, and NaCl could be used to separate water from acetone in produce samples extracts.

6.2.1.3 Ethyl Acetate Extraction

Ethyl acetate is only slightly miscible with water, which simplifies the problem of separating water from the sample extract. Since ethyl acetate is more nonpolar than the other solvents discussed, the polar pesticides do not readily partition into ethyl acetate. Large amounts of sodium sulfate are usually added to bind the coextracted water and force the polar pesticides into the organic phase. Most methods using ethyl acetate extraction entail two extractions of the sample matrix rather than the single extraction commonly used with acetonitrile and acetone. Polar solvents like ethanol may be added to the extraction solvent to increase the recovery of polar compounds [39]. Ethyl acetate extraction will result in a cleaner extract as it will extract less of the polar plant matrix compounds, but more lipids and waxes [12,40].

6.2.1.4 Methanol Extraction

Methanol has not been as commonly used as an extraction solvent. Krause [41] found methanol to be the most effective solvent for extracting ^{14}C-labeled *N*-methylcarbamate insecticides from produce samples. Produce samples were extracted with methanol, and methanol was removed from the methanol–water extract using a vacuum rotary evaporator. Pang et al. [42] used methanol to extract pyrethroids from produce samples. Water and NaCl were added to the methanol–water extract, and the nonpolar pesticides were partitioned into toluene. Klein and Alder [43] and Alder et al. [44] extracted produce samples with methanol, added NaCl to the methanol–water extract, transferred the extract to a diatomaceous earth column, and eluted the pesticide residues with dichloromethane. Granby et al. [45] extracted produce and dry samples with a methanol–acetate buffer mixture. The extracts were filtered and determined directly by HPLC with tandem mass spectrometry (HPLC-MS/MS) with no further cleanup.

6.2.2 Matrix Solid-Phase Dispersion Extraction

Matrix solid-phase dispersion (MSPD) extraction was demonstrated in 1989, when Barker et al. [46] homogeneously dispersed biological matrices with C_{18} (40 μm octadecylsilyl-derivatized silica) solid-phase extraction (SPE) sorbent. MSPD involves mixing the sample with a sorbent using a mortar and pestle. The resulting mixture is then packed into a small column, and the adsorbed residues were then selectively eluted from the column with solvent. MSPD methods for pesticides in fresh fruits and vegetables have been reported, using C_{18} [47–50], octyl-derivatized silica (C_8) [51–55], aminopropyl-derivatized silica [56], and Florisil (synthetic magnesium silicate) [57–60] SPE sorbents or diatomaceous earth [61–64]. The disadvantage of the MSPD applications using SPE sorbents is that the small sample sizes used may not be representative of the whole sample. A comprehensive review of the analytical applications of MSPD for the analysis of various contaminants and analytes in food and biological samples is presented by Barker [65].

6.2.3 SUPERCRITICAL FLUID EXTRACTION

Supercritical fluid extraction (SFE) is a process by which a solvent is obtained at its supercritical point, thus possessing the properties of a gas and a liquid, and used to extract solutes from the sample matrix. The advantages of using supercritical fluids over conventional liquid solvents are their liquid-like densities offering higher solubilities and their gas-like low viscosities and high diffusivities enabling for efficient extraction. Pesticides have been extracted from several food types of vegetal origin such as benzimidazole fungicides from apple, banana, and potato [66]; OH, OP, and ON pesticides from grapes, carrots, potatoes, and broccoli [67], and 300 pesticides in soybean, spinach, and orange [68]. A 17 laboratory AOAC collaborative study of SFE for produce samples was conducted in 2002 [69]. SFE applications for pesticide analysis in foods are presented in further detail in reviews [70–73]. Despite many studies with SFE, its major drawback is that it requires a tremendous amount of method development. Method development is not an easy task because several parameters such as the temperature and pressure of the supercritical fluid, choice and concentration of cosolvent, flow rate, and collection mode need to be optimized for different matrices [74,75].

6.2.4 PRESSURIZED FLUID EXTRACTION

Pressurized fluid extraction (PFE), also called accelerated solvent extraction (ASE), is similar to SFE because liquid solvents are used at elevated temperatures and pressures to increase the rate of solvent solubilization, allowing for increased extraction of the solutes and preventing the solvent from boiling. Since it is not necessary for the solvent to reach its supercritical point, PFE is much simpler than SFE. The major disadvantage of PFE (as well as for SFE) is that initial equipment costs are relatively expensive compared with conventional extraction blending or homogenization.

There are numerous studies using PFE for the analysis of pesticides in vegetal matrixes. PFE has been used to analyze OPs in orange juice, grapefruit, broccoli, and flour [76] and herbal supplements [77,78]; eight pesticide classes (or 28 pesticides) in pear, cantaloupe, potato, and cabbage [79]; thiabendazole, carbendazim, and phenyl urea herbicides in oranges [80]; carbamates in banana, green bean, broccoli, melon, and carrot [81]; and pesticides in 15 vegetable samples by GC-ion trap-mass spectrometry and HPLC-MS/MS [82]. The analysis of 405 pesticides in grains (maize, wheat, oat, rice, barley) was developed using PFE, SPE for cleanup, and GC-MS and HPLC-MS/MS for analysis [83]. Grain samples (10 g) were mixed with Celite 545 (10 g) and the mixture was placed in a 34 mL cell and extracted with acetonitrile in the static state at 1500 psig and 80°C. Recoveries of 382 pesticides ranged from 60% to 120% at fortification levels ranging from 0.0125 to 0.100 mg/kg. A comprehensive review of applications of PFE for the analysis of various contaminants and analytes besides pesticides in food and biological samples is presented by Carabias-Martínez et al. [84].

6.2.5 MICROWAVE-ASSISTED EXTRACTION

Microwave-assisted extraction (MAE) is a process that uses microwaves to heat the solvent in a closed and pressurized vessel, further enhancing and improving the extraction of target compounds from the matrix samples. The advantages of MAE are less extraction time and solvent use compared with conventional extraction procedure. However, the disadvantages of MAE are that extensive cleanup is usually required, the organic solvents used should be able to absorb microwave energy, and some organic solvents at high temperatures and pressures can corrode the equipment [85]. Eskilsson and Björklund [86] present a comprehensive review of the MAE process and applications. MAE has been successfully applied to analyze many different pesticides in a variety of fruits and vegetables [87–90].

6.2.6 SOLID-PHASE MICROEXTRACTION

Solid-phase microextraction (SPME) was developed by Pawliszyn and coworkers in the 1990s [91,92]. The major features of SPME are that it uses a fused silica fiber that is coated with a stationary phase and the fiber is retrofitted into a needle of a syringe device. The stationary phase can vary in film thickness and coating types. This syringe device can be inserted in a gas (headspace) or liquid (direct immersion) sample and the target analytes are extracted and concentrated in the stationary phase of the fiber. A similar procedure to SPME is stir bar sorptive extraction (SBSE), in which extraction takes place with the sample and a stir bar coated with a stationary phase [93–95]. The advantages of SPME are that little solvent is needed and the technique can be automated and coupled to a chromatographic instrument, while its major disadvantage is the lack of sensitivity due to the small sorbent size, which affects the amount of analyte that can be adsorbed onto the sorbent [96]. Sample preparation for direct immersion SPME usually involves processing the vegetal matrix into a liquid extract, which may contain added water and salt such as sodium chloride to enhance retention of the pesticide to the stationary phase of the fiber. SPME has been used to analyze phenyl urea herbicides and their aniline homologs, OPs, pyrethroid insecticides, and strobilurin fungicides in fresh fruits and vegetables [97–102].

6.3 SAMPLE CLEANUP

Sample extracts, resulting from solvent extraction and liquid–liquid partition, will contain a large number of sample matrix coextractants. The amount of cleanup required will depend on the type of chromatographic system used, the specificity of the detector used, and the desired limit of detection. In the 1960s, when gas chromatographs (GCs) equipped with electron capture detectors (ECD) were used primarily for OCs, an extensive amount of sample cleanup was required due to the lack of selectivity. Later, with the development of more selective GC detectors, such as the Hall electrolytic conductivity detector (ELCD), the flame photometric detector (FPD), and the nitrogen phosphorus detector (NPD), the only cleanup required for produce sample extracts was the removal of the coextracted water [35]. The large amounts of coextractants did not present a problem with these highly selective detectors and they did not have much effect on the packed open tubular GC columns used. Today, with the widespread use of capillary columns, and lower detection levels, a greater amount of cleanup is desirable for GC analysis. Historically, more cleanup of sample extracts was required when using HPLC, since less selective detectors such as UV, diode array, and fluorescence detectors were used. Today, HPLC-MS/MS provides excellent selectivity and sensitivity, so that in some cases a very minimal cleanup or even no cleanup may be required for sample extracts [103].

6.3.1 GEL PERMEATION CHROMATOGRAPHY

Gel permeation chromatography (GPC) or size exclusion chromatography separates particles on the basis of size. GPC is best suited for removing high-molecular-weight pigments and lipidic materials like waxes and fats [104]. A number of methods use GPC for the cleanup of solvent extracts of foods of vegetal origin [37,39,104–108].

6.3.2 SOLID-PHASE EXTRACTION

SPE is a cleanup technique where HPLC sorbents are used to separate analytes from the sample matrix. While the sorbents used for SPE are in many cases chemically similar to those used for HPLC, the particle size is much larger. SPE differs from the HPLC process, in that the analyte is eluted stepwise, rather than a gradient increase (or decrease) of the solvent strength in the mobile phase used in gradient HPLC. While the resolution of HPLC columns is typically measured as thousands of theoretical plates, SPE columns are typically in the 10–50 theoretical plate range [109]. The SPE process is usually conducted by eluting sample extracts through SPE columns (open syringe barrels or cartridges) or disks. Recently, dispersive SPE has been introduced, which involves mixing the sample extracts directly with SPE sorbents [12]. The most common types of sorbents commonly used for SPE cleanup include: C_{18}, styrene–divinyl benzene (SDVB), aminopropyl (NH_2), primary secondary amine (PSA), trimethyl ammonium strong anion exchange (SAX), graphitized carbon black (GCB), Florisil, silica, and alumina.

Two strategies have been used for SPE cleanup of sample extracts, which can be referred to as *analyte isolation* and *matrix isolation*. Analyte isolation typically entails analytes that are adsorbed from an organic solvent–water mixture onto a nonpolar SPE sorbent. The SPE column may be rinsed with an aqueous solution to remove interfering coextractives, followed by elution of the adsorbed analytes with an organic solvent. Matrix isolation entails using the SPE columns which act as "chemical filters" and trapping matrix components while the pesticides are eluted through the column. Analyte isolation using nonpolar reversed phase SPE columns (C_{18} or SDVB) works very well with nonpolar pesticides such as the OCs and the synthetic pyrethroids but not with polar, water-soluble pesticides, as they may not be consistently adsorbed onto nonpolar reversed phase SPE sorbents from an organic solvent–water mixture. Therefore, few pesticide multiclass MRMs use analyte isolation [110–114].

Matrix isolation is more widely used in pesticide residue cleanup. Generally, matrix isolation SPE cleanup procedures are devised so that both polar and nonpolar pesticides will be eluted through the SPE column(s) while certain matrix components will be retained. The following section provides examples of how various SPE sorbents have been used for matrix isolation SPE cleanup. Further information on matrix isolation cleanups of sample extracts is provided in Table 6.1.

TABLE 6.1
SPE Column and Dispersive SPE Cleanups Used for the Analysis of Pesticides in Produce Samples

SPE Sorbent(s)[a]	Pesticides[b]	Elution Solvent[c]	References
SPE Columns			
C_{18}	Multiclass	Acetone–water (7:3)	[3,38,115]
		Acetonitrile–water (7:3)	[11,25]
		Acetonitrile	[27,116–118]
Florisil	OH	Acetone–hexane (1:7)	[11]
	OH	EtAc–hexane (1:9)	[39]
	OH	Ethyl ether–PE (3:7)	[119]
	OH, PYR	EtAc–hexane (3:7)	[120]
	OH, OP, PYR	EtAc–hexane (1:1)	[108]
Florisil/silica	ON, OP	Acetone–PE (3:7)	[121]
GCB	Multiclass	Acetone–EtAc (1:1)	[122,123]
		Acetonitrile–toluene (3:1)	[118]
	OP	Acetonitrile–toluene (3:1)	[78]
GCB/Florisil	OH, PYR	Hexane–DCM–acetone (10:60:30)	[79]
GCB/NH2	Multiclass	Acetonitrile–toluene (3:1)	[27,116,117,124]
	OH, OP	Acetone–toluene (3:1)	[125]
GCB/PSA	Multiclass	Acetonitrile–toluene (3:1)	[21]
		Hexane–acetone (1:1)	[126]
	OH, OP	Acetone–toluene (3:1)	[127]
	OP	Acetone–toluene (3:1)	[128]
GCB/SAX/PSA	Multiclass	Acetonitrile–toluene (3:1)	[28]
NH_2	Multiclass	Acetone	[129–131]
	NMC	Methanol–DCM (1:99 or 2:98)	[25,132–135]
		Methanol–acetone (3:7)	[11]
		EtAc	[39]
PSA	Multiclass	Acetone	[136]
SAX	OH, OP	DCM	[63]
SAX/PSA	Multiclass	PE–acetone (2:1)	[115]
	OH, ON	PE–acetone (2:1)	[3,4]
	OH, PYR	Acetone–hexane (3:7)	[137]

(continued)

TABLE 6.1 (continued)
SPE Column and Dispersive SPE Cleanups Used for the Analysis
of Pesticides in Produce Samples

SPE Sorbent(s)[a]	Pesticides[b]	Elution Solvent[c]	References
Silica	Fungicides	EtAc	[50]
	Fungicides	Acetone–DCM (1:1)	[138]
	OH, OP, NMC	EtAc–hexane (3:7)	[39]
Dispersive SPE			
PSA	Multiclass	Acetonitrile	[12,13,15,16,17,139,140]
	Fungicides		[18]
	OH, OP		[8]
	NMC, OP		[14]
GCB/PSA	OH	Acetonitrile–toluene (3:1)	[141]

[a] C_{18}, Octadecyl-derivatized silica; GCB, graphitized carbon black; NH_2, aminopropyl; PSA, primary secondary amine; SAX, strong anion exchanger.

[b] OH, Organohalogens; ON, organonitrogens; OP, organophosphates; PYR, synthetic pyrethroids; NMC, *N*-methylcarbamates.

[c] DCM, dichloromethane; EtAc, ethyl acetate; PE, petroleum ether.

6.3.2.1 C_{18} SPE Cleanup

Octadecyl (C_{18}) SPE columns have been used to filter out planar nonpolar sample matrix components, primarily pigments and waxes, from sample extracts. When acetonitrile–water [11,25], acetone–water [4,5,115], or acetonitrile [37,116] extracts are eluted through C_{18} SPE columns, nonpolar pigments and lipids will be retained in the column, whereas both nonpolar and polar pesticides will be eluted through the column. Dispersive SPE with C_{18} may also be used for the cleanup of plant pigments in acetone–water produce sample extracts (F.J. Schenck, unpublished data, 2006).

6.3.2.2 NH_2 and PSA SPE Cleanup

The most effective SPE sorbents for the matrix isolation cleanup of plant extracts are aminopropyl (NH_2) and ethylenediamine-*N*-propyl (PSA) (see Figure 6.1) [12,21,126,142]. These two sorbents are weak anion exchangers and will interact strongly with acidic matrix components, such as fatty acids, to the extent that strong organic solvents, such as acetone and acetonitrile, are incapable of desorbing them. This is important since high concentrations of fatty acids (e.g., oleic, linoleic, and palmitic acids) in food extracts will produce ions that can interfere with the GC-MS selected ion monitoring (SIM) determination of coeluting pesticides [126,129]. NH_2 and PSA are the most effective sorbents for reducing the effects of sample matrix enhancement [143]. Dispersive SPE with PSA sorbent, which greatly reduces the time and effort required for an SPE cleanup, has been widely used (see Table 6.1). Dispersive SPE cleanup with PSA is not as effective as PSA SPE columns [129], especially for the HPLC-postcolumn derivatization fluorescence determination of *N*-methylcarbamates (see Figure 6.2) (F.J. Schenck, unpublished data, 2006).

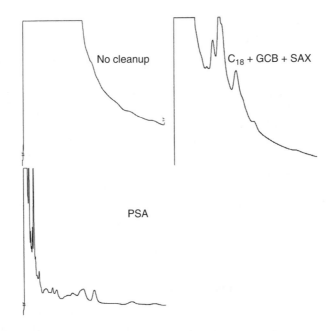

FIGURE 6.1 GC/ECD chromatograms before and after SPE cleanup of sample extracts from a blank asparagus sample. Asparagus was extracted with acetone using the method of Luke et al. [35].

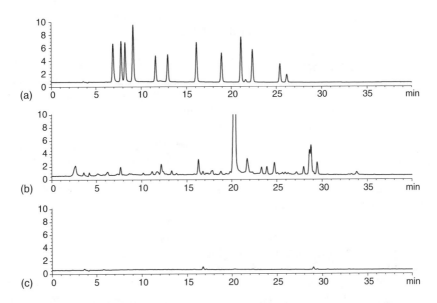

FIGURE 6.2 LC-FLD chromatograms of (a) mixed carbamate standard 1.0 ng/μL in methanol (peak identities are aldicarb sulfoxide, aldicarb sulfone, oxamyl, methomyl, 3-OH carbofuran, aldicarb, propoxur, carbofuran, carbaryl, isoprocarb, BDMC, methiocarb); (b) blank green bean-QuEChERS acetonitrile extraction with PSA-dispersive SPE cleanup; and (c) blank green bean-QuEChERS acetonitrile extraction with PSA SPE column cleanup.

6.3.2.3　Tandem GCB/NH$_2$, GCB/PSA, and GCB/SAX/PSA SPE Column Cleanup

GCB adsorbs planar compounds, and has been widely used for the removal planar molecules like pigments and sterols from produce sample extracts. GCB will not remove fatty acids and other polar compounds, thus it is usually used in combination with NH$_2$ or PSA SPE sorbents. These two sorbents when used together provide an excellent cleanup. The problem with GCB is that it may retain planar aromatic pesticides such as hexachlorobenzene (HCB) [12,16,27]. Thus, solvent mixtures containing toluene are commonly used to elute pesticides through GCB columns. Acetonitrile–toluene (3:1) mixtures have been used for the multiclass MRM elution of pesticides through tandem GCB-NH$_2$ [27,116,124], GCB-PSA [21], and GCB-SAX-PSA [28]. Schenck and Howard-King [125,127] found acetone–toluene (3:1) to be a better elution solvent than acetonitrile–toluene (3:1) since much smaller volumes of solvent were needed to elute HCB through tandem GCB-NH$_2$ or GCB-PSA SPE columns.

6.3.2.4　Alumina, Florisil, and Silica SPE Column Cleanup

Alumina (Al$_2$O$_3$), silica (SiO$_2$), and Florisil are very polar sorbents that have been used for the cleanup of sample extracts for many years. While these three sorbents provide a very effective cleanup, they will oxidize, decompose, or adsorb polar OPs pesticides [40]. They are suitable for the cleanup of nonpolar analytes such as OCs and pyrethroids. Some multiclass MRMs entail splitting the sample extract, cleaning up part of the extract on Florisil for OH analysis, and using the part that had not been cleaned up on Florisil for OP analysis [11].

6.4　DETERMINATION

GC with element-selective detection has been the primary method for the determination of pesticides for many years. These systems are used to screen a wide variety of pesticides containing heteroatoms such as halogens, phosphorus, sulfur, and nitrogen. These detectors provide not only sensitivity, but also great specificity, such that very little sample cleanup is required. Confirmation of identity of the pesticide usually required the use of a second GC with a different type of column. While element-selective detectors are still widely used, they are being replaced or complemented by mass spectrometric (MS) detectors.

The most popular GC-MS procedure for pesticide analysis of vegetal matrices is an instrument equipped with electron impact ionization and a single quadrupole mass spectrometer in the SIM mode (GC-MS/SIM). Multiresidue pesticide screening of fresh produce using GC-MS/SIM is widely used because these methods are economically efficient because any pesticide present can be screened, quantitated, and confirmed. The FDA uses GC-MS/SIM methods to detect and confirm >100 pesticides in a single injection [4,5]. Other mass analyzers used for GC analysis of pesticides besides a single quadrupole are triple quadrupole [82,144], ion trap [115,145], and time-of-flight (TOF) [146–149]. The advantage of a triple quadrupole mass spectrometer is that it offers enhanced specificity and sensitivity. The first and third quadrupoles are mass filters for the parent and product, respectively, whereas the second quadrupole is a collision chamber where collisional dissociation of the parent is used to form the product ions. Ion trap mass spectrometers can provide superior sensitivity and qualitative information, since MS/MS and MS^n experiments can be conducted, but there are problems with quantitation. TOF mass spectrometers provide the capability of nontarget identification, since these techniques provide either mass resolution or full-scan spectra at all times of the analyte, which is not possible with single ion or multiple reaction monitoring [56]. The major drawback of triple quadrupole and TOF techniques is that the instruments are expensive, compared with the cost of a single quadrupole mass spectrometer.

Methods based on HPLC were not as common in the past, because the traditional UV and fluorescence detectors were much less selective and sensitive than the GC detectors. This has changed with the commercial availability of HPLC-mass spectrometry (HPLC-MS), which has increased sensitivity and selectivity. The HPLC can be coupled to various types of mass spectrometers, such as ion trap, single quadrupole, triple quadrupole (or tandem), TOF, or hybrid instruments such as quadrupole–ion trap and quadrupole–TOF. For small molecule analysis, such as pesticides, these instruments operate in atmospheric pressure ionization (API), either in electrospray ionization (ESI) or in atmospheric pressure chemical ionization (APCI) in the positive or negative mode. A number of multiclass MRMs using HPLC-MS/MS have been reported for pesticide screening in vegetal matrices [43,48,150–157]. These multiclass MRMs have the advantage of excellent sensitivity, quantitation, and the capabilities of screening a large number and a wide variety of pesticides in a single HPLC run. The disadvantage of tandem instruments is that they are limited to targeted screening, which only permits screening of masses or transitions in the selected ion recording or multiple reaction monitoring mode, but does not allow for the discovery of unknown pesticides in the sample matrix. HPLC-TOF/MS benefits from the high resolving power of m/z signals, which enables for the accurate masses of ions for identity of the analyte [158,159]. These instruments are also sensitive, quantitative, and provide accurate mass measurements for confirmation. These characteristics allow the analysis to be conducted under full-scan conditions.

6.5 APPLICATION TO REAL SAMPLES

6.5.1 ACETONE EXTRACTION AND LIQUID–LIQUID PARTITION

Using the Luke et al. method [35], samples are extracted with acetone, partitioned once with petroleum ether–dichloromethane (1:1), partitioned twice with sodium chloride–dichloromethane, and solvent exchanged to acetone. Schenck et al. [136] eluted the Luke acetone extracts through PSA SPE columns with acetone for a quick multiresidue cleanup. Podhorniak et al. [128] eluted the Luke acetone extracts through tandem GCB/PSA SPE columns with acetone–toluene (3:1) for the GC determination of OPs at one part per billion (1.0 ppb) using GC with pulsed flame photometric detection. In another study by Podhorniak et al. [132], NH$_2$ SPE cleanup was followed by HPLC with postcolumn derivatization and fluorescence detection for the determination of *N*-methylcarbamates at 1.0 ppb level.

Cairns et al. [115] first presented an excellent and lengthy cleanup for acetone extracts of produce samples, that has been incorporated into the FDA PAM as the 302 E7C6 [3] method. Acetone–water extracts are eluted through C_{18} SPE columns to remove pigments and lipids, following which the eluant is subjected to two liquid–liquid partition cleanups with dichloromethane–sodium chloride. After azeotroping the extract with petroleum ether and acetone, the extract was eluted through tandem SAX/PSA columns with a petroleum ether–acetone mixture. Luke et al. [38] modified this method to eliminate dichloromethane partitioning, by removing water from the acetone–water eluant by salting out with a mixture of fructose, NaCl, and $MgSO_4$.

6.5.2 ACETONITRILE EXTRACTION AND SALTING OUT FOLLOWED BY MULTIPLE SPE COLUMN CLEANUPS

Using the method used by government agencies in Canada and China, produce samples are extracted with acetonitrile. Water is removed from the extract by salting out, and the acetonitrile is eluted through a C_{18} SPE column. The extract is eluted through tandem GCB/NH_2 SPE columns with an acetonitrile–toluene mixture. The analysis is performed by GC-MS and HPLC fluorescence [27] or by GC-MS and HPLC-MS/MS [116].

6.5.3 ACETONITRILE EXTRACTION AND LIQUID–LIQUID PARTITION WITH SALT AND MAGNESIUM SULFATE

Using the QuEChERS method, produce samples are extracted by mixing vigorously with acetonitrile. Magnesium sulfate and NaCl are added and the resulting mixture is vigorously shaken and centrifuged. Aliquots of the resulting supernatants are cleaned up by dispersive SPE, mixing PSA SPE sorbent, and $MgSO_4$ [12]. A modification of the method using a buffer was employed to improve the recoveries of base-sensitive compounds such as captan, folpet, dichlofluanid, and chlorothalonil. Samples are extracted with 1.0% acetic acid–acetonitrile, and the salt-out partitioning was performed using sodium acetate and $MgSO_4$ [16].

6.5.4 SOLVENT EXTRACTION WITH MINIMAL OR NO CLEANUP AND HPLC-MS/MS DETERMINATION

HPLC-MS/MS provides both excellent sensitivity and selectivity, making it possible to analyze produce extracts that had been subjected to a very minimal cleanup. Jansson et al. [153] extracted produce samples with ethyl acetate in the presence of sodium sulfate. An aliquot of the extract was evaporated to dryness, dissolved in methanol before injection on HPLC-MS/MS. Klein and Alder [43] extracted produce samples with methanol rather than ethyl acetate, since ethyl acetate would not work well with fatty matrices such as avocado. The methanol–water extracts were mixed with NaCl, and applied to a diatomaceous earth column, and the pesticides were eluted from the column with dichloromethane.

6.5.5 MATRIX ENHANCEMENT AND SUPPRESSION

The matrix effect is a nontrivial issue in the analysis of pesticides using GC and HPLC-MS. In GC and GC-MS, the sample matrix usually causes enhancement in the chromatographic response of the pesticide compared with the same pesticide at the same concentration in the matrix-free solvent. This enhancement is a design problem observed with capillary columns and liners in GC instruments. Components in the matrix block active sites in the GC injection liner and column which reduce the loss and protect the analyte from thermal degradation and adsorption sites [143,160]. Several possible solutions are used to minimize or reduce enhancement effects, such as higher column flow rates to reduce the residence times the pesticides spend in the injection liner; difficult matrix or direct sample introduction (DMI, DSI) techniques to reduce matrix accumulation in the liner; isotope-labeled internal standards for GC-MS applications that compensate for any enhancement effects from the native analytes; matrix-matched standards to correct for the equivalent response of matrix-free calibration solutions; and the use of analyte protectants or standard addition for quantitation in the matrix sample [13,146,161–163]. However, these techniques can be expensive, time-consuming, impractical, and involve additional resources and effort.

Matrix suppression is a major issue in HPLC-MS because when API (electrospray or atmospheric chemical ionization) procedures are used, the coextractives from the matrix compete with the analyte during the ionization process, reducing the number of analyte molecules from getting ionized. Possible solutions are the use of isotope-labeled standards, matrix-matched standards, improved cleanup procedures, and dilution of the final extract. Alder et al. [44] developed a procedure known as the "ECHO effect," based on simultaneous injections of the reference standard and the sample in a single run. The first and second injections are made ahead and behind a HPLC precolumn, respectively, which results in the separation of the standard and sample peaks. The peaks are compared with compensate for any difference in the peak responses of the standard and the sample.

6.6 FUTURE TRENDS

Multiresidue analysis must be continually updated to not only include the environmentally persistent or commonly used pesticides, but also the newly registered ones. In addition, the cost is a major factor in determining the pesticide types to analyze, sample preparation procedures, and the type of instrumentation to use. More laboratories are adopting HPLC-MS methods because these procedures are becoming more cost-effective, efficient, sensitive, and available. Alder et al. [164] evaluated the applicability of HPLC-MS to determine various pesticide classes, including those pesticides (organophosphorus, triazine, and pyrethroid) that have been traditionally screened by GC and GC-MS procedures. Their assessment revealed that only one class, organochlorine pesticides, achieved better performance with GC-MS than with HPLC-MS.

Real-time analysis or faster analysis time is another area which can increase throughput and productivity in multiresidue analysis. Faster analysis times can be achieved through new developments in column technology, such as smaller particle size (1.8 or 1.7 μm) or monolithic packing materials and HPLC pumps, that can deliver high flow rates and yet withstand high pressures can reduce the chromatographic run times significantly [165–167].

HPLC-TOF/MS instruments, because of its high acquisition speed and the ability to provide accurate mass measurements and full spectra, could be used for untargeted pesticide analysis [158,167]. In combination with a quadrupole filter, a hybrid instrument HPLC-QTOF/MS can provide not only accurate mass measurements but also selectivity and identification [168]. However, it would be ideal to determine the presence of pesticides in real time, in which the vegetal matrix requires no or little sample preparation. Three new ionization techniques, direct analysis in real time (DART), desorption ESI (DESI), and desorption APCI (DAPCI) coupled to a TOF mass spectrometer, have been shown to require no sample preparation or extraction before the analysis of various gases, liquids, and solid surfaces, including fruits and vegetables [169–171].

REFERENCES

1. Anonymous, European Commission (EC), status of active substances under EU review, http://europa.eu.int/comm/food/plant/protection/evaluation/stat_active_subs_3010 _en.xls, 2006.
2. Anonymous, What is a pesticide? www.epa.gov/pesticides, U.S. Environmental Protection Agency, Washington, D.C., 1999.
3. McMahon, B.M. and Wagner, R.F., *Pesticide Analytical Manual Volume 1*, 3rd edn, http://vm.cfsan.fda.gov/~frf/pami3.html, U.S. Food and Drug Administration, Washington, D.C., 1999, chap. 3.
4. Mercer, G.E. and Hurlbut, J.A., A multiresidue pesticide monitoring procedure using gas chromatography/mass spectrometry and selected ion monitoring for the determination of pesticides containing nitrogen, sulfur, and/or oxygen in fruits and vegetables, *J. AOAC Int.*, 87, 1224, 2004.
5. Mercer, G.E., Determination of 112 halogenated pesticides using gas chromatography/mass spectrometry with selected ion monitoring, *J. AOAC Int.*, 88, 1452, 2005.
6. McMahon, B.M. and Wagner, R.F., *Pesticide Analytical Manual Volume 1*, 3rd edn, U.S. Food and Drug Administration, Washington, D.C., http://vm.cfsan.fda .gov/~frf/pami3.html, 1994, updated October 1999, chap. 2.
7. Young, S.J.V. et al., Homogeneity of fruits and vegetables comminuted in a vertical cutter mixer, *J. AOAC Int.*, 79, 976, 1996.
8. Schenck, F.J. and Hobbs, J., Evaluation of the quick, easy, cheap, effective, rugged, and safe (QuEChERS) approach to pesticide residue analysis, *Bull. Environ. Contam. Toxicol.*, 73, 24, 2004.
9. Bertuzzi, P.F., Kamps, L., and Miles, C.I., Extraction of chlorinated pesticides from nonfatty samples of low moisture content, *J. Assoc. Off. Anal. Chem.*, 50, 623, 1967.
10. Luke, M.A. and Doose, G.M., A modification of the Luke multiresidue procedure for low moisture, nonfatty products, *Bull. Environ. Contam. Toxicol.*, 30, 110, 1983.

11. Cook, J. et al., Multiresidue analysis of pesticides in fresh fruits and vegetables using procedures developed by the Florida Department of Agriculture and Consumer Services, *J. AOAC Int.*, 82, 1419, 1999.

12. Anastassiades, M. et al., Fast and easy multiresidue method employing acetonitrile extraction/partitioning and "dispersive solid phase extraction" for the determination of pesticide residues in produce, *J. AOAC Int.*, 86, 412, 2003.

13. Čajka, T. et al., Use of automated direct sample introduction with analyte protectants in the GC-MS analysis of pesticide residues, *J. Sep. Sci.*, 28, 1048, 2005.

14. Liu, M. et al., Simultaneous determination of carbamate and organophosphorus pesticides in fruits and vegetables by liquid chromatography-mass spectrometry, *J. Chromatogr. A*, 1097, 183, 2005.

15. Leandro, C.C., Fussell, R.J., and Keely, B.J., Determination of priority pesticides in baby foods by gas chromatography tandem quadrupole mass spectrometry, *J. Chromatogr. A*, 1085, 207, 2005.

16. Lehotay, S.J., Maštovská, K., and Lightfeld, A.R., Use of buffering and other means to improve results of problematic pesticides in a fast and easy method for residue analysis of fruits and vegetables, *J. AOAC Int.*, 88, 615, 2005.

17. Lehotay, S.J. et al., Validation of a fast and easy method for the determination of residues from 229 pesticides in fruits and vegetables using gas and liquid chromatography and mass spectrometric detection, *J. AOAC Int.*, 88, 595, 2005.

18. Thurman, E.M. et al., Discovering metabolites of post-harvest fungicides in citrus with liquid chromatography/time-of-flight mass spectrometry and ion trap mass spectrometry, *J. Chromatogr. A*, 1082, 71, 2005.

19. Ortelli, D., Edder, P., and Corvi, C., Pesticide residues survey in citrus fruits, *Anal. Chim. Acta*, 22, 423, 2005.

20. Pérez-Ruiz, T. et al., High performance liquid chromatographic assay of phosphate and organophosphorus pesticides using post column photochemical reaction and fluorimetric detection, *Anal. Chim. Acta*, 540, 383, 2005.

21. Okihashi, M. et al., Rapid method for the determination of 180 pesticide residues in foods by gas chromatography/mass spectrometry and flame photometric detection, *J. Pestic. Sci.*, 30, 368, 2005.

22. Mills, P.A., Onley, J.H., and Gaither, R.A., Rapid method for chlorinated pesticides in nonfatty foods, *J. Assoc. Off. Anal. Chem.*, 46, 186, 1963.

23. Horowitz, W., *Official Methods of Analysis of AOAC International*, 18th edn, AOAC International, Gaithersburg, MD, 2005, chap. 10.

24. Mumma, R.O. et al., Dieldrin: extraction of accumulations by root uptake, *Science*, 152, 530, 1966.

25. Lee, S.M. et al., Multipesticide residue method for fruits and vegetables: California Department of Food and Agriculture, *Frezenius J. Anal. Chem.*, 339, 376, 1991.

26. Liao, W., Joe, T., and Cusick, W.G., Multiresidue screening method for fresh fruits and vegetables with gas chromatography/mass spectrometric detection, *J. AOAC Int.*, 74, 554, 1991.

27. Fillion, J., Sauvé, F., and Selwyn, J., Multiresidue method for the determination of 251 pesticides in fruits and vegetables by gas chromatography/mass spectrometry and liquid chromatography with fluorescence detection, *J. AOAC Int.*, 83, 698, 2000.

28. Sheridan, R.S. and Meola, J.R., Analysis of pesticide residues in fruits, vegetables, and milk by gas chromatography/tandem mass spectrometry, *J. AOAC Int.*, 82, 982, 1999.

29. Jones, L.R. and Riddick, J.A., Separation of organic insecticides from plant and animal tissues, *Anal. Chem.*, 24, 569, 1952.

30. Cabras, P. et al., Simplified multiresidue method for the determination of organophosphorus insecticides in olive oil, *J. Chromatogr. A*, 761, 327, 1997.

31. Garcia Sánchez, A. et al., Multiresidue analysis of pesticides in olive oil by gelpermeation chromatography followed by gas chromatography-tandem mass-spectrometric detection, *Anal. Chim. Acta*, 558, 53, 2006.

32. Di Muccio, A.D. et al., Selective clean-up applicable to aqueous acetone extracts for the determination of carbendazim and thiabendazole in fruits and vegetables by high performance liquid chromatography with UV detection, *J. Chromatogr. A*, 833, 61, 1999.

33. Leoni, V., Caricchia, A.M., and Chiavarini, S., Multiresidue method for quantitation of organophosphorus pesticides in vegetable and animal foods, *J. AOAC Int.*, 75, 511, 1992.

34. Luke, M.A., Froberg, J.E., and Masumoto, H.T., Extraction and cleanup of organochlorine, organophosphate, organonitrogen, and hydrocarbon pesticides in produce for determination by gas–liquid chromatography, *J. Assoc. Off. Anal. Chem.*, 58, 1020, 1975.

35. Luke, M.A. et al., Improved multiresidue gas chromatographic determination of organophosphorus, organonitrogen and organohalogen pesticides in produce, using flame photometric and electrolytic conductivity detectors, *J. Assoc. Off. Anal. Chem.*, 64, 1187, 1981.

36. Hopper, M., Improved method for partition of organophosphate pesticide residues on a solid phase partition column, *J. Assoc. Off. Anal. Chem.*, 71, 131, 1988.

37. Specht, W., Pelz, S., and Gilsbach, W., Gas chromatographic determination of pesticide residues after cleanup by gel-permeation chromatography and mini-silica gel-chromatography, *Frezenius J. Anal. Chem.*, 353, 183, 1995.

38. Luke, M.A., Cassias, I., and Yee, S., A multiresidue analytical method using solid phase extraction without methylene chloride, Laboratory Information Bulletin 4178, U.S. Food and Drug Administration, Division of Field Science, HFC-140, Rockville, MD, 1999.

39. Holstedge, D.M. et al., A rapid multiresidue screen for organophosphorus, organochlorine and *N*-methyl carbamate insecticides in plant and animal tissues, *J. AOAC Int.*, 77, 1263, 1994.

40. Ambrus, A. and Their, H.-P., Applications of multiresidue procedures in pesticide residues analysis, *Pure Appl. Chem.*, 58, 1035, 1986.

41. Krause, R.T., Multiresidue method for determining *N*-methylcarbamate insecticides in crops using high performance liquid chromatography, *J. Assoc. Off. Anal. Chem.*, 63, 1114, 1980.

42. Pang, G.F. et al., Multiresidue liquid chromatographic method for simultaneous determination of pyrethroid insecticides in fruits and vegetables, *J. AOAC Int.*, 78, 1474, 1995.

43. Klein, J. and Alder, L., Applicability of gradient liquid chromatography with tandem mass spectrometry to the simultaneous screening for about 100 pesticides in crops, *J. AOAC Int.*, 86, 1015, 2003.

44. Alder, L. et al., The ECHO technique—the more effective way of data evaluation in liquid chromatography-tandem mass spectrometry analysis, *J. Chromatogr. A*, 1058, 67, 2004.

45. Granby, K., Andersen, J.H., and Christensen, H.B., Analysis of pesticides in fruit, vegetables and cereals using methanolic extraction and detection by liquid chromatography-tandem mass spectrometry, *Anal. Chim. Acta*, 520, 165, 2004.

46. Barker, S.A. et al., Isolation of drug residues from tissues by solid phase dispersion, *J. Chromatogr. A*, 475, 353, 1989.

47. Soler, C., Manes, J., and Picó, Y., Routine application using single quadrupole liquid chromatography-mass spectrometry to pesticides analysis in citrus fruits, *J. Chromatogr. A*, 1088, 224, 2005.

48. Soler, C., Manes, J., and Picó, Y., Liquid chromatography-electrospray quadrupole ion-trap mass spectrometry of nine pesticides in fruits, *J. Chromatogr. A*, 1048, 41, 2004.

49. Valenzuela, A.I. et al., Comparison of various liquid chromatographic methods for the analysis of avermectin residues in citrus fruits, *J. Chromatogr. A*, 918, 59, 2001.

50. Navarro, M. et al., Application of matrix solid phase dispersion to a new generation of fungicides in fruits and vegetables, *J. Chromatogr. A*, 968, 201, 2002.

51. Takino, M., Yamaguchi, K., and Nakahara, T., Determination of carbamate pesticide residues in vegetables and fruits by liquid chromatography-atmospheric pressure photoionization and -atmospheric pressure chemical ionization-mass spectrometry, *J. Agric. Food Chem.*, 52, 727, 2004.

52. Torres, C.M., Picó, Y., and Manes, J., Comparison of octadecylsilica and graphitized carbon black as materials for solid phase extraction of fungicide and insecticide residues from fruit and vegetables, *J. Chromatogr. A*, 778, 127, 1997.

53. Blasco, C., Font, G., and Picó, Y., Comparison of microextraction procedures to determine pesticides in oranges by liquid chromatography-mass spectrometry, *J. Chromatogr. A*, 970, 201, 2002.

54. Blasco, C. et al., Determination of fungicide residues in fruits and vegetables by liquid chromatography-atmospheric pressure chemical ionization mass spectrometry, *J. Chromatogr. A*, 947, 227, 2002.

55. Kristenson, E.M. et al., Miniaturized automated matrix solid phase dispersion extraction of pesticides in fruit followed by gas chromatographic-mass spectrometric analysis, *J. Chromatogr. A*, 917, 277, 2001.

56. Ferrer, C. et al., Determination of pesticide residues in olives and olive oil by matrix solid-phase dispersion followed by gas chromatography-mass spectrometry and liquid chromatography/tandem mass spectrometry, *J. Chromatogr. A*, 1069, 183, 2005.

57. Hu, Y.-Y. et al., Response surface optimization for determination of pesticide multi-residues by matrix-solid phase dispersion and gas chromatography, *J. Chromatogr. A*, 1098, 188, 2005.

58. Albero, B. et al., Determination of herbicide residues in juice by matrix solid-phase dispersion and gas chromatography-mass spectrometry, *J. Chromatogr. A*, 1043, 127, 2004.

59. Albero, B., Sánchez-Brunette, C., and Tadeo, J.L., Determination of organophosphorus pesticides in fruit juices by matrix solid-phase dispersion and gas chromatography, *J. Agric. Food Chem.*, 51, 6915, 2003.

60. Sanchez-Brunete, C., Albero, B., and Tadeo, J.L., High-performance liquid chromatographic multiresidue method for determination of *N*-methylcarbamates in fruit and vegetable juices, *J. Food Protect.*, 67, 2565, 2004.

61. Chu, X.G.O., Hu, X.Z., and Yao, H.Y., Determination of 266 pesticide residues in apple juice by matrix solid phase dispersion and gas chromatography-mass selective detection, *J. Chromatogr. A*, 1063, 201, 2005.

62. Sicbaldi, F., Sarra, A., and Copeta, G.L., Diatomaceous earth-assisted extraction for the multiresidue determination of pesticides, *J. Chromatogr. A*, 765, 23, 1997.

63. Bicchi, C., D'Amato, A., and Balbo, C., Multiresidue method for quantitative gas chromatographic determination of pesticide residues in sweet cherries, *J. AOAC Int.*, 80, 1281, 1997.

64. Perret, D. et al., Validation of a method for multiclass pesticide residues in fruit juices by liquid chromatography tandem mass spectrometry after extraction by matrix solid phase dispersion, *J. AOAC Int.*, 85, 724, 2002.

65. Barker, S.A., Applications of matrix solid phase dispersion in food analysis, *J. Chromatogr. A*, 880, 63, 2000.

66. Aharonson, N., Lehotay, S.J., and Ibrahim, M.A., Supercritical fluid extraction and HPLC analysis of benzimidazole fungicides in potato, apple, and banana, *J. Agric. Food Chem.*, 42, 2817, 1994.

67. Lehotay, S.J. and Eller, K.I., Development of a method of analysis for 46 pesticides in fruits and vegetables by supercritical fluid extraction and gas chromatography/ion trap mass spectrometry, *J. AOAC Int.*, 78, 821, 1995.

68. Ono, Y. et al., Pesticide multiresidue analysis of 303 compounds using supercritical fluid extraction, *Anal. Sci.*, 22, 1473, 2006.

69. Lehotay, S.J., Determination of pesticide residues in nonfatty foods by supercritical fluid extraction and gas chromatography/mass spectrometry: collaborative study, *J. AOAC Int.*, 85, 1148, 2002.

70. Camel, V., Supercritical fluid extraction as a useful method for pesticides determination, *Analysis*, 26, M99, 1998.

71. Lehotay, S.J., Supercritical fluid extraction of pesticides in foods, *J. Chromatogr. A*, 785, 289, 1997.

72. Motohashi, N., Nagashima, H., and Párkányi, C., Supercritical fluid extraction for the analysis of pesticide residues in miscellaneous samples, *J. Biochem. Biophys. Methods*, 43, 313, 2000.

73. Valcárcel, M. and Tena, M.T., Applications of supercritical fluid extraction in food analysis, *Fresenius J. Anal. Chem.*, 358, 561, 1997.

74. Torres, C.M., Picó, Y., and Mañes, J., Determination of pesticide residues in fruit and vegetables, *J. Chromatogr. A*, 754, 301, 1996.

75. Schantz, M., Pressurized liquid extraction in environmental analysis, *Anal. Bioanal. Chem.*, 386, 1043, 2006.

76. Obana, H. et al., Determination of organophosphorus pesticides in foods using an accelerated solvent extraction system, *Analyst*, 122, 217, 1997.

77. Yi, X. and Lu, Y., Multiresidue determination of organophosphorus pesticides in ginkgo leaves by accelerated solvent extraction and gas chromatography with flame photometric detection, *J. AOAC Int.*, 88, 729, 2005.

78. Yi, X., Hua, Q., and Lu, Y., Determination of organophosphorus pesticide residues in the roots of *Platycodon grandiflorum* by solid-phase extraction and gas chromatography with flame photometric detection, *J. AOAC Int.*, 89, 225, 2006.

79. Adou, K., Bontoyan, W., and Sweeney, P.J., Multiresidue method for the analysis of pesticide residues in fruits and vegetables by accelerated solvent extraction and capillary gas chromatography, *J. Agric. Food Chem.*, 49, 4153, 2001.

80. Bester, K. et al., How to overcome matrix effects in the determination of pesticides in fruit by HPLC-ESI-MS-MS, *Fresenius J. Anal. Chem.*, 371, 550, 2001.

81. Okihashi, M., Obana, H., and Hori, S., Determination of *N*-methylcarbamate pesticides in foods using an accelerated solvent extraction with a mini-column cleanup, *Analyst*, 123, 711, 1998.

82. Garrido Frenich, A. et al., Determination of multiclass pesticides in food commodities by pressurized liquid extraction using GC-MS/MS and LC-MS/MS, *Anal. Bioanal. Chem.*, 383, 1106, 2005.

83. Pang, G.-F. et al., Simultaneous determination of 405 pesticide residues in grain by accelerated solvent extraction then gas chromatography-mass spectrometry or liquid chromatography-tandem mass spectrometry, *Anal. Bioanal. Chem.*, 384, 1366, 2006.

84. Carabias-Martínez, R. et al., Pressurized liquid extraction in the analysis of food and biological samples, *J. Chromatogr. A*, 1089, 1, 2005.

85. Eskilsson, C. et al., Fast and selective analytical procedures for determination of persistent organic pollutants in food and feed using recent extraction techniques. In: *Modern Extraction Techniques. Food and Agricultural Samples*, C. Turner, ed., ACS Symposium Series 926, American Chemical Society, Washington, D.C., 2006.

86. Eskilsson, C. and Björklund, E., Analytical-scale microwave-assisted extraction, *J. Chromatogr. A*, 902, 227, 2000.

87. Pylypiw, H.M. et al., Suitability of microwave-assisted extraction for multiresidue pesticide analysis of produce, *J. Agric. Food Chem.*, 45, 3522, 1997.

88. Bouaid, A. et al., Microwave-assisted extraction method for the determination of atrazine and four organophosphorus pesticides in oranges by gas chromatography, *Frezenius J. Anal. Chem.*, 367, 291, 2000.

89. Barriada-Pereira, M. et al., Microwave-assisted extraction versus Soxhlet extraction in the analysis of 21 organochlorine pesticides in plants, *J. Chromatogr. A*, 1008, 115, 2003.

90. Singh, S.B., Foster, G.D., and Khan, S.U., Microwave-assisted extraction for the simultaneous determination of thiamethoxam, imidacloprid, and carbendazim residues in fresh and cooked vegetable samples, *J. Agric. Food Chem.*, 52, 105, 2004.

91. Pawliszyn, J., *Solid Phase Microextraction: Theory and Practice*, Wiley-VCH, New York, 1997.

92. Arthur, C.L. and Pawliszyn, J., Solid phase microextraction with thermal desorption using fused silica optical fibers, *Anal. Chem.*, 62, 2145, 1990.

93. Baltussen, E. et al., Stir bar sorptive extraction (SBSE), a novel extraction technique for aqueous samples: theory and principles, *J. Microcol. Sep.*, 11, 737, 1999.

94. Baltussen, E., Cramers, C.A., and Sandra, P., Sorptive sample preparation—a review, *Anal. Bioanal. Chem.* 373, 3, 2002.

95. Sandra, P., Tienpont, B., and David, F., Multi-residue screening of pesticides in vegetables, fruits and baby food by stir bar sorptive extraction-thermal desorption-capillary gas chromatography-mass spectrometry, *J. Chromatogr. A*, 1000, 299, 2003.

96. Turner, C., Overview of modern extraction techniques for food and agricultural samples. In: *Modern Extraction Techniques: Food and Agricultural Samples*, ACS Symposium Series 926, American Chemical Society, Washington, D.C., 2006.

97. Simplicio, A.L. and Boas, L.V., Validation of a solid-phase microextraction method for the determination of organophosphorus pesticides in fruits and fruit juice, *J. Chromatogr. A*, 833, 35, 1999.

98. Berrada, H., Font, G., and Moltó, J.C., Application of solid-phase microextraction for determining phenylurea herbicides and their homologous anilines from vegetables, *J. Chromatogr. A*, 1042, 9, 2004.

99. Fytianos, K. et al., Solid phase microextraction applied to the analysis of organophosphorus insecticides in fruits, *Chemosphere*, 65, 2090, 2006.

100. Beltran, J. et al., Application of solid-phase microextraction for the determination of pyrethroid residues in vegetable samples by GC-MS, *Anal. Bioanal. Chem.*, 376, 502, 2003.

101. De Melo Abreu, S. et al., Screening of grapes and wine for azoxystrobin, kresoxim-methyl and trifloxystrobin fungicides by HPLC with diode array detection, *Food Addit. Contam.*, 22, 549, 2005.

102. Blasco, C. et al., Solid-phase microextraction liquid chromatography/tandem mass spectrometry to determine postharvest fungicides in fruits, *Anal. Chem.*, 75, 3606, 2003.

103. Hogenboom, A.C. et al., Determination of pesticides in vegetables using large volume injection column liquid chromatography-electrospray tandem mass spectrometry, *J. Chromatogr. A*, 892, 379, 2000.

104. Grob, K. and Kälin, I., Attempt for an on-line size exclusion chromatography-gas chromatography method for analyzing pesticide residues in foods, *J. Agric. Food Chem.*, 39, 1950, 1991.

105. Andersson, A. and Pålsheden, H., Comparison of the efficiency of different GLC multi-residue methods on crops containing pesticide residues, *Frezenius J. Anal. Chem.*, 339, 365, 1991.

106. Roos, A.H. et al., Universal extraction/clean-up procedure for screening of pesticides by extraction with ethyl acetate and size exclusion chromatography, *Anal. Chim. Acta*, 196, 95, 1987.

107. Zrostliková, J., Hajšlová, J., and Čajka, T., Evaluation of two-dimensional gas chromatography-time-of-flight mass spectrometry for the determination of multiple pesticides in fruit, *J. Chromatogr. A*, 1019, 173, 2003.

108. Zhang, W.-G. et al., Simultaneous determination of 109 pesticides in unpolished rice by a combination of gel permeation chromatography and Florisil column purification, and gas chromatography/mass spectrometry, *Rapid Comm. Mass Spectrom.*, 20, 609, 2006.

109. Poole, C.F., Gunatilleka, A.D., and Sethuraman, R., Contributions of theory to method development in solid-phase extraction, *J. Chromatogr. A*, 885, 17, 2000.

110. Casanova, J., Use of solid phase extraction disks for analysis of moderately polar and nonpolar pesticides in high moisture foods, *J. AOAC Int.*, 79, 936, 1996.

111. Nordmeyer, K. and Their, H.-P., Solid-phase extraction for replacing dichloromethane partitioning in pesticide multiresidue analysis, *Z. Lebensm Unters. Forsch A*, 208, 259, 1999.

112. Wong, J.W. et al., Multiresidue determination of pesticides in malt beverages by capillary gas chromatography with mass spectrometry and selected ion monitoring, *J. Agric. Food Chem.*, 52, 6361, 2004.

113. Wong, J.W. et al., Multiresidue pesticide analysis in wines by solid-phase extraction and capillary gas chromatography-mass spectrometric detection with selective ion monitoring, *J. Agric. Food Chem.*, 51, 1148, 2003.

114. Štajnbaher, D. and Zupančič-Kralj, L., Multiresidue method for determination of 90 pesticides in fresh fruits and vegetables using solid phase extraction and gas chromato-graphy-mass spectrometry, *J. Chromatogr. A*, 1015, 185, 2003.

115. Cairns, T. et al., Multiresidue pesticide analysis by ion-trap technology: a cleanup approach for mass spectral analysis, *Rapid Commun. Mass Spectrom.*, 7, 1070, 1993.

116. Pang, G.F. et al., Simultaneous determination of 446 pesticide residues in fruits and vegetables by three cartridge solid phase extraction/gas chromatography-mass spectrometry and liquid chromatography-tandem mass spectrometry, *J. AOAC Int.*, 89, 740, 2005.

117. Kondo, H. et al., Multiresidue analysis of pesticides in agricultural products by GC/MS, GC/ECD and GC/FTD using acetonitrile extraction and clean-up with mini-columns, *J. Food Hyg. Soc. Jpn.*, 45, 161, 2003.

118. Chun, O.K., Kang, H.G., and Kim, M.H., Multiresidue method for the determination of pesticides in Korean domestic crops by gas chromatography/mass selective detection, *J. AOAC Int.*, 86, 823, 2003.

119. Valverde García, A. et al., Simple efficient multiresidue screening method for analysis of nine halogen containing pesticides on peppers and cucumbers by GLC-ECD, *J. Agric. Food Chem.*, 39, 2188, 1991.

120. Nakamura, Y. et al., Multiresidue analysis of 48 pesticides in agricultural products by capillary gas chromatography, *J. Agric. Food Chem.*, 42, 2508, 1994.

121. Ueno, E. et al., Determination of nitrogen- and phosphorus-containing pesticide residues in vegetables by gas chromatography with nitrogen–phosphorus and flame photometric detection after gel permeation chromatography and a two-step minicolumn cleanup, *J. AOAC Int.*, 86, 1241, 2003.

122. Ling, T., Xiadong, M., and Chongjiu, L., Application of gas chromatography-tandem mass spectrometry (GC-MS-MS) with pulsed splitless injection for the determination of multiclass pesticides in vegetables, *Anal. Lett.*, 39, 985, 2006.

123. Shuling, S., Xiadong, M., and Chongjiu, L., Multi-residue determination method of pesticides in leek by gel permeation chromatography and solid phase extraction followed by gas chromatography with mass spectrometric detector, *Food Control*, 18, 448, 2006.

124. Pang, G.F. et al., Multi-residue determination of 450 pesticide residues in honey, fruit juice and wine by double-cartridge solid-phase extraction gas chromatography-mass spectrometry and liquid chromatography-mass spectrometry, *Food Addit. Contam.*, 23, 777, 2006.

125. Schenck, F.J. and Howard-King, V., Determination of organochlorine and organophosphorus pesticides in low-moisture nonfatty products using a solid phase extraction cleanup, *J. Environ. Sci. Health B*, 35, 1, 2000.

126. Saito, Y. et al., Multiresidue determination of pesticides in agricultural products by gas chromatography/mass spectrometry with large volume injection, *J. AOAC Int.*, 87, 1356, 2004.

127. Schenck, F.J. and Howard-King, V., Rapid solid phase extraction cleanup for pesticide residues in fresh fruits and vegetables, *Bull. Environ. Contam. Toxicol.*, 63, 277, 1999.

128. Podhorniak, L.V., Negron, J.F., and Griffith, F.D. Jr., Gas chromatography with pulsed flame photometric detection multiresidue method for organophosphate pesticide and metabolite residues at parts-per-billion level in representative commodities of fruit and vegetable crop groups, *J. AOAC Int.*, 84, 873, 1999.

129. Hercegová, A. et al., Comparison of sample preparation methods combined with fast gas chromatography-mass spectrometry for ultratrace analysis of pesticide residues in baby food, *J. Sep. Sci.*, 29, 1102, 2006.

130. Tsumura, Y. et al., Simultaneous determination of 13 synthetic pyrethroids and their metabolite, 3-phenoxybenzoic acid, in tea by gas chromatography, *J. Agric. Food Chem.*, 42, 2922, 1994.

131. Hercegová, A. et al., Comparison of sample preparation methods combined with fast gas chromatography-mass spectrometry for ultratrace analysis of pesticide residues in baby food, *J. Sep. Sci.*, 29, 1102, 2006.

132. Podhorniak, L.V. et al., A multiresidue method (MRM) for *N*-methyl carbamate and metabolite pesticide residues at ppb levels in selected representative commodities of fruit and vegetable crop groups, *J. AOAC Int.*, 87, 1237, 2004.

133. De Kok, A. and Hiemstra, M., Optimization, automation and validation of the solid phase extraction cleanup and on-line liquid chromatographic determination of *N*-methylcarbamate pesticides in fruits and vegetables, *J. AOAC Int.*, 75, 1063, 1992.

134. De Kok, A., Hiemstra, M., and Vreeker, C.P., Optimization of the postcolumn hydrolysis reaction on solid phases for the routine high-performance liquid chromatographic determination of *N*-methylcarbamate pesticides in food products, *J. Chromatogr. A*, 507, 459, 1990.

135. De Kok, A., Hiemstra, M., and Vreeker, C.P., Improved cleanup method for the multiresidue analysis of *N*-methylcarbamates in grains, fruits and vegetables by means of HPLC with postcolumn fluorescence detection, *Chromatographia*, 24, 469, 1987.

136. Schenck, F.J., Hobbs, J., and Parker, A., A rapid cleanup for Luke extracts, Laboratory Information Bulletin 4320, U.S. Food and Drug Administration, Division of Field Science, HFC-140, Rockville, MD, 2004.

137. Sharif, Z. et al., Determination of organochlorine and pyrethroid pesticides in fruit and vegetables using solid phase extraction cleanup cartridges, *J. Chromatogr. A*, 1127, 254, 2006.

138. Rial Otero, R. et al., Multiresidue method for fourteen fungicides in white grapes by liquid–liquid and solid phase extraction followed by liquid chromatography-diode array detection, *J. Chromatogr. A*, 992, 121, 2003.

139. Díez, C. et al., Comparison of acetonitrile extraction/partitioning and "dispersive solid-phase extraction" method with classic multi-residue methods for extraction of herbicides in barley samples, *J. Chromatogr. A*, 1131, 11, 2006.

140. Leandro, C.C., Fussell, R.J., and Keely, B.J., Determination of priority pesticides in baby foods by gas chromatography tandem quadrupole mass spectrometry, *J. Chromatogr. A*, 1085, 207, 2005.

141. Wong, J. et al., Multiresidue analysis of organohalogen pesticides in fresh produce by the QuEChERS procedure and GC-MS/SIM, Poster presentation at the California Pesticide Workshop, Sacramento, CA, 2006.

142. Schenck, F.J., Lehotay, S.J., and Vega, V.A., Comparison of solid phase extraction sorbents for cleanup in pesticide residue analysis of fresh fruits and vegetables, *J. Sep. Sci.*, 25, 883, 2002.

143. Schenck, F.J. and Lehotay, S.L., Does further clean-up reduce the matrix enhancement effect in gas chromatographic analysis of pesticide residues in food? *J. Chromatogr. A*, 868, 51, 2000.

144. Gamón, M., Lleó, C., and Ten, A., Multiresidue determination of pesticides in fruit and vegetables by gas chromatography/tandem mass spectrometry, *J. AOAC Int.*, 84, 1209, 2001.

145. Agüera, A. et al., Multiresidue method for the analysis of multiclass pesticides in agricultural products by gas chromatography-tandem mass spectrometry, *Analyst*, 127, 347, 2002.

146. de Koning, S. et al., Trace level determination of pesticides in food using difficult matrix introduction-gas chromatography-time of flight mass spectrometry, *J. Chromatogr. A*, 1008, 247, 2003.

147. Dallüge, J. et al., Comprehensive two-dimensional gas chromatography with time-of-flight mass spectrometric detection applied to the determination of pesticides in food extracts, *J. Chromatogr. A*, 965, 207, 2002.

148. Patel, K. et al., Evaluation of large volume-difficult matrix introduction-gas chromatography-time of flight-mass spectrometry (LVI-DMI-GC-TOF-MS) for the determination of pesticides in fruit-based baby foods, *Food Addit. Contam.*, 21, 658, 2004.

149. Zrostliková, J. et al., Performance of programmed temperature vaporizer, pulsed splitless on column injection techniques in analysis of pesticide residues in plant matrices, *J. Chromatogr. A*, 937, 73, 2001.

150. Taylor, M.J. et al., Multi-residue method for rapid screening and confirmation of pesticides in crude extracts of fruits and vegetables using isocratic liquid chromatography with electrospray tandem mass spectrometry, *J. Chromatogr. A*, 982, 225, 2002.
151. Ortelli, D., Edder, P., and Corvi, C., Multiresidue analysis of 74 pesticides in fruits and vegetables by liquid chromatography-electrospray-tandem mass spectrometry, *Anal. Chim. Acta*, 520, 33, 2004.
152. Garrido Frenich, A. et al., Potentiality of gas chromatography-triple quadrupole mass spectrometry in vanguard and rearguard methods of pesticide residues in vegetables, *Anal. Chem.*, 77, 4640, 2005.
153. Jansson, C. et al., A new multi-residue method for analysis of pesticide residues in fruit and vegetables using liquid chromatography with tandem mass spectrometric detection, *J. Chromatogr. A*, 1023, 93, 2004.
154. Sannino, A., Bolzoni, L., and Bandini, M., Application of liquid chromatography with electrospray tandem mass spectrometry to the determination of a new generation of pesticides in processed fruits and vegetables, *J. Chromatogr. A*, 1036, 161, 2004.
155. Tanizawa, H. et al., Multi-residue method for screening of pesticides in crops by liquid chromatography with tandem mass spectrometry, *J. Food Hyg. Soc. Jpn.*, 46, 185, 2005.
156. Hernández, F. et al., Multiresidue liquid chromatography tandem mass spectrometry determination of 52 non gas chromatography-amenable pesticides and metabolites in different food commodities, *J. Chromatogr. A.*, 1109, 242, 2006.
157. Garrido Frenich, A.G. et al., Monitoring multi-class pesticide residues in fresh fruits and vegetables by liquid chromatography with tandem mass spectrometry, *J. Chromatogr. A.*, 1048, 199, 2004.
158. Ferrer, I., Thurman, E.M., and Fernández-Alba, A.R., Quantitation and accurate mass analysis of pesticides in vegetables by LC/TOF-MS, *Anal. Chem.*, 77, 2818, 2005.
159. García-Reyes, J.F., Molina-Díaz, A., and Fernández-Alba, A.R., Identification of pesticide transformation products in food by liquid chromatography/time-of-flight mass spectrometry via "fragmentation-degradation" relationships, *Anal. Chem.*, 79, 307, 2007.
160. Erney, D.R. et al., Explanation of the matrix-induced chromatographic response enhancement of organophosphorus pesticides during open tubular column gas chromatography with splitless or hot on-column injection and flame photometric detection, *J. Chromatogr. A*, 638, 57, 1993.
161. Lehotay, S.J., Analysis of pesticide residues in mixed fruit and vegetable extracts by direct sample introduction/gas chromatography/tandem mass spectrometry, *J. AOAC Int.*, 83, 680, 2000.
162. Patel, K. et al., Evaluation of large volume-difficult matrix introduction-gas chromatography-time of flight-mass spectrometry (LVI-DMI-GC-TOF-MS) for the determination of pesticides in fruit-based baby foods, *Food Addit. Contam.*, 21, 658, 2004.
163. Anastassiades, M., Maštovská, K., and Lehotay, S.J., Evaluation of analyte protectants to improve gas chromatographic analysis of pesticides, *J. Chromatogr. A*, 1015, 163, 2003.
164. Alder, L. et al., Residue analysis of 500 high priority pesticides: better by GC-MS or LC-MS/MS? *Mass Spectrom. Rev.*, 25, 838, 2006.
165. Guillarme, D. et al., Method transfer for fast liquid chromatography in pharmaceutical analysis: application to short columns packed with small particle. Part I. Isocratic separation, *Eur. J. Pharmaceut. Biopharmaceut*, 66, 475, 2007.

166. Churchwell, M.I. et al., Improving LC–MS sensitivity through increases in chromato-graphic performance: comparisons of UPLC–ES/MS/MS to HPLC–ES/MS/MS, *J. Chromatogr. B*, 825, 134, 2005.

167. Lacorte, S. and Fernandez-Alba, A.R., Time of flight mass spectrometry applied to the liquid chromatographic analysis of pesticides in water and food, *Mass Spectrom. Rev.*, 25, 866, 2006.

168. Picó, Y., Blasco, C., and Font, G., Environmental and food applications of LC-tandem mass spectrometry in pesticide-residue analysis. An overview, *Mass Spectrom. Rev.*, 23, 45, 2004.

169. Cody, R.B., Laramée, J.A., and Durst, H.D., Versatile new ion source for the analysis of materials in open air under ambient conditions, *Anal. Chem.*, 77, 2297, 2005.

170. Cooks, R.G. et al., Ambient mass spectrometry, *Science*, 311, 15, 2006.

171. Williams, J.P. et al., The use of recently described ionization techniques for the rapid analysis of some common drugs and samples of biological origin, *Rapid Comm. Mass Spectrom.*, 20, 1447, 2006.

7 Determination of Pesticides in Food of Animal Origin

Antonia Garrido Frenich, Jose Luis Martinez, and Adrian Covaci

CONTENTS

7.1   INTRODUCTION

Organic pollutants present in foods of animal origin can be classified into two large categories: contaminants and residues. Contaminants (e.g., dioxins or polychlorinated biphenyls) are substances that are not added deliberately to the foods and that can enter into foods during their production process, transformation, storage, packed, transport, or fraudulent practices. Residues are compounds that can occur in food-stuffs as a consequence of intentional usage of phytosanitary or veterinary products during plant or animal production. Therefore, pesticides found incorporated or onto the surface of foods belong to the residues' group.

Pesticides are a group of chemicals used either to directly control pest populations or to prevent or reduce pest damage on crops, landscape, or animals. In consequence, pesticides may be present in fresh or processed animal foods, if animals have been fed with contaminated feed or water, or from practices involving pesticides in places where animals are living (stables, beehives) or in food-processing factories. Pesticides are classified into several classes according to the target organisms they are designed to control: insecticides (to control insects including acari); fungicides (to control fungus diseases); herbicides (to control weeds); and any other product used to control pests except medication. Pesticide residues in foods of animal origin of great public and regulatory concern have been *insecticides*, such as organo-chlorines (OCs), organophosphates (OPs), carbamates, and pyrethroids (Py), and *fungicides*, such as benzimidazoles and OPs.

Knowledge of the physicochemical properties of pesticides is very important in environmental risk assessment because these influence the distribution, persistence, and fate of either parent compounds or metabolites. The most important properties are vapor pressure, water solubility, and the octanol–water partition coefficient (K_{ow}), which dictates how a pesticide will distribute among animal fatty tissues.

Chronologically, OCs have been the first class of chemicals used to control pests in agriculture and public health. They are persistent, mobile in the environment, lipophilic (high K_{ow} values), bioaccumulate through the food chain, and present potential for adverse effects in humans and the environment. As a consequence of their restrictions in usage and production since the 1970s and 1980s, newer classes of pesticides, such as OPs and carbamates, have been used on a large scale as alternatives to OCs. Despite their lower environmental persistence compared with OCs and their wider range of lipophilicity (log K_{ow} ranging between -0.9 and 5.7), OPs and carbamates can also accumulate in fatty matrices of animal origin. However, unlike most OCs, OPs and carbamates are stored in the animal fatty tissues only for short periods of time (e.g., days). Nevertheless, these compounds present a high acute toxicity. For these reasons, OPs and carbamates are replaced by Py pesticides. Py pesticides are nonpersistent compounds, have low water solubility, are lipophilic compounds (log K_{ow} ranging between 4.2 and 7), and present low toxicity to mammals and birds. Lastly, benzimidazole fungicides, effective as antihelmintics for the treatment of parasitism in both human and domestic animals, present relatively low water solubility and vapor pressure, and are moderately lipophilic substances (log K_{ow} between 1.4 and 2.7).

Despite several benefits, one disadvantage of pesticides is their toxicity to humans. This makes necessary the monitoring of food safety to avoid risks to consumers, as well as to regulate international trade. To address this issue, the European Union and the U.S. Environmental Protection Agency (EPA) have established maximum residue levels (MRLs), for pesticides in some matrices of animal origin. It is possible to roughly classify pesticides into several categories according to their mode of action. Pesticides in the same category do not need to be chemically related, and one substance might act through several mechanisms.[1] The most common modes of toxic action are as follows:

- *Enzyme inhibition*: This is by far the most important mechanism of toxicity. The toxicant reacts with an enzyme or a transport protein and inhibits its normal function. Typical pesticides of this group are OPs and carbamates that inhibit acetyl cholinesterase. However, some enzyme inhibitors have little specificity and as a consequence, many different enzymes may be targeted.
- *Disturbance of the chemical signal systems*: The toxicant imitates the true signal substances, and thus transmits a signal too strongly, too long lasting, or at a wrong time. Typical examples are nicotine that gives signals similar to acetylcholine in the nervous system but it is not eliminated, or phenoxy herbicides, such as 2,4-D, that mimic the plant hormone auxin.
- *Generation of very reactive molecules*: The toxicant forms easily reactive radicals of intermediates which are very aggressive and nonselectively attack biomolecules. The classical example of a free radical producing poison is the herbicide paraquat, which delivers an electron to molecular oxygen, which further produces a reactive hydroxyl radical.
- *Disturbance of the physical properties of membranes and cells*: These compounds react with proteins which are needed for cellular processes, such as cell division. Benzimidazoles are the most representative class of pesticides with this mode of action.

The aim of this chapter is to critically review the literature methods for pesticide analysis in foods of animal origin, including practical aspects on sample preparation, analytical techniques, and quality control. Finally, future perspectives with regard to the pesticide analysis are also discussed. The reviewed categories include (1) milk and milk-derived products (cheese, butter, cream); (2) meat and meat-derived products (including fat); (3) fish (fillet and organs) and shellfish (shrimps, mussels, etc.); (4) honey; and (5) eggs.

7.2 SAMPLE PREPARATION

Despite the tremendous growth during recent years in number of work dealing with the determination of the pesticide levels in various food of animal origin, no standard analytical procedures have yet been set for these compounds. This has resulted in a variety of analytical approaches for both sample preparation and instrumental analysis. Because of the low levels at which these compounds may be present in complex animal matrices, sample treatments include a number of steps for exhaustive extraction and preconcentration of the target compounds, followed by purification before final chromatographic separation and detection. In most instances, the need for additional fractionation usually depends on the selected chromatographic and detection systems and/or on the specific study goal. Tables 7.1 and 7.2 summarize relevant data on selected analytical procedures used for the determination of pesticides in a wide variety of food samples of animal origin. Furthermore, due to

TABLE 7.1

Overview of Typical Analytical Procedures Used for the Determination of Pesticides by GC-MS or LC-MS Techniques in Food of Animal Origin

Pesticide	Sample Type (g, mL)	Extraction Procedure	Cleanup	Analytical Column	Instrumental Analysis	Recovery (%)	References
368 Multiclass	Tissues of beef, mutton, chicken, pork, and rabbit (10)	Blender homogenization Cyclohex:EtOAc (1:1)	GPC	DB-1701 (30 m × 0.25 mm × 0.25 μm) dC$_{18}$ (150 mm × 2.1 mm × 3 μm)	GC-MS (SIM) LC-MS/MS	40–120	[2]
69 Multiclass							
Atrazine	Beef kidney (0.5)	MSPD (XAD-7 HP) + PLE (EtOH:water) + SPME	—	DB-5 (30 m × 0.25 mm × 0.10 μm)	GC-MS (SIM)	104–111	[3]
OCs + Py	Ground beef (100)	Isooctane-ACN partition	Florisil	DB-5 (30 m × 0.25 mm × 0.25 μm)	GC-IT-EI-MS (SIM)	40–82	[4]
OCs, OPs	Muscle of chicken, pork, and lamb (5)	Polytron (Na$_2$SO$_4$ + EtOAc)	GPC	VF-5MS (30 m × 0.25 mm × 0.25 μm)	GC-EI-MS/MS	70–90	[5]
OCs, OPs	Liver of chicken, pork, and lamb (5)	Polytron (Na$_2$SO$_4$ + EtOAc)	GPC	VF-5MS (30 m × 0.25 mm × 0.25 μm)	GC-EI-MS/MS	70–115	[6]
Benomyl metabolites	Goat liver (5)	Raney Ni reduction + acidic dehydration + defatting + EtOAc	—	SB-PH (150 mm × 4.6 mm) + SB-PH (250 mm × 4.6 mm)	LC-TSP-MS/MS	—	[7]
Carbamates	Beef and chicken tissues (5)	SFE (CO$_2$)	—	SB-PH-30 (1.5 m × 0.1 mm + 0.25 mm)	SFC-MS (SIM)	19–146	[8]
OCs	Pork fat (1.25)	EtOAc:Cyclohex 1:1	GPC	VF-5MS (30 m × 0.25 mm × 0.25 μm)	GC-EI-MS/MS	66–101	[9]
Triazine herbicides	Bovine milk (50)	HFM-SPME	—	DB-5 (30 m × 0.32 mm × 0.25 μm)	GC-MS (SIM)	57–107	[10]

40 Multiclass	Milk (whole or powdered) (1)	Formic acid + TEA + Shake (water) + SPME (PDMS-DVB)	—	CP-Sil 8 (10 m × 0.53 mm × 0.25 μm)	LP-GC-EI-MS/MS	81–110	[11]
Carbamates	Bovine milk (3)	MSPD (sand) + hot water (pH = 4.6) extraction	—	Alltima C$_{18}$ (250 mm × 4.6 mm × 5 μm)	LC-ESI-MS/MS (SIM)	76–104	[12]
Herbicides and fungicides	Bovine milk (4)	Carbograph 4 + back-flushing with MeOH and DCM	—	Alltima C$_{18}$ (250 mm × 4.6 mm × 5 μm)	LC-ESI-MS/MS (MRM)	85–102	[13]
OCs	Butter (1.7)	Hexane	Acidified silica (Hex) + activated silica (Hex:DCM 1:1)	CP Sil 8 (50 m × 0.25 mm)	GC-MS (SIM)	n.r.	[14]
OCs	Eggs (2–5)	Shake (MeOH), sonication (Hex, Na$_2$SO$_4$)	Fractionation by HPLC	DB-XLB (60 m × 0.25 mm × 0.25 μm)	GC-MS (SIM)	51–108	[15]
OCs, OPs, Py, ONs	Honey (10)	Shake (water) + EtOAc extraction	Florisil (Hex: EtOAc 1:1)	LM-5 (35 m × 0.25 mm × 0.25 μm)	GC-MS (SIM)	76–119	[16]
OCs, OPs, triazines	Honey (10)	Shake (water:MeOH, 70:30) + SPE C$_{18}$ (Hex:EtOAc 1:1)	—	ZB-5MS (30 m × 0.25 mm × 0.25 μm)	GC-MS (SIM)	75–107	[17]
51 Multiclass	Honey (10)	Shake (water:MeOH, 70:30) + SPE C$_{18}$ (Hex:EtOAc 1:1)	—	ZB-5MS (30 m × 0.25 mm × 0.25 μm)	GC-MS (SIM)	86–101	[18]
OCs, OPs, Py, ON	Honey (10[a], 5[b])	[a]Shake (water) + Hex:Acet (60:40), [b]shake (water, 40°C) + SFE (CO$_2$ with 10% AcN)	Florisil (DCM: Hex 8:2 + Hex: Act 6:4)	LM-5 (35 m × 0.25 mm × 0.25 μm)	GC-EI-MS (SIM)	75–94	[19]
OCs	Honey (5)	Shake (water) + SPE C$_{18}$ (EtOAc + MeOH + DCM)	—	DB-5 (30 m × 0.25 mm × 0.25 μm)	GC-MS (SIM)	79–98	[20]
Fluvalinate	Honey (5–10)	Shake (water:MeOH, 2:8) + EtOAc extraction	C$_{18}$ cartridge (Hex, 8 mL)	SE-54 (12 m × 0.20 mm × 0.30 μm)	GC-MS (SIM)	94–100	[21]
14 Multiclass	Honey (0.5)	Shake (water) + SPME (PDMS)	—	RTX-5MS (30 m × 0.25 mm × 0.25 μm)	GC-EI-MS (SIM)	36–127	[22]

(continued)

TABLE 7.1 (continued)

Overview of Typical Analytical Procedures Used for the Determination of Pesticides by GC-MS or LC-MS Techniques in Food of Animal Origin

Pesticide	Sample Type (g, mL)	Extraction Procedure	Cleanup	Analytical Column	Instrumental Analysis	Recovery (%)	References
Chlordimeform and degradation products	Honey (1)	Shake (water) + Hex:Acet 8:2, extraction	—	DB-17 (30 m × 0.25 mm × 0.25 μm)	GC-EI-MS (SIM)	n.r.	[23]
Vinclozolin	Honey (1)	Shake (water) + aHex:Acet 7:3, extraction, bSPE C18 (Acet)	aFlorisil (Hex: DCM 1:1 v/v)	DB-17 (30 m × 0.25 mm × 0.25 μm)	GC-EI-MS (SIM)	>90	[24]
Several pesticides	Honey (1)	Shake (water) + SPE C18 (Hex:DCM 1:1)	—	BP-1 (12 m × 0.22 mm × 0.25 μm)	GC-EI-MS (SIM)	74–96	[25]
OCs	Honey (10)	Shake (water) + PE:EtOAc 8:2 extraction	Florisil (20% DE in Hex)	DB-5 (30 m × 0.25 mm × 0.25 μm)	GC-full-scan-MS	78–108	[26]
Triazole fungicides	Honey (50–60)	Hydromatrix + Soxhlet (DE)	GPC + Florisil	DB-5MS (30 m × 0.25 mm × 0.25 μm)	GC-EI-MS/MS	51–99	[27]
OPs	Honey (0.5)	MSPD (C18)	Florisil (DCM: MeOH 85:15)	Spherisorb C18 (150 mm × 4.6 mm × 5 μm)	LC-APCI-MS (SIM)	14–102	[28]
OPs	Honey (2.5)	aShake (water) + SPME (PDMS) bShake (water) + SBSE (PDMS)	—	Luna C18 (250 mm × 4.6 mm × 5 μm)	LC-APCI-MS (SIM)	a<10 b40–66	[29]
OCs	Fish (10)	Ultrasonic (Acet:Hex 5:2)	Cool extracts (−24°C), Florisil (Acet:Hex 1:9, 13 mL)	DB-5 (30 m × 0.25 mm × 0.25 μm)	GC-EI-MS (SIM)	78–115	[30]

OCs	Fish (1–10), fish liver (0.50)	Soxhlet (Hex:Acet 3:1)	Acidified silica (Hex + DCM)	HT-8 (25 m × 0.22 mm × 0.25 μm)	GC-ECNI-MS (SIM)	75–90	[31]
OCs	Fish (5–10)	Soxhlet (PE)	Sulfuric acid in Hex + silica column with Na$_2$SO$_4$	ZB-5 (30 m × 0.25 mm × 0.25 μm)	GC-MS (SIM)	n.r.	[32]
OCs and fipronil	Fish (10–12)	Polytron (DCM:Hex 1:1)	GPC	BGB-172 Chiral column (30 m)	GC-MS (SIM)	n.r.	[33]
OCs	Fish (2)	Soxhlet (Hex:Acet 3:1, 60 mL)	a Acidified silica (Hex:DCM 3:1) b Alumina + silica + Florisil	HP-5 (30 m × 0.25 mm × 0.25 μm)	GC-EI-MS (SIM)	65–107	[34]
OCs	Fish (10), fish liver (5)	Soxhlet (Hex:DCM:Acet 3:1:1)	Acid silica + neutral silica + deactivated basic silica (Hex)	HT-8 (50 m × 0.22 mm × 0.25 μm)	GC-EI-MS (SIM), GC-ECNI-MS (SIM)	72–80	[35,36]
OCs	Fish/shellfish (1)	Al$_2$O$_3$-activated basic + acidic silica gel + SFE (CO$_2$) + SPME (PDMS)	—	HP-5 (30 m × 0.25 mm × 0.25 μm)	GC-EI-MS/MS	65–89	[37]
OCs	Shellfish (1)	Al$_2$O$_3$-activated basic + acidic silica gel + SFE (CO$_2$)	—	HP-5 (30 m × 0.25 mm × 0.25 μm)	GC-EI-MS/MS	78–119	[38]
OCs	Fish (8)	Na$_2$SO$_4$ + Hex extraction	NPLC	DB-5 (30 m × 0.25 mm × 0.25 μm)	GC-EI-MS/MS	75–112	[39]

n.r., not reported. Solvents: Acet, acetone; AcN, acetonitrile; Hex, hexane; DCM, dichloromethane; EtOAc, ethyl acetate; *i*-PrOH, 2-propanol; DE, diethyl ether; MeOH, methanol; EtOH, ethanol; Cyclohex, cyclohexane; PE, petroleum ether. Solvent mixtures: proportions as v/v; TEA, triethylamine; Py, pyrethroids; ON, organonitrogen; HFM, hollow fiber membrane; SPME, solid-phase microextraction; PDMS-DVB, polydimethylsiloxane–divinylbenzene; LP, low pressure; SFE, supercritical fluid extraction; SBSE, stir bar sorptive extraction; SFC, supercritical fluid chromatography; NPLC, normal phase liquid chromatography; SIM, selected ion monitoring; IT, ion trap; EI, electron impact; ECNI, electron capture negative ionization; APCI, atmospheric pressure chemical ionization; ESI, electrospray; TSP, thermospray.

TABLE 7.2

Overview of Typical Analytical Procedures Used for the Determination of Pesticides by GC or LC with Classical Detectors in Foods of Animal Origin

Pesticide	Sample Type (g, mL)	Extraction Procedure	Cleanup	Analytical Column	Instrumental Analysis	Recovery (%)	References
OPs	Cow milk (25), liver, and muscle boar (10)	Acet:AcN 1:4 + DCM partition + Na$_2$SO$_4$ (AcN)	SPE C$_{18}$	ZB-5 (30 m × 0.32 mm × 0.25 μm) + ZB-50 (30 m × 0.32 mm × 0.25 μm)	GC-NPD	59–117	[40]
OPs	Meat (1)	Hydromatrix + SFE (CO$_2$)	Florisil (heptane + Acet)	DB-1701 (30 m × 0.32 mm × 0.25 μm)	GC-NPD	78–95	[41]
OPs	Fat (0.75)	Warm (hydromatrix + water) + SFE (CO$_2$ + 3% AcN)	Florisil	DB-1 (30 m × 0.53 mm × 1.5 μm)	GC-FPD	74–95	[42]
OPs	[a]Milk (5) [b]Butter (5)	[a]MSPD (silica gel + Na$_2$SO$_4$); Acet:DCM 2:1 [b]Melt (50°C) and filter + Hex + AcN partition + DCM partition + Na$_2$SO$_4$	—	Pyrex glass column (1 m × 2 mm i.d.)	GC-NPD	92–96	[43]
OPs, Herbicides	Milk, cheese	AcN + Hex partition	Florisil	DB-1 (30 m × 0.32 mm × 0.25 μm)	GC-FID		[44]
OPs	Milk (50)	Shake (EtOAc + Na$_2$SO$_4$), dryness (Hex) + AcN saturated in Hex extraction	—	HP-1 (25 m × 0.20 mm × 0.25 μm)	GC-FPD	43–99	[45]
OPs	Milk (10)	AcN + EtOH in Ultra Turrax + dispersion on solid matrix diatomaceous	—	SPB-608 (15 m × 0.53 mm × 0.83 μm)	GC-FPD	72–109	[46]

Analyte	Sample	Extraction	Cleanup	Column	Detection	Recovery	Ref.
OPs	Milk (10)[a] Butterfat (1)[b]	Acet:AcN + DCM[a] Hex + Extrelux + C18 columns + MeOH[b]	SPE C18[a]	DB-17 (15 m × 0.53 mm × 1.0 μm)	GC-FPD	n.r.	[47]
Py (acrinathrine and metabolites)	Honey (1)	Shake (water) + benzene: i-PrOH 1:1 Shake (MeOH)	Florisil cartridge (Hex:DCM 1:1)	DB-17 (25 m × 0.25 mm × 0.25 μm)	GC-FID	91–101	[48]
Py	Milk (10)	AcN:EtOH (5:1) + solid matrix diatomaceous	GPC	DB-17 (15 m × 0.53 mm × 1.5 μm)	GC-ECD	86–93	[49]
OCs	Chicken eggs (1)	MSPD–Florisil (DCM)		DB-5 MS (30 m × 0.25 mm × 0.25 μm)	GC-ECD	80–115	[50]
Carbamates	Egg (25)	Acetonitrile–polytron	Aminopropyl SPE	Hypersil C18 (250 mm × 4.6 mm × 5 μm)	LC-fluorescence	70–106	[51]
OPs, Py	Honey (1–5)	Shake (water/MeOH) + SPME (PDMS)	Florisil column (Hex:DCM 1:1)	007–17 (60 m × 0.25 mm × 0.25 μm)	GC-ECD; GC-NPD	88–115	[52,53]
OPs, Py	Honey (1.5)	Shake (water + phosphate buffer) + SPME (PDMS)	—	HP-5 (30 m × 0.32 mm × 0.25 μm)	GC-AED	>91	[54]
OPs, Py	Honey (20)	Hex:i-PrOH 6:3 + water with ammonia (0.28)	—	RP-18 (250 mm × 4 mm × 5 μm)	LC-DAD	91–109	[55]
OPs	Honey (1)	Shake (water) + SPME (sol–gel crown ether fiber)	—	HP-5 (30 m × 0.32 mm × 0.25 μm)	GC-FPD	74–105	[56]
Misc. insecticides	Honey (2)	Shake (water) + SPE C18	—	Hypersil (250 mm × 4.6 mm)	LC-UV	78–98	[57]
Benomyl, carbendazim	Honey (1)	Shake (HCl, 0.05 M + EtOAc) + EtOAc + NaOH 0.1 M + EtOAc	—	Novapack C18 (150 mm × 3.9 mm)	LC-fluorescence	91–99	[58]
Atrazine and simazine	Honey (5)	Ultrasonic extraction (benzene:water 1:1)		HPTLC Silica gel (10 × 20 cm)	TLC	84	[59]

(continued)

TABLE 7.2 (continued)

Overview of Typical Analytical Procedures Used for the Determination of Pesticides by GC or LC with Classical Detectors in Foods of Animal Origin

Pesticide	Sample Type (g, mL)	Extraction Procedure	Cleanup	Analytical Column	Instrumental Analysis	Recovery (%)	References
OCs	Fish and shellfish	Soxhlet, Hex	SPE-SPE (C_{18} + Florisil or alumina)	SPB-608 (30 m $\times$ 0.53 mm $\times$ 0.5 μm)	GC-ECD	63–129	[60]
OCs, OPs, Py, and carbamates	Fish and shellfish (10)	Blending with water, Acet, and AcN	SPE–SPE (C_{18} and aminopropyl) + AcN	n.r.	GC-ECD; GC-FPD; LC-UV	70–140	[61]
Benomyl, thiabendazole	Table-ready food (100)	Acetone—blender	SCX cartridge	Platinum C_8 (150 mm $\times$ 4.6 mm $\times$ 5 μm)	LC-fluorescence + LC-UV	74–122	[62]

n.r., not reported. Solvents: Acet, acetone; AcN, acetonitrile; Hex, hexane; DCM, dichloromethane; EtOAc, ethyl acetate; *i*-PrOH, 2-propanol; MeOH, methanol; EtOH, ethanol. Solvent mixtures: proportions as *v*/*v*; Py, pyrethroids; SPME, solid-phase microextraction; SPMD, solid-phase matrix dispersion; SFE, supercritical fluid extraction; TLC, thin layer chromatography; GPC, gel permeation chromatography; ECD, electron capture detection; FID, flame ionization detection; FPD, photometric detection; NPD, nitrogen–phosphorus detection; UV, ultraviolet detection; DAD, diode array detection.

particular physicochemical properties of various classes of pesticides, the determination of some pesticides may require specific analytical approaches.

7.2.1 SAMPLE PRETREATMENT

Only drying and homogenization are usually carried out before extraction of biological samples. Alternatively, (semi)liquid (e.g., eggs or milk) samples may be freeze-dried and then treated as any other solid biological sample. Losses of volatile compounds might also occur during freeze-drying and there is a greater potential for contamination in the laboratory. The samples should be stored in an inert material (e.g., glass containers) to avoid possible sorption of the pesticides into the storage medium or to prevent contamination of the sample from the storage medium (e.g., with phthalates). Most commonly, samples are kept in glass or Teflon containers. Screw caps should be lined with solvent-rinsed aluminum foil or with Teflon inserts. Polyethylene or other plastics should be avoided unless a thorough validation has been previously carried out and contamination has been ruled out. Alternatively, solid samples can be wrapped in aluminum foil and then inserted into plastic bags. Preferably, samples should be frozen (−20°C or −80°C) as soon as possible after sampling. Freezing and storage of multiple small samples suitable for analysis, rather than larger masses, is recommended to avoid multiple freezing and thawing of tissue and to reduce sample handling, which in turn reduces the potential for contamination.

In general, maintaining sample tissues in their original wet state is regarded as the most appropriate approach for preparing samples for pesticide analysis. Instead, homogenized samples should be mixed with a desiccant such as sodium sulfate, Celite, or Hydromatrix to bind water. The desiccant must be free of analytes, for example, by heating at high temperature in the case of sodium sulfate or by preextraction (Celite or Hydromatrix).

7.2.2 EXTRACTION TECHNIQUES

For food samples of animal origin, the selection of the extraction technique depends on the nature of the matrix investigated; different procedures are used for solid and liquid samples. The amount of sample required varies largely depending on the contamination level anticipated in the sample and on the sensitivity provided by the detection technique. Table 7.3 summarizes relevant extraction methods related to pesticides in food samples of animal origin.

Solid samples can be extracted by any one of a number of techniques (Tables 7.1 and 7.2). The main points to consider here are the use of adequate solvent systems (e.g., low boiling solvents to facilitate concentration), adequate exposure time between solvents and the sample matrix, and limitation of sample handling steps, that is, avoid filtration steps by using Soxhlet, extraction columns (sample matrix eluted after soaking in solvent), or semiautomated extraction systems (e.g., pressurized liquid extraction, PLE). Cross-contamination from residues left behind by high levels in previous samples is a concern at this stage and equipment must be thoroughly cleaned and checked from batch to batch. Purity of extraction solvents

TABLE 7.3

Description of the Most Common Techniques Used for the Extraction of Pesticides from Solid Food Samples of Animal Origin

Technique	Overview
Conventional Soxhlet	Sample + desiccant mixture in glass or paper thimble is extracted with condensed (cold) solvent for 4–12 h
Automated Soxhlet (e.g., "Soxtec")	Sample + desiccant mixture in extraction thimble is immersed in boiling solvent (30–60 min), then raised for Soxhlet extraction. Solvent can also be evaporated
Supercritical fluid extraction (SFE)	Sample + desiccant mixture is placed in high-pressure cartridge and CO_2 at 150–450 atm at temperature of 40°C–150°C passed through. After depressurization, analytes are collected in solvent trap
Column extraction	Sample + desiccant mixture is placed in large column with filter. Eluted with large volume of extraction solvent
Sonication-assisted extraction	Sample (+desiccant) in open or closed vessel immersed in solvent and heated with ultrasonic radiation using ultrasonic bath or probe
Pressurized liquid extraction (PLE)	Sample + desiccant mixture is placed in extraction cartridge and solvent (heated, pressurized) passed through, then dispensed in extraction vial
Microwave-assisted extraction (MAE)	Sample (+desiccant) is placed in open or closed vessel immersed in solvent and heated with microwave energy

is also a major consideration here. Only pesticide-grade or high-purity glass-distilled solvents should be used because evaporation steps will later concentrate any contaminants.

A generally used method is homogenization of the solid sample with a wide variety of solvents, such as dichloromethane–hexane[33] or ethyl acetate.[5] Such technique allows quantitative extraction of pesticides directly from matrices or after drying with anhydrous Na_2SO_4, but also uses large volumes of solvents. Alternatively, the use of matrix solid-phase dispersion (MSPD) in different variants is a suitable choice, which results in an intimate contact between the sample components and the sorbent particles and therefore in a more efficient retention of impurities.[28] Moreover, lower solvent consumption and cleaner extracts can be expected using MSPD compared with the column extraction technique, in which the sample is packed above the sorbent.

Binary solvent mixtures typically containing acetone–*n*-hexane[31] or petroleum ether[32] have been preferred for Soxhlet-based extractions. In general, extraction with a polar–nonpolar binary mixture has been found to be more efficient for recovering pesticides from fish tissues of low lipid content than a nonpolar solvent.[31,35] This technique has a number of advantages, such as minimum sample pretreatment required, simplicity, and high recoveries obtained for most pesticides. The time- and solvent-consuming nature of Soxhlet extraction (or related techniques involving percolation of a solvent through the sample) is generally thought to be related to the slow diffusion and desorption of the analytes from the sample matrix. Semivolatile pesticides can also be lost from Soxhlet apparatus via volatilization.

Therefore, more automated extraction techniques, such as PLE or microwave-assisted extraction (MAE), are an alternative to Soxhlet and column extraction methods. The use of microwave energy (for MAE) or elevated temperatures and pressure (as in PLE) increases the rates of diffusion and desorption and thus speeds up extraction. Although these techniques use less solvent, they suffer the disadvantage of initial high cost. Moreover, the preparation of a homogeneous dry sample from wet tissue for PLE can be a challenge due to the limited size of PLE vials, typically <100 mL. The use of PLE-based extraction methods for organic pollutants, including pesticides, has recently been reviewed by Björklund et al.[63] and Carabias-Martinez et al.[64]

Supercritical fluid extraction (SFE) has also been used for the extraction of pesticides from biological samples. Due to the wide polarity range of pesticides, a polar modifier was typically used during extraction with supercritical CO_2.[8,19,37,38,41] However, SFE is less popular due to the high number of variables which have to be optimized and to its lesser ability to accommodate a wide variety of analytes and matrices.[41]

For liquid samples, such as milk, solid-phase extraction (SPE) has been established as a robust extraction method for a wide variety of pesticides and has been increasingly used compared with the classical liquid–liquid extraction technique. Various adsorbents, for example, silica-based C_{18}[40,47] or Carbograph 4[13] have shown high recoveries of investigated pesticides from milk.

Another sampling device, solid-phase microextraction (SPME) involves immersing a polydimethylsiloxane-coated syringe into or above liquid samples.[10,11] Hydrophobic compounds are adsorbed onto the coated fiber and the syringe is then placed into a hot injection liner which desorbs these compounds into the GC. The benefit of this approach is that it requires no solvent or multistepped cleanup/concentration procedures. A drawback is that the adsorption efficiency can be affected by complex matrices, especially when the fiber is directly immersed in the sample.

Homogenization of a liquid sample with solvents, for example, acetonitrile–ethanol[46,49] is less used than for solid samples, while MSPD has found some applications for milk,[12,43] butter,[43] and eggs.[50,65,66]

For honey, the first step consisted of transferring the analytes into an aqueous phase by shaking the samples with water. This phase can subsequently be extracted as described earlier for milk samples, with SPE[21,24,57] and SPME[29,52,54,56] as the most used techniques.

Lipid contents of biological samples should be determined during the pesticide analysis in food samples of animal origin. Most studies have determined total extractable lipid gravimetrically by drying a fraction of the sample extract to constant weight. However, results can vary widely among laboratories due to different extraction efficiencies of various combinations of solvent and extraction systems. The benchmark method for total lipid is that of Bligh and Dyer.[67] Recently, Smedes[68] demonstrated that mixtures of isopropanol–cyclohexane–water (8:10:11) are an effective substitute for the Bligh and Dyer mixture of chloroform–methanol–water. The Smedes method gave more consistent results for extractable lipids in fish tissue with low lipid content (<1% lipid).

7.2.3 CLEANUP AND FRACTIONATION

The nonselective nature of the exhaustive extraction procedures and the complexity of sample matrices result in complex extracts that require further purification. Despite the inherent advantages derived from partial or complete integration of this tedious and time-consuming purification with the extraction, up to now, the development in this field has been rather limited and the analytical steps involved in cleanup protocols for pesticides have usually been carried out off-line. Typical purification and fractionation procedures have been summarized in Tables 7.1 and 7.2. For biotic samples, lipid elimination should be accomplished before chromatographic analysis. Lipid elimination can be accomplished by destructive or nondestructive methods. Otherwise, similar protocols can be used for purification of the extracts almost irrespective of the matrix nature.

Nondestructive methods for lipid removal. Gel permeation chromatography (GPC) and adsorption chromatography on selected sorbents are nondestructive treatments applied for lipid elimination. GPC is mainly carried out either in automated systems or by gravity flow columns. The current use of prepacked polystyrene–divinylbenzene-based high-performance GPC columns has resulted in higher separation efficiencies, improved reproducibility, and lower solvent consumption as compared with manually packed columns. Satisfactory isolation of the target compounds from the coextracted organic material after single GPC analysis has been achieved for samples containing limited amounts of lipids.[5] However, for more complex matrices, GPC followed by further cleanup by adsorption chromatography may be required to remove remaining low-molecular-weight lipids, waxes, and pigments.

Silica gel, alumina, and Florisil with different degrees of activation have been widely used for lipid/pigment removal by adsorption chromatography under atmospheric conditions (Tables 7.1 and 7.2). Basic alumina and silica gel columns have been effective for the separation of OCs from fish lipids, although there is the possibility of minor losses due to dehydrochlorination of some OCs, for example, *p,p'*-DDT, on the alumina.[69] Alumina and Florisil have been preferred as fat retainers because of their higher lipid-retaining capacity in procedures involving MSPD[28] or SFE.[19,41] The effectiveness of these adsorption columns depends on the mass and the water content of the adsorbent together with the polarity of the solvent. In general, 3–8 g of absorbent is used in a 0.5–1.0 cm diameter column with silica gel or Florisil deactivated with a low percentage of water (0%–5%). Alumina and Florisil have the capacity to retain about 100 mg lipid per 10 g of adsorbent.

Destructive methods for lipid removal. Drastic treatments of the extracts are usually conducted when analytes of interest are stable in acidic or basic conditions. Lipid removal using a sulfuric acid wash or elution of the extract through sulfuric acid-impregnated silica has been described as an effective cleanup for the analysis of OCs, such as hexachlorocyclohexanes, DDT, or chlordanes.[31,36] The dispersion of sulfuric acid onto the surface of activated silica results in a sorbent which can be easily loaded into a column. The use of acidified silica avoids the emulsion problems, reduces the sample handling and solvent consumption, and increases sample throughput. However, the acidic treatment destroys all other classes of pesticides, including several other OCs, such as dieldrin. KOH-treated silica columns or KOH wash of extracts is less effective for the removal of lipids and can be used for matrices with low lipid content.[34] Similar to acidic treatment, basic treatment also results in degradation of some analytes, for example, DDT is converted to DDE.

Fractionation. For specific applications, isolation of the target analytes from other compounds present in the extract can be mandatory to avoid interferences during final determination. Due to their wide range of polarity, pesticides can be isolated in a separate fraction using classical adsorbents, such as silica gel,[70,71] while alumina and Florisil show less selectivity for pesticides. The fractionation is usually done by applying the extract in a small volume of nonpolar solvent to the adsorbent and by eluting with various volumes of solvents with increasing polarity.[19]

7.3 ANALYTICAL TECHNIQUES

The choice of analytical technique used for the detection of pesticides is strongly dependent on the analyte's polarity. Compounds with high log K_{ow}, such as OCs, Py, and most OPs, are nonpolar and are preferably analyzed by gas chromatography (GC), while polar compounds such as carbamates, benzimidazoles, and some OPs are amenable by liquid chromatography (LC). An overview of the advantages and limitations of various chromatographic detectors is given in Table 7.4. Other methods using sensors or immunoassay and electrochemical techniques have also been applied.

7.3.1 GAS CHROMATOGRAPHY

Capillary gas chromatography (GC), coupled to selective detectors, such as electron capture (ECD), nitrogen–phosphorus (NPD), atomic emission (AED), flame photometric (FPD), or to nonspecific detectors, such as flame ionization (FID), is still one of the most used techniques for the determination of pesticide residues in foods (Table 7.2). Several applications describe the use of such detectors. However, more and more methods are using mass spectrometric (MS) detection, because it allows to identify, quantify, and confirm the compounds present in the sample on basis of their structure in one single run.

Nowadays, GC-MS is the primary analytical technique used for confirmation of results obtained with classical detectors. In addition, EU requirements indicate that all confirmatory methods for pesticide residues in animal foods must use MS detection.[72] There are three modes of GC-MS available, electron impact (EI), positive chemical ionization, and negative chemical ionization; the first one is the most widely used in this field. Due to its adequate sensitivity and selectivity, GC-MS in selected ion monitoring (SIM) is commonly used in the determination of different classes of pesticides in animal tissues,[2–4] milk,[10] butter,[14] egg,[15] honey,[16–22] and fish and shellfish.[30–39] On the contrary, the less sensitive method of mass scanning, full-scan mode, has been applied only when concentrations of investigated pesticides were high enough.[26]

GC coupled to tandem mass spectrometry (MS/MS) commonly provides higher sensitivity and selectivity, as well as degree of certainty, than GC-MS in SIM mode,

TABLE 7.4

Advantages and Drawbacks of Different Detection Techniques for Chromatographic Determination of Pesticides

Detection	Advantages	Drawbacks
ECD	Fair sensitivity for OCs	Interferences from other halogen-containing species
	Purchase cost	Limited linear range
	Maintenance cost	Very low selectivity
	Easy-to-use	
NPD	Fair sensitivity for OPs and ONs	Interferences from other halogen-containing species
	Purchase cost	Limited linear range
	Maintenance cost	Very low selectivity
	Easy-to-use	
FPD	Fair sensitivity for OPs	Interferences from other halogen-containing species
	Purchase cost	Limited linear range
	Maintenance cost	Very low selectivity
	Easy-to-use	
EI-LRMS	Good selectivity	Low sensitivity
	Relatively cheap and easy-to-use	
ECNI-LRMS	Relatively cheap and easy-to-use	Frequent source maintenance required
	Good sensitivity for organohalogens	Limited number of applications
	Good selectivity	
QTrap-MS	Relatively cheap	Needs adequate optimization
	Good sensitivity	Consistent but sometimes unpredictable
	Very good selectivity	fragmentation
IT-MS/MS	Relatively cheap	Needs adequate optimization
	Good selectivity	Limited linear range
	No isobaric interferences	Variable sensitivity for different pesticide classes
HR-TOFMS	Full-scan spectra and fast scanning rate	Limited dynamic range
		Matrix can saturate detection system
	Spectral deconvolution and identification of unknown pesticides or metabolites	Quantitation can be difficult
	Benchtop high resolution easy-to-use system	
	Excellent screening tool	
	Can also be used in ECNI mode	
HRMS	Good sensitivity	Purchase and maintenance cost
	Very good selectivity	Experienced analyst required
		Exclusive use in EI mode
LC-MS		
Q-MS	Relatively cheap	Low sensitivity
	Easy-to-use system	Relatively low selectivity

TABLE 7.4 (continued)

Advantages and Drawbacks of Different Detection Techniques for Chromatographic Determination of Pesticides

Detection	Advantages	Drawbacks
QqQ-MS	Good selectivity	Needs adequate optimization
	Good sensitivity	Purchase cost
IT-MS/MS	Good selectivity	Needs adequate optimization
	Fair sensitivity	Cutoff limitations for daughter ions
		Limited linear range

Abbreviations: LR, low resolution; HR, high resolution; ECNI, electron capture negative ion; TOF, time-of-flight; QqQ, triple quadrupole.

because it involves at least two stages of mass analysis, separated by a fragmentation step. The most common tandem mass spectrometers for GC, ion trap (IT), and triple quadrupole (QqQ) are important tools in food analysis. However, a limited number of examples are presented in the literature on application of GC-MS/MS in the area of pesticide residue determination in foods of animal origin. GC-IT-MS/MS has been employed in the determination of multiclass pesticide in different kinds of milk[11] and in the analysis of OCs in fish samples.[37–39] However, GC-QqQ-MS/MS has been less frequently used to determine pesticide residues in animal tissues such as muscle,[5] liver,[6] and fat (Figure 7.1).[9]

In all the previously summarized GC-MS applications, fused capillary columns with bonded phases of different polarities (nonpolar BP-1; low polar VF-5, DB-5, LM-5, ZB-5MS, HP-5MS, RTX-5MS, DB-XLB, SE-54, HT-8, or CP Sil 8; low-/midpolarity (DB-1701); and medium polar DB-17), various lengths (10–60 m), internal diameters (0.20–0.53 mm), and film thickness (0.10–0.30 μm) have been used (Tables 7.1 and 7.2).

7.3.1.1   Enantioselective Gas Chromatography

Several pesticides have optically active or chiral isomers (e.g., α-HCH, *o,p'*-DDT, *cis/trans*-chlordane, or heptachlor).[73] As a consequence, biotransformation reactions in biological samples can result in nonracemic patterns in environmental samples. Crucial for chiral analysis is the availability of chiral capillary GC columns such as those with various cyclodextrins chemically bonded to a polysiloxane. These phases are relatively heat stable and have low bleed.[73] Current methods range from the simple use of 30 m chiral columns to a two-dimensional "heart-cutting" technique, providing higher peak capacity and generally further separation of chiral compounds.[73] While use of chiral GC separations is not part of routine pesticide analysis, it is a well-developed technology that is relatively easy to implement in existing GC-ECD and GC-MS instruments.

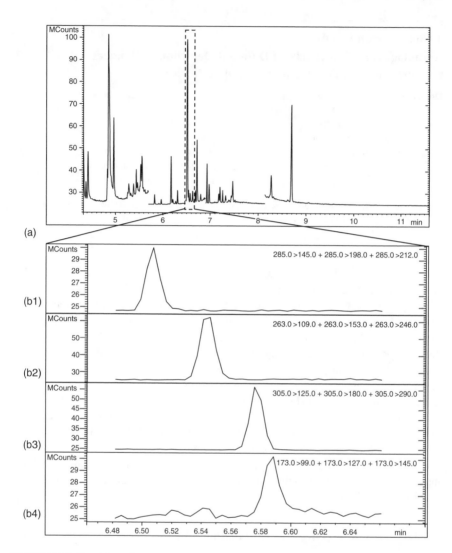

FIGURE 7.1 Total ion chromatogram obtained by GC-QqQ-MS/MS of a spiked sample of chicken liver with 34 pesticides (organochlorine + organophosphorus) at 50 µg/kg, and detail of extracted ion chromatograms of (b1) vinclozoline, (b2) parathion methyl, (b3) pirimiphos methyl, and (b4) malathion. (Modified from Garrido Frenich et al., 2007, *J. Chromatogr.*, 1153, 194.)

7.3.2 LIQUID CHROMATOGRAPHY

Although LC is the method of choice for thermally labile, ionic, and polar compounds, such as carbamates, urea, and phenoxy herbicides or benzimidazoles, it has been less applied than GC. LC coupled to universal detectors, such as UV or fluorescence detector, was successfully used for the determination of OPs and Py in honey[55] and carbamates in eggs,[51] respectively. However, in the last years, LC-MS has emerged as a prominent tool for the monitoring pesticide residues in foods of animal origin. This technique meets the EU legislation[72] to ensure appropriate selectivity and sensitivity. For that, LC-MS has become the common method to determine pesticides, although LC-MS/MS is the recommended mode due to its high selectivity and sensitivity.[74–76] From the different LC-MS methodologies, the atmospheric pressure interfacing system, using electrospray (ESI) or atmospheric pressure chemical ionization (APCI) in either positive or negative ionization modes, is the most widely used ionization technique. The thermospray interface has been less applied, for example, the determination of benomyl metabolites in goat liver.[7]

LC-MS using ESI in positive mode has been applied to the analysis of carbamate insecticides in bovine milk,[12] whereas APCI in positive and negative modes[28] or negative mode[29] has been employed in the determination of OPs in honey. All of these applications involve the use of the C_{18} reversed-phase column (250 mm × 4.6 mm i.d., 5 μm particle size) and elution using a gradient methanol–water or methanol–aqueous formic acid. LC-MS/MS using ESI in positive mode has been applied to the analysis of multiclass pesticide residues in animal tissues[77] coupled with a C_{18} column and elution using a gradient acetonitrile–water. An overview of the applications of MS/MS in food analysis has recently been published.[76]

On the other hand, SFE coupled to supercritical fluid chromatography (SFC) using MS detection has also been used to analyze carbamate pesticides in beef and chicken tissues.[8] Supercritical fluids can be used as mobile phases because they are liquid-like in some aspects and gas-like in others, thus forming complementary aids to LC and GC techniques.

7.3.3 IMMUNOASSAYS

Currently, chromatographic techniques are generally used to determine pesticides in foods due to their reliable analyte identification and the simultaneous determination of several compounds. However, they require expensive equipment and experienced personnel. Screening approaches, which can differentiate potentially positive samples from hundreds of negative samples in one simple test and thus can increase the sample throughput, have therefore emerged as an alternative. Immunoassays are inexpensive, easy and rapid to perform, sensitive, and reliable to determine low concentrations of pesticides. They are also useful for screening analysis. The most commonly used immunoassay for pesticide residue analysis is enzyme-linked immunosorbent assay (ELISA). This technique is based on competitive binding in which the binder molecule, an excess amount of labeled analyte or coating antigen, and the target analyte are allowed to approach equilibrium. The sample antigen competes with the coated antigen for binding sites on the labeled antibody. After a wash step, detection is performed by adding substrate and chromophore.

An ELISA method using two monoclonal antibodies was used to determine fenbendazole and its metabolites in bovine milk.[78] Immunoassay techniques were also used for the detection of the simazine[79] and atrazine herbicides[80] in milk. A magnetic particle-based ELISA was employed for the determination of the pesticide spinosad and its metabolites in milk and beef tissue samples.[81] The presence of alachlor, carbofuran, atrazine, benomyl, and 2,4-D in extracts of bovine liver, ground beef, and lard was quantitatively determined by the magnetic particle-based ELISA.[82] The pesticides were extracted from samples by SFE using supercritical CO_2. A commercial ELISA kit was applied for screening the presence of carbofuran and aldicarb sulfone in meat and liver extracts of pig and chicken.[83] Samples were extracted with water or acetonitrile, although the final extract must be in aqueous solution. This ELISA kit was also applied for the determination of aldicarb sulfone directly in bovine milk, that is, no previous extraction was required. Results obtained by an ELISA method used for the determination of permethrin in fish extracts significantly correlated ($R^2 = 0.985$) with those obtained by GC-MS.[84] ELISA technique has also been used to quantify DDT and metabolites in fatty food samples.[85] In general, fatty coextractive materials must be removed before application of ELISA for OCs.

7.3.4 ELECTROANALYTICAL TECHNIQUES

Different electroanalytical techniques have also been developed for determining pesticides in foods of animal origin. An amperometric cholinesterase biosensor has been used for the detection of 2,4-dichlorophenoxyacetic acid and 2,4,5-trichloro-phenoxyacetic acid pesticides in diluted milk samples using electrochemical detection.[86] A direct electrochemical method based on cyclic voltammetry was used for the determination of amitraz and its final breakdown product (2,4 dimethylaniline) in milk and honey samples.[87] An electroanalytical method for determination of the pesticide dichlorvos in milk using gold-disk microelectrodes and square-wave voltammetry was applied.[88] Neither extraction nor preparation of the samples was necessary in the method. A triazine herbicide minisensor based on surface-stabilized bilayer lipid membranes has been applied to the rapid screening of triazine herbicides in protein-free milk,[89] whereas a disposable biosensor has been developed for the detection of OPs and carbamates in milk.[90]

7.4 INTERNAL QUALITY CONTROL

Internal quality control (IQC) is defined as the set of procedures undertaken by laboratory staff for the continuous monitoring of an operation and the measurements to decide whether results are reliable enough to be released.[91] It should not be confused with external quality control (EQC). In the framework of using validated analytical methods by internationally accepted criteria,[92-95] IQC activities must be sufficient to ensure that the measurement chemical process is under statistical control.[96] This goal is achieved when the quality level is good enough to detect whether unexpected or unwanted changes have occurred during analysis of samples.

7.4.1 BASIC ACTIVITIES OF IQC IN PESTICIDE ANALYSIS

1. The laboratory must be divided in well-defined working areas to avoid contamination of standards, samples, and extracts.
2. All equipments (measurement instruments, balances, flasks, pipettes) must be regularly calibrated.

3. Personnel must be qualified, trained, and motivated. Personnel must not carry out a sample analysis without demonstrating previously their ability to give good results.
4. Sampling requires highly specialized analytical expertise. Samples must be transported to the laboratory and processed as soon as possible. If not possible, they must be stored in adequate conditions to assure their stability.
5. The quality of reagents used must be appropriate for the concerned test. Individual primary calibration solutions, generally 100–1000 mg/L, should be prepared by weighing not less than about 10 mg of the pesticide.
6. The method validation process has to include the estimation of uncertainty, indispensable in establishing the comparability of the measurement.
7. Other questions related on safety, chemical hygiene, or clean environment have also to be addressed.

7.4.2 Internal Quality Control Measures

IQC measures[97] must be included in the analytical batches to enable the decision whether the batches satisfy the preset quality criteria and that a set of the results can be accepted.

1. Reagent blank eliminates false positives by contamination in the extraction process, instruments, or chemicals used. Moreover, a matrix blank has to be analyzed to detect interferences of sample matrix. If a matrix blank is not available, the use of a simulated homemade matrix is allowed.
2. Certified Reference Materials (CRMs) or homemade reference materials (checked for stability and homogeneity), prepared by spiking a blank sample matrix with 5%–10% of the target pesticides, can be used to study the variation between batches of samples. This reference must be analyzed everyday by applying the analytical method, providing additional information about instrument performance (instrument sensitivity, column performance, etc.). The variation in the data obtained from the analysis of the quality control sample is normally monitored on a quality control chart. However, CRMs are expensive and, although several materials have been produced for the analysis of OCs, most of them show limitations, such as a limited number of certified pesticides, wide uncertainty ranges, concentrations well above the current values of interest, or a physical state not matching routine samples (e.g., freeze-dried materials).
3. A calibration curve must be carried out for every sample batch. A minimum of three standard concentrations, prepared in solvent or in extract of blank matrix, has to be used for each pesticide. The first of them has to be equal or preferably lower than the MRL allowed for each pesticide in the target matrix, but must be still higher than or equal to the limit of quantification. The fit (linear, quadratic, etc.) of the calibration function must be inspected to ensure that it is satisfactory. Individual points of the calibration curve must not differ more than $\pm 20\%$ ($\pm 10\%$ if the MRL is exceeded or approached).

TABLE 7.5

Acceptable Recovery Ranges of Quantitative Methods

Mass Fraction (µg/kg)	Range of Acceptable Recovery (%)
≤ 1	50–120
>1–10	70–110
≥ 10	80–110

4. Recoveries measured from spiked matrix blanks must be used to check the extraction efficiency in each sample batch, by adding all analytes to a matrix blank at a concentration level about 30% above the LOQ. This level of addition may be varied to have information over a range of concentrations. The Council Directive 96/23/EC[98] has established acceptable recovery ranges in function of the amount of the analyte(s) added to the blank matrix (Table 7.5).

5. Replicated samples provide a less formal means of checking for drift than quality control samples. The results obtained from the analysis of these samples must be comparable, taking into account the uncertainty of the method. They are located in the batch every certain number of samples (e.g., each 10 samples) and their presence is known by the analyst.

6. Blind samples are replicated samples placed in the analytical batch without known by the analyst. They are complementary to replicated samples, providing information about the analyst's proficiency.

7.5 APPLICATION TO REAL SAMPLES

7.5.1 ORGANOCHLORINE PESTICIDES

From the historical point of view, OCs have been the first pesticides introduced in the market and as a consequence, numerous methods have been developed for their determination in food of animal origin. A recent review by Muir and Sverko[99] has emphasized that, while the use of GC-MS will be essential for most laboratories, new analytical techniques with low costs and low environmental impact (e.g., immunoassays of "fast" GC) may be well-suited for broader use in routine analysis of OCs.

OCs are a class of compounds which have been heavily used in the past as insecticides, but which are nowadays obsolete and, with a few exception, banned in most industrialized countries. Moreover, the environmental levels of various OCs are on a decreasing trend which in most cases has lead to the development of sensitive methods for their detection. Several published articles are based on the use of GC coupled to ECD, which has a very good sensitivity toward halogenated compounds. GC-ECD methods have been applied for the determination of OCs in fish and fish-derived products[60,61] or in eggs.[50] However, due to the stricter quality control criteria, the analysis method of choice is GC-MS (Table 7.1). For all studied matrices, a cleanup step (SPE, MSPD, or solvent partition) was necessary to eliminate lipids (Tables 7.1 and 7.2). Due to their good thermal stability, OCs have been analyzed in food of animal origin exclusively by GC and there are virtually no LC methods reported.

7.5.2 ORGANOPHOSPHORUS PESTICIDES

OPs represent a group of compounds with a wide range of physicochemical properties, but all having the same mode of action which has enabled to develop compounds against ecto- and endoparasites, active in the vapor phase or in the soil, and used against a wide range of crop and public health pests. Although the majority of the OPs are insecticides, a small part of them also have fungicidal or herbicidal activities. Many OPs present high acute toxicity and are suspected of carcinogenic, mutagenic, and endocrine disruptive effects. The majority of the published articles are based on the use of GC coupled to various detectors. GC-ECD[52,53] or GC-AED[23,54] has been applied for the determination of OPs in honey. GC-NPD was applied to the analysis of methyl parathion in milk and cheese[44] and the analysis of OPs in cow milk and boar tissues,[40] milk and butter,[43] meat,[41] and honey.[53] However, the most used detectors for analyzing OPs are FPD and MS (Table 7.1). GC-FPD has been also used in the determination of OPs in honey,[56] fatty matrices,[42] and milk.[45–47] In most of these studies, a cleanup step (SPE, MSPD, or solvent partition) was necessary due to the complexity of the samples, despite the selectivity of the used detectors. In addition, the applicability of the methods to the analysis of real samples revealed the absence of the target OP compounds or their presence at lower concentrations than the established levels. Although the determination of OPs has been successfully performed in foods of animal origin by GC, only a few LC methods have been reported.[28,29,55]

7.5.3 OTHER PESTICIDES

The carbamates and Py represent chronologically the third and fourth major groups of insecticides. Py are important in veterinary medicine where they are used as ectoparasiticides. A rapid procedure has been developed for determining 13 Py compounds in milk after MSPD and GC-ECD.[49] Acrinathrine and its main metabolite, 3-phenoxybenzaldehyde, have been analyzed in honey by GC-FID, after liquid–liquid extraction and SPE,[24] whereas residues of fluvalinate have been determined in honey by LC-UV detection.[55]

Although they are also suspected carcinogens and mutagens, carbamates are increasingly used due to their lower persistence. LC with ultraviolet detection has also been applied to the determination of miscellaneous carbamates in honey,[57] whereas an LC method with fluorescence detection has been reported for analysis of the benomyl and carbendazin fungicides in honey.[48] Other methods[7,12] for carbamate compounds are reported in Table 7.1.

On the other hand, triazines, pre- and postemergence herbicides, are widely used worldwide. Both parent compounds and their degradation products have caused concern because they are toxic and persistent in organisms, water, and soil. Although triazines (atrazine or simazine) and their degradation products are one of the most extensively investigated group of pesticides in environmental matrices, up-to-date, they have been determined only in honey after ultrasonic solvent extraction and thin layer chromatography.[59]

7.6 EMERGING ISSUES IN ANALYTICAL METHODS

Although pesticide analysis is a mature area within environmental and food analytical chemistry, analytical methods are constantly evolving and improving and undoubtedly new technologies will emerge in the future. Some procedures with low environmental impact (SPME, microscale glassware, low solvent use, etc.) may be particularly suited when analytical budgets are small. A major development in the pesticide analysis has been the introduction of semiautomated extraction instruments for PLE, which can further be combined with solid-phase adsorbents to extract and isolate analytes of interest. In general, automated and semiautomated systems are available for cleanup and isolation of samples for pesticides using disposable solid-phase cartridges, as well as HPLC columns.

Advances in GC that are potentially applicable to pesticide analysis are the commercial availability of multidimensional GC (2D-GC) and "fast GC." In GC × GC, or comprehensive two-dimensional GC, all of the analyte's mass is transferred to a second column, and thus resolving power and detection limits are increased by up to an order of magnitude. Rapid GC separation of pesticides by "fast GC" was obtained on narrow-bore (0.1 mm) columns, which reduced peak widths and shortened total run times to minutes. Both techniques can be run with ECD, NPD, or FPD and may thus be relatively inexpensive to implement. Additionally, enantioselective GC and LC may find interesting applications for studying degradation or metabolism.

Another important area expected to be of increasing importance in the future is the use of LC-MS-based techniques due to the introduction of robust LC-MS interfaces, as well as of new mass analyzers, such as IT- or QqQ-MS/MS instruments, which allow a suitable identification, confirmation, and analyte quantification. LC-MS/MS is also adequate for the determination of metabolites or degradation products of pesticides. Moreover, the recent introduction of time-of-flight (TOF) mass analyzers is a powerful tool for the detection of target/nontarget pesticides and degradation products in food. This technique gives accurate masses for both parent and fragment ions and enables the measurement of the elemental formula of a compound achieving compound identification. In addition, the combination of Q-TOF allows MS/MS, provides more structural information, and enhances selectivity.

Finally, it will also be important improving strategies for the confirmation of analyte's identity, for instance, based on the use of identification points indicated in international guidelines. For this, the potentiality of TOF-MS will be of great importance, mainly to avoid false positive results.

ACKNOWLEDGMENTS

Dr. Adrian Covaci was financially supported by a postdoctoral fellowship from the Research Scientific Foundation-Flanders (FWO). Dr. Martínez Vidal and Dr. Garrido Frenich would like to thank to the Instituto Nacional de Investigación y Tecnología Agraria y Alimentaria (INIA-FEDER) of Spain (Research Project CAL03-087-C2-2) for financial support.

REFERENCES

1. Stenersen, J., *Chemical Pesticides. Modes of Action and Toxicology*, CRC Press, Boca Raton, FL, USA, 2004.
2. Guo-Fang, P. et al., Validation study on 660 pesticide residues in animal tissues by gel permeation chromatography cleanup and GC-MS or LC-MS. *J. Chromatogr. A*, 1125, 1, 2006.
3. Current, M.S.S. and King, J.W., Ethanol-modified subcritical water extraction combined with solid-phase microextraction for determining atrazine in beef kidney. *J. Agric. Food Chem.*, 49, 2175, 2001.
4. Argauer, R.J., Lehotay, S.J., and Brown, R.T., Determining lipophilic pyrethoids and chlorinated hydrocarbons in fortified ground beef using ion-trap mass spectrometry. *J. Agric. Food Chem.*, 45, 3936, 1997.
5. Garrido Frenich, A. et al., Multiresidue analysis of organochlorine and organophosphorus pesticides in muscle of chicken, pork and lamb by gas chromatography-triple quadrupole mass spectrometry. *Anal. Chim. Acta*, 558, 42, 2006.
6. Garrido Frenich, A., Plaza Bolaños, P., and Martínez Vidal, J.L., Multiresidue analysis of pesticides in animal liver by gas chromatography using triple quadrupole tandem mass spectrometry. *J. Chromatogr. A*, 1153, 194, 2007.
7. Moghaddam, M.F., Trubey, R.K., and Anderson, J.J., An LC/MS/MS method for improved quantitation of the bound residues in the tissues of animals orally dosed with [^{14}C]-benomyl. *J. Agric. Food Chem.*, 42, 1469, 1994.
8. Voorhees, K.J., Gharaibeh, A.A., and Murugavery, B., Integrated SFE/SFC/MS system for the analysis of pesticides in animal tissues. *J. Agric. Food Chem.*, 46, 2353, 1998.
9. Patel, K. et al., Evaluation of gas chromatography-tandem quadrupole mass spectrometry for the determination of organochlorine pesticides in fats and oils. *J. Chromatogr. A*, 1068, 289, 2005.
10. Basheer, C. and Lee, H.K., Hollow fibre membrane-protected solid-phase microextraction of triazine herbicides in bovine milk and sewage sludge samples. *J. Chromatogr. A*, 1047, 189, 2004.
11. González-Rodríguez, M.J. et al., Determination of pesticides and some metabolites in different kinds of milk by solid-phase microextraction and low-pressure gas chromatography-tandem mass spectrometry. *Anal. Bioanal. Chem.*, 382, 164, 2005.
12. Bogialli, S. et al., Simple and rapid assay for analyzing residues of carbamate insecticides in bovine milk: hot water extraction followed by LC-MS. *J. Chromatogr. A*, 1054, 351, 2004.
13. Bogialli, S. et al., Development of a multiresidue method for analyzing herbicide and fungicide residues in bovine milk based on SPE and LC-MS-MS. *J. Chromatogr. A*, 1102, 1, 2006.
14. Kalantzi, O.I. et al., The global distribution of PCBs and organochlorine pesticides in butter. *Environ. Sci. Technol.*, 35, 1013, 2001.

15. Chu, S.G. et al., Methodological refinements in the determination of 146 polychlorinated biphenyls, including non-ortho and mono-ortho substituted PCBs, and 26 organochlorine pesticides as demonstrated in heron eggs. *Anal. Chem.*, 75, 1058, 2003.
16. Rissato, S.R. et al., Multiresidue determination of pesticides in honey samples by gas GC-MS and application in environmental contamination. *Food Chem.*, 101, 1719, 2007.
17. Sánchez-Brunete, C. et al., Determination of pesticide residues by GC-MS using analyte protectants to counteract the matrix effect. *Anal. Sci.*, 21, 1291, 2005.
18. Albero, B., Sánchez-Brunete, C., and Tadeo, J.L., Analysis of pesticides in honey by solid-phase extraction and gas chromatography-mass spectrometry. *J. Agric. Food Chem.*, 52, 5828, 2004.
19. Rissato, S.R. et al., Supercritical fluid extraction for pesticide multiresidue analysis in honey: determination by gas chromatography with electron-capture and mass spectrometry detection. *J. Chromatogr. A*, 1048, 153, 2004.
20. Blasco, C. et al., Assessment of pesticide residues in honey samples from Portugal and Spain. *J. Agric. Food Chem.*, 51, 8132, 2003.
21. Russo, M.V. and Neri, B., Fluvalinate residues in honey by capillary gas chromatography-electron capture detection-mass spectrometry. *Chromatographia*, 55, 607, 2002.
22. Volante, M. et al., A SPME-GC-MS approach for antivarroa and pesticide residues analysis in honey. *Chromatographia*, 54, 241, 2001.
23. Jiménez, J.J. et al., Capillary gas chromatography with mass spectrometric and atomic emission detection for characterization and monitoring chlordimeform degradation in honey. *J. Chromatogr. A*, 946, 247, 2002.
24. Bernal, J.L. et al., Determination of the fungicide vinclozolin in honey and bee larvae by solid-phase and solvent extraction with gas chromatography and electron-capture and mass spectrometric detection. *J. Chromatogr. A*, 754, 507, 1996.
25. Fernández Muiño, M.A. and Simal Lozano, J., Gas chromatographic-mass spectrometric method for the simultaneous determination of amitraz, bromopropylate, coumaphos, cymiazole and fluvalinate residues in honey. *Analyst*, 118, 1519, 1993.
26. Tahboub, Y.R., Zaater, M.F., and Barri, T.A., Simultaneous identification and quantitation of selected organochlorine pesticide residues in honey by full-scan gas chromatography-mass spectrometry. *Anal. Chim. Acta*, 558, 62, 2006.
27. Charlton, A.J.A. and Jones, A., Determination of imidazole and triazole fungicide residues in honeybees using gas chromatography–mass spectrometry. *J. Chromatogr. A*, 1141, 117, 2007.
28. Fernández, M., Picó, Y., and Mañes, J., Rapid screening of organophosphorus pesticides in honey and bees by liquid chromatography-mass spectrometry. *Chromatographia*, 56, 577, 2002.
29. Blasco, C. et al., Comparison of solid-phase microextraction and stir bar sorptive extraction for determining six organophosphorus insecticides in honey by liquid chromatography-mass spectrometry. *J. Chromatogr. A*, 1030, 77, 2004.
30. Hong, J. et al., Rapid determination of chlorinated pesticides in fish by freezing-lipid filtration, solid-phase extraction and GC-MS. *J. Chromatogr. A*, 1038, 27, 2004.
31. Voorspoels, S. et al., Levels and profiles of PCBs and OCPs in marine benthic species from the Belgian North Sea and the Western Scheldt Estuary. *Mar. Pollut. Bull.*, 49, 391, 2004.
32. Perugini, M. et al., Levels of polychlorinated biphenyls and organochlorine pesticides in some edible marine organisms from the Central Adriatic Sea. *Chemosphere*, 57, 391, 2004.

33. Konwick, B.J. et al., Bioaccumulation, biotransformation, and metabolite formation of fipronil and chiral legacy pesticides in rainbow trout. *Environ. Sci. Technol.*, 40, 2930, 2006.

34. Manirakiza, P. et al., Persistent chlorinated pesticides and polychlorinated biphenyls in selected fish species from Lake Tanganyika, Burundi, Africa. *Environ. Pollut.*, 117, 447, 2002.

35. Jacobs, M.N., Covaci, A., and Schepens, P., Investigation of selected persistent organic pollutants in farmed Atlantic Salmon (*Salmo salar*), salmon aquaculture feed, and fish oil components of the feed. *Environ. Sci. Technol.*, 36, 2797, 2002.

36. Covaci, A. et al., Determination of organohalogenated contaminants in liver of harbour porpoises (*Phocoena phocoena*) stranded on the Belgian North Sea coast. *Mar. Pollut. Bull.*, 44, 1156, 2002.

37. Rodil, R. et al., Selective extraction of trace levels of polychlorinated and polybrominated contaminants by supercritical fluid-solid-phase microextraction and determination by GC-MS. Application to aquaculture fish feed and cultured marine species. *Anal. Chem.*, 77, 2259, 2005.

38. Rodil, R. et al., Multicriteria optimisation of a simultaneous supercritical fluid extraction and clean-up procedure for the determination of persistent organohalogenated pollutants in aquaculture samples. *Chemosphere*, 67, 1453, 2007.

39. Serrano, R. et al., Determination of low concentrations of organochlorine pesticides and PCBs in fish feed and fish tissues from aquaculture activities by GC-MS-MS. *J. Sep. Sci.*, 26, 75, 2003.

40. Pagliuca, G. et al., Residue analysis of organophosphorus pesticides in animal matrices by dual column capillary gas chromatography with nitrogen-phosphorus detection. *J. Chromatogr. A*, 1071, 67, 2005.

41. Juhler, R.K., Supercritical fluid extraction of pesticides from meat: a systematic approach for optimisation. *Analyst*, 123, 1551, 1998.

42. Hopper, M.L., Automated one-step supercritical fluid extraction and clean-up system for the analysis of pesticide residues in fatty matrices. *J. Chromatogr. A*, 840, 93, 1999.

43. Battu, T.S., Singh, B., and Kang, B.K., Contamination of liquid milk and butter with pesticide residues in the Ludhiana district of Punjab state, India. *Ecotoxicol. Environ. Saf.*, 59, 324, 2004.

44. Mallatou, H. et al., Pesticide residues in milk and cheese from Greece. *Sci. Total Environ.*, 196, 111, 1997.

45. Salas, J.H. et al., Organophosphorus pesticide residues in Mexican commercial pasteurized milk. *J. Agric. Food Chem.*, 51, 4468, 2003.

46. Di Muccio, A. et al., Selective solid-matrix dispersion extraction of organophosphate pesticide residues from milk. *J. Chromatogr. A*, 754, 497, 1996.

47. Erney, D.R. et al., Explanation of the matrix-induced chromatographic response enhancement of organophosphorus pesticides during open-tubular column gas-chromatography with splitless or hot on-column injection and flame photometric detection. *J. Chromatogr. A*, 638, 57, 1993.

48. Bernal, J.L. et al., Gas chromatographic determination of acrinathrine and 3-phenoxy-benzaldehyde residues in honey. *J. Chromatogr. A*, 882, 239, 2000.

49. Di Muccio, A. et al., Selective, extraction of pyrethroid pesticide residues from milk by solid-matrix dispersion. *J. Chromatogr. A*, 765, 51, 1997.

50. Valsamaki, V.I. et al., Determination of organochlorine pesticides and polychlorinated biphenyls in chicken eggs by matrix solid phase dispersion. *Anal Chim. Acta*, 573, 195, 2006.

51. Schenck, F.J. et al., Liquid chromatographic determination of *N*-methyl carbamate pesticide residues at low parts-per-billion levels in eggs. *J. AOAC Int.*, 89, 196, 2006.
52. Jiménez, J.J. et al., Solid phase microextraction applied to the analysis of pesticide residues in honey using GC-ECD. *J. Chromatogr. A*, 829, 269, 1998.
53. Jiménez, J.J. et al., Gas chromatography with electron-capture detection and nitrogen-phosphorus detection in the analysis of pesticides in honey after elution from Florisil column. Influence of the honey matrix on the quantitative results. *J. Chromatogr. A*, 823, 381, 1998.
54. Campillo, N. et al., Solid phase microextraction and gas chromatography with atomic emission detection for multiresidue determination of pesticides in honey. *Anal. Chim. Acta*, 562, 9, 2006.
55. Martel, A.C. and Zeggane, S., Determination of acaricides in honey by high-performance liquid chromatography with photodiode array detection. *J. Chromatogr. A*, 954, 173, 2002.
56. Yu, J., Wu, C., and Xing, J., Development of new solid-phase microextraction fibers by sol-gel technology for the determination of organophosphorus pesticide multiresidues in food. *J. Chromatogr. A*, 1026, 101, 2004.
57. Jiménez, J.J. et al., Determination of rotenone residues in raw honey by solid-phase extraction and high-performance liquid chromatography. *J. Chromatogr. A*, 871, 67, 2000.
58. Bernal, J.L. et al., High performance liquid chromatographic determination of benomyl and carbendazim residues in apiarian samples. *J. Chromatogr. A*, 787, 129, 1997.
59. Rezic, I. et al., Determination of pesticides in honey by ultrasonic solvent extraction and thin-layer chromatography. *Ultrason. Sonochem.*, 12, 477, 2005.
60. Doong, R. and Lee, C., Determination of organochlorine pesticide residues in foods using solid-phase extraction clean-up cartridges. *Analyst*, 124, 1287, 1999.
61. Sun, F. et al., Multiresidue determination of pesticide in fishery products by a tandem solid-phase extraction technique. *J. Food Drug Anal.*, 13, 151, 2005.
62. Levine, R.A. et al., Automated method for cleanup and determination of benomyl and thiabendazole in table-ready foods. *J. AOAC Int.*, 81, 1217, 1998.
63. Björklund, E. et al., New strategies for extraction and clean-up of persistent organic pollutants from food and feed samples using selective pressurized liquid extraction. *Trends Anal. Chem.*, 25, 318, 2006.
64. Carabias-Martinez, R. et al., Pressurized liquid extraction in the analysis of food and biological samples. *J. Chromatogr. A*, 1089, 1, 2005.
65. Lehotay, S.J. et al., Analysis of pesticide residues in eggs by direct sample introduction/gas chromatography/tandem mass spectrometry. *J. AOAC Int.*, 49, 4589, 2001.
66. Lehotay, S.J., Mastovska, K., and Yun, S.J., Evaluation of two fast and easy methods for pesticide residue analysis in fatty food matrices. *J. AOAC Int.*, 88, 630, 2005.
67. Bligh, E.G. and Dyer, W.J., A rapid method of total lipid extraction and purification. *Can. J. Biochem. Physiol.*, 37, 911, 1959.
68. Smedes, F., Determination of total lipid using non-chlorinated solvents. *Analyst*, 124, 1711, 1999.
69. Van der Valk, F. and Wester, P.G., Determination of toxaphene and other organochlorine pesticides in fish from Northern Europe. *Chemosphere*, 22, 57, 1991.
70. Van der Hoff, G.R. et al., Determination of organochlorine compounds in fatty matrices: application of normal-phase LC clean-up coupled on-line to GC/ECD. *J. High Resolut. Chromatogr.*, 20, 222, 1997.
71. Serrano, R. et al., Sample clean-up and fractionation of organophosphorus pesticide residues in mussels using normal-phase LC. *Intern. J. Environ. Anal. Chem.*, 70, 3, 1998.

72. Commission Decision, 2002/657/EC. Implementing Council Directive 96/23/EC concerning the performance of analytical methods and the interpretation of results, 2002.

73. Vetter, W., Enantioselective fate of chiral chlorinated hydrocarbons and their metabolites in environmental samples. *Food Rev. Intern.*, 17, 113, 2001.

74. Soler, C. and Pico, Y., Recent trends in liquid chromatography-tandem mass spectrometry to determine pesticides and their metabolites in food. *Trends Anal. Chem.*, 26, 103, 2007.

75. Pico, Y. et al., Control of pesticide residues by LC-MS to ensure food safety. *Mass Spectrom. Rev.*, 25, 917, 2006.

76. Kotretsou, S.I. and Koutsodimou, A., Overview of the applications of tandem mass spectrometry (MS/MS) in food analysis of nutritionally harmful compounds. *Food Rev. Intern.*, 22, 125, 2006.

77. Pang, G.F. et al., Validation study on 660 pesticide residues in animal tissues by gel permeation chromatography cleanup and GC-MS or LC-MS-MS. *J. Chromatogr. A*, 1125, 1, 2006.

78. Brandom, D.L. et al., Analysis of fenbendazole residues in bovine milk by ELISA. *J. Agric. Food Chem.*, 50, 5791, 2002.

79. Yazymina, E.V. et al., Immunoassay techniques for detection of the herbicide simazine based on use of oppositely charged water-soluble polyelectrolytes. *Anal. Chem.*, 71, 3538, 1999.

80. Bushway, R.J., Perkins, L.B., and Hurst, H.L., Determination of atrazine in milk by immunoassay. *Food Chem.*, 43, 283, 1992.

81. Young, D.L. et al., Determination of spinosad and its metabolites in food and environmental matrices. 3. Immunoassay methods. *J. Agric. Food Chem.*, 48, 5146, 2000.

82. Nam, K.S. and King, J.W., Supercritical fluid extraction and enzyme immunoassay for pesticide detection in meat products. *J. Agric. Food Chem.*, 48, 5195, 2000.

83. Lehotay, S.J. and Argauer, R.J., Detection of aldicarb sulfone and carbofuran in fortified meat and liver with commercial ELISA kits after rapid extraction. *J. Agric. Food Chem.*, 41, 2006, 1993.

84. Bonwick, G.A. et al., Synthetic pyrethroid insecticides in fish: analysis by GC-MS operated in the negative ion chemical ionization mode and ELISA. *Food Agric. Immunol.*, 8, 185, 1996.

85. Botchkareva, A.E. et al., Development of chemiluminescent ELISA to DDT and its metabolites in food and environmental samples. *J. Immunol. Meth.*, 283, 45–57, 2003.

86. Medyantseva, E.P. et al., The specific immunochemical detection of 2,4-dichlorophenoxyacetic acid and 2,4,5-trichlorophenoxyacetic acid pesticides by amperometric cholinesterase biosensors. *Anal. Chim. Acta*, 347, 71, 1997.

87. Brimecombe, R. and Limson, J., Voltammetric analysis of the acaricide amitraz and its degradant, 2,4-dimethylaniline. *Talanta*, 71, 1298, 2007.

88. De Souza, D. and Machado, S.A.A., Electroanalytical method for determination of the pesticide dichlorvos using gold-disk microelectrodes. *Anal. Bioanal. Chem.*, 382, 1720, 2005.

89. Siontorou, C.G. et al., A triazine herbicide minisensor based on surface-stabilized bilayer lipid membranes. *Anal. Chem.*, 69, 3109, 1997.

90. Zhang, Y.D. et al., Disposable biosensor test for organophosphate and carbamate insecticides in milk. *J. Agric. Food Chem.*, 53, 5110, 2005.

91. Inczedy, J., Lengyel, T., and Ure, A.M., *Compendium of Analytical Nomenclature. Definitive Rules*, 3rd edn, Blackwell Science, Oxford, UK, 1997.

92. Massart, D.L. et al., *Handbook of Chemometrics and Qualimetrics: Part A*, Elsevier, Amsterdam, 1997.

93. Cuadros-Rodríguez, L. et al., Assessment of uncertainty in pesticide multiresidue analytical methods: main sources and estimation. *Anal. Chim. Acta*, 454, 297, 2002.
94. Thompson, M., Ellison, S.L.R., and Wood, R., Harmonized guidelines for single laboratory validation of methods of analysis (IUPAC technical report). *Pure Appl. Chem.*, 74, 835, 2002.
95. EURACHEM, http://www.eurachem.ul.pt/guides/CITAC%20EURACHEM%20GUIDE.pdf, 2002.
96. Analytical Methods Committee, Principles of data quality control in chemical analysis. *Analyst*, 114, 1497, 1989.
97. Martínez Vidal, J.L., Garrido Frenich, A., and Egea González, F.J., Internal quality control criteria for environmental monitoring of organic micro-contaminants in water. *Trends Anal. Chem.*, 22, 34, 2003.
98. Council Directive, 96/23/EC. Performance of analytical methods and interpretation of results, 1996.
99. Muir, D. and Sverko, E., Analytical methods for PCBs and organochlorine pesticides in environmental monitoring and surveillance: a critical appraisal. *Trends Anal. Chem.*, 386, 769, 2006.

8 Determination of Pesticides in Soil

Consuelo Sánchez-Brunete, Beatriz Albero,
and José L. Tadeo

CONTENTS

8.1 INTRODUCTION

Pesticides may reach the soil compartment by different ways. Direct soil application is normally employed for the control of weeds, insects, or microorganisms, the use of herbicides being a typical example. Pesticides may also reach the soil indirectly, when the pesticide fractions applied to the aerial part of plants (to control weeds, crop pests, or diseases) drop to the soil during application, or lixiviate from the crops. Other ways the pesticides reach the soil are by transportation from a different compartment, e.g., with the irrigation water, or by atmospheric deposition.

Once in the soil, pesticides may undergo a series of transformation and distribution processes. These transformation processes may have a biotic or abiotic origin and cause the degradation of pesticides through several mechanisms, such as oxidation, reduction, or hydrolysis. The distribution of pesticides can be originated by various processes, such as volatilization, leaching, runoff, and absorption by plants. In these processes, the physical–chemical properties of pesticides and the adsorption–desorption equilibrium in soil are the main factors involved. Figure 8.1 shows the most important pathways of pesticide distribution and transformation in soil.

The fate of pesticides and their degradation products in soil will depend on different factors, such as the agricultural practices, the climate, and the type of soil. Pesticides and their degradation or transformation products may cause toxic effects to man and the environment, making necessary to evaluate if their application may cause an unacceptable risk. Consequently, many developed countries have regulated the pesticide use in agriculture [1,2].

8.2 SAMPLE PREPARATION

8.2.1 Sampling and Preparation of Soil Samples

The plough layer of soil (0–20 cm) is generally sampled for the determination of pesticides in this compartment. Nevertheless, other layers may be sampled at

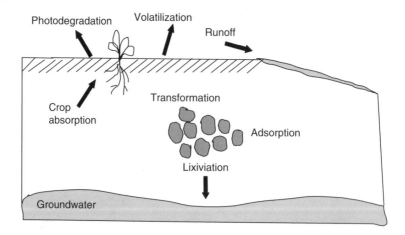

FIGURE 8.1 Distribution and transformation pathways of pesticides in soil.

different depths to study the distribution of these compounds in soil and, in addition, soil solution may be sometimes sampled to know the bioavailability of pesticides.

After field sampling, soil is usually air dried and sieved through a 2 mm mesh in the laboratory. Then, soil samples are placed in closed glass flasks and stored frozen until the analysis of pesticides.

The addition of known amounts of pesticides to blank soil samples is a normal practice to study the recovery of these compounds. However, the recovery of pesticides from soil may be different in freshly spiked than in aged soil samples. Pesticides in soil may undergo transformation processes that lead to the formation of bound residues, which cannot be extracted even after exhaustive extraction with organic solvents. The use of reference soil samples with certified concentrations of the studied pesticides is recommended for the validation of the analytical methods, but these reference materials are difficult to prepare and maintain and are available only for a few pesticides.

8.2.2 EXTRACTION

The liquid–solid extraction (LSE) of pesticides from soil is generally carried out by organic solvents. Two techniques have been widely used, the shaking and filter method and the Soxhlet extraction method. These classical analytical techniques have the advantage of being simple and low cost methods, but they are time consuming, laborious, difficult to automate, and nonselective methods. In addition, they suffer from various disadvantages, such as the use of large volume of organic solvents and the need of cleanup steps.

Several modern analytical techniques have been developed to overcome these problems. Accelerated solvent extraction (ASE), also named pressurized liquid extraction (PLE), is a fast technique that uses low volumes of solvents and can be automated, although the high temperatures used to accelerate the process may degrade some pesticides. Supercritical fluid extraction (SFE) uses fluids above their critical temperature and pressure. In these conditions, supercritical fluids behave similar to liquids, CO_2 being widely employed because of its reduced cost and low critical temperature (31°C) and pressure (73 atm). Microwave-assisted extraction (MAE) is also a fast technique that is able to extract multiple samples at the same time, but the extraction vessels are expensive and must be cooled at room temperature before opening. Ultrasonic or sonication assisted extraction with various organic solvents has also been employed to extract pesticides from soil. A miniaturized technique based on the sonication assisted extraction in small columns (SAESC) has been recently developed in our laboratory. In this method, the soil sample located in a small column is placed in an ultrasonic water bath, wherein pesticides are extracted with a low solvent volume, assisted by sonication. Tables 8.1 through 8.3 summarize representative published papers on the analysis of pesticides in soil using those extraction techniques.

8.2.2.1 Herbicides

Analyses of herbicide residues in soil have been frequently performed because of the wide application of these compounds. Initially, polar herbicides, such as benzonitriles and phenoxy acids, were extracted from soil with organic solvents of

TABLE 8.1

Extraction Methods of Herbicides from Soil

Technique	Class	Solvent	References
Shaking	Benzonitriles, phenoxy acids	Low–medium polarity, acidic pH	[3–6]
	Dinitroanilines	Acetonitrile–water (99:1, v/v)	[7]
	Phenoxy acids, glyphosate	Water, basic pH	[8–10]
	Phenylureas, triazines	Methanol	[11–16]
	Sulfonylureas	Methanol, acidic pH	[17]
	Multiclass	Ethyl acetate	[18–20]
		Acetonitrile	[21]
		Acetone	[22]
Soxhlet	Triazines, benzonitriles	Methanol	[23–25]
Sonication	Phenoxy acids, pyrimidines	Water, basic pH	[26,27]
	Triazines	Hexane–acetone (2:1, v/v)	[28]
	Multiclass	Cyclohexane–acetone (3:1, v/v)	[29]
SAESC		Ethyl acetate	[30,31]
PLE	Phenoxy acids	Water	[32]
	Multiclass	Acetone	[33]
MAE	Phenoxy acids	Water–methanol, pH 7	[34]
	Triazines	Water–methanol (1:1, v/v) pH 7	[35]
	Multiclass	Acetonitrile	[36,37]
SPME	Triazines		[36]

SAESC, sonication assisted extraction in small columns; PLE, pressurized liquid extraction; MAE, microwave-assisted extraction; SPME, solid-phase microextraction.

low–medium polarity at acidic pH, using manual or mechanical shaking or sonication. For less polar herbicides, such as triazines, chloroacetamides, and dinitroanilines, organic solvents such as acetone, ethyl acetate, methanol, and acetonitrile, alone or in mixtures with water, were commonly used.

More recently, a considerable reduction in solvent consumption has been achieved by miniaturizing the scale of sample extraction. In addition, MAE and SPME have been successfully applied to the extraction of various herbicides from soil. MAE is a technique with a reduced consumption of solvent, which is normally acetonitrile or methanol, alone or in mixtures with water, and solid-phase microextraction (SPME) eliminates the need of solvent and an ulterior cleanup step is not needed.

In multiclass herbicide analysis, soil samples were generally extracted with a polar or medium polarity solvent, such as acetone or acetonitrile. PLE is a new technique used successfully for the extraction of herbicides, such as triazines and phenoxy acids, using water and acetone as solvents.

8.2.2.2 Insecticides and Fungicides

Conventional methods have been widely used in the extraction of organochlorine (OC) insecticides from soil, although the use of new extraction techniques has

TABLE 8.2
Extraction Methods of Insecticides and Fungicides from Soil

Technique	Class	Solvent	References
Shaking	Organophosphorus	Methanol	[38]
	Strobilurins	Acetone	[39]
	Benzimidazoles	Ethyl acetate	[40,41]
	Multiclass-fungicides	Acetone	[42]
Soxhlet	Multiclass-insecticides	Dichloromethane	[43]
Sonication	Organochlorines	Petroleum ether–acetone (1.1, v/v)	[44]
	Organophosphorus	Acetonitrile	[45]
		Water, acetone	[46]
	Pyrethroids	Isooctane–Dichloromethane (15:85, v/v)	[47]
	Multiclass-fungicides	Water, acetone	[48]
SAESC	Carbamates	Methanol	[49]
	Multiclass-insecticides	Ethyl acetate	[50]
	Multiclass-fungicides	Ethyl acetate	[51]
SFE	Carbamates, Pyrethroids	CO_2–3%methanol	[52,53]
	Organochlorines	CO_2	[54]
	Multiclass-insecticides	CO_2–3%methanol	[55]
PLE	Organochlorines	Acetone–hexane (1:1, v/v)	[56–58]
MAE	Carbamates	Methanol	[52]
	Organochlorines	Acetone–hexane (1:1, v/v)	[59]
	Pyrethroids	Toluene	[60,61]
SPME	Organochlorines		[62,63]
	Organophosphorus		[64,65]
	Multiclass-fungicides		[66,67]

SAESC, sonication assisted extraction in small columns; SFE, solid-phase extraction; PLE, pressurized liquid extraction; MAE, microwave-assisted extraction; SPME, solid-phase microextraction.

increased during the last years. In the PLE, the soil sample is placed in a cartridge and extracted with mixtures of acetone and hexane. The use of MAE has also increased because of the good recoveries obtained. Moreover, headspace SPME has been successfully used to determine OC insecticides in soil with limits of detection (LOD) similar to other extraction techniques.

Organophosphorus (OP) pesticides are compounds highly polar and soluble in water that have been extracted from soil by shaking with organic solvents such as methanol. Other new techniques, such as SPME, are now frequently used for the extraction of these compounds in soil samples.

Carbamates were initially extracted from soil by conventional methods using mechanical shaking with different solvents. SFE and MAE were afterwards successfully applied to soil as a practical alternative to traditional methods. In recent years, analysis by means of SAESC has obtained good results.

TABLE 8.3
Multiresidue Methods of Pesticide Extraction from Soil

Technique	Class	Solvent	References
Shaking	H, I, F	Acetonitrile–water (70:30, v/v)	[68]
		Ethyl acetate	[69]
Soxhlet	I, A	Hexane–acetone (1:1, v/v)	[70]
	H, I	Acetone	[71]
	H, I	Methylene chloride–acetone (1:1, v/v)	[72]
Sonication	F, I	Acetonitrile–water (2:1, v/v)	[73]
	H, F, I, A	Methanol–water (4:1, v/v)	[74]
	H, I, A	Ethyl acetate	[75]
SAESC	H, I, F, A	Ethyl acetate	[76,77]
SFE	H, I, F	CO_2–3%methanol	[78,79]
PLE	H, I	Water	[73]
SPME	H, I		[80]

H, herbicides; I, insecticides; F, fungicides; A, acaricides; SAESC, sonication assisted extraction small columns; SFE, solid-phase extraction; PLE, pressurized liquid extraction; SPME, solid-phase microextraction.

Pyrethroid insecticides are a class of natural and synthetic compounds that are retained in soils because of their high lipophility and low water solubility and extracted from soil samples by sonication with organic solvents, alone or in binary mixtures. Investigations with fortified samples showed that good and similar recoveries of these compounds were obtained with MAE and SFE.

The analysis of multiclass mixtures of insecticides was initially carried out by Soxhlet or shaking methods with low or medium polarity solvents. SFE with CO_2 modified with methanol and SAESC with ethyl acetate are other techniques used more recently.

The analysis of fungicides in soil was initially accomplished by classical extraction methods, such as the shaking and filter method using acetone or ethyl acetate. The ultrasonic assisted extraction and SPME have been other techniques used more recently for the determination of fungicides in soil samples.

8.2.2.3 Multiresidue

Reliable multiresidue analytical methods are needed for monitoring programs of pesticide residues in soil. The classical procedure for pesticide extraction from soil was to shake soil samples with an organic solvent, ethyl acetate or acetonitrile, alone or in mixtures with water, being the most widely used solvents.

SFE with carbon dioxide containing 3% methanol, as a modifier used to improve recoveries of polar pesticides, has been employed for the multiresidue extraction of pesticides having a wide range of polarities and molecular weights. SFE using CO_2 is essentially a solvent-free extraction wherein the carbon dioxide is easily removed at atmospheric pressure.

Recently, a modification of the SAESC has been used for the simultaneous determination of different classes of pesticides. The good reproducibility and detection limits achieved with this method allow its application to the monitoring of pesticide residues in soil [76].

SPME has been mainly used for the extraction of pesticides from aqueous samples; however, head space SPME has been recently used for the determination of pesticides volatilized from soil. The application of MAE for the extraction of pesticide residues is increasing in the last years and together with other modern techniques, such as sonication and PLE, are the most widely used methods at present.

8.2.3 CLEANUP

Soil sample extracts, obtained with any of the methods described earlier, generally contain a considerable amount of other components that may interfere in the subsequent analysis. Therefore, the determination of pesticides at residue level frequently requires a further cleanup of soil extracts. Liquid–liquid partition (LLP) between an aqueous and an organic phase, at modulated pH in some cases, has been the most common first step in the cleanup of extracts. An alternative cleanup technique is column chromatography, using reverse or normal phases, in which pesticides are separated from interferences by elution with a solvent of adequate polarity. Tables 8.4 through 8.6 summarize the cleanup procedures employed in the determination of pesticides in soil.

8.2.3.1 Herbicides

Phenoxy acid herbicides are normally formulated as amine salts or esters, which are rapidly hydrolyzed in soil to the acidic form. Cleanup techniques for the

TABLE 8.4

Cleanup Techniques Used in the Analysis of Herbicides

Class	Technique	Solvent	References
Phenoxy acids	LLP, pH 8–9	Methylene chloride	[3]
	LLP, SPE-florisil	Diethyl ether	[5]
	LLP-pH 2	Ether:hexane	[32]
	SPE-silica gel	Dichloromethane	[4,26]
	SPE-polymer	Benzene–hexane (1:9, v/v)	[8,10]
	SPE-C8	Methanol	[17]
Phenylureas	SPE-florisil	Ethyl ether–n-hexane (1:1, v/v)	[23,24]
Pyrimidines	SPE-alumina	Ethyl ether–n-hexane (1:2, v/v)	[15]
Triazines	SPE-polymer	Methanol–ethyl acetate (7:3, v/v)	[35]
Multiclass	LLP-SPE-florisil-alumina	Dichloromethane–diethyl ether	[21]

LLP, liquid–liquid partition; SPE, solid-phase extraction.

TABLE 8.5

Cleanup Techniques Used in the Analysis of Insecticides and Fungicides

Class	Technique	Solvent	References
Insecticides			
Organochlorines	SPE-alumina	Hexane–ethyl acetate (7:3, v/v)	[44]
	SPE-carbon	Hexane–ethyl acetate (80:20, v/v)	[57]
	SPE-florisil	Heptane–ethyl acetate (1:1, v/v)	[58]
Organophosphorus	LLP	Dichloromethane	[46]
	SPE-MISPE	Water	[46]
Pyrethroids	SPE-florisil	Hexane–ethyl acetate (2:1, v/v)	[60,61]
Multiclass	LLP	Methylene chloride	[42]
	SPE-C18	Methanol	[43]
Fungicides			
Strobilurins	SPE-florisil	Toluene-ethyl acetate (20:1, v/v)	[39]

LLP, liquid–liquid partition; SPE, solid-phase extraction; MISPE, molecularly imprinted solid-phase extraction.

purification of soil extracts include liquid–liquid partitioning, at basic or acidic pH, and column chromatography using various adsorbents (Florisil, alumina, or silica gel). These cleanup processes are time consuming and large quantities of solvents are generally required. Therefore, minicolumns and cartridges, which reduce the solvent consumption and the analysis time, have replaced conventional chromatographic columns. Various organic solvents with different polarity, such as methanol, dichloromethane, or other intermediate polarity solvents, have been used to elute phenoxy acid herbicides from cleanup columns. In recent years, new polymeric packing materials have been developed.

The cleanup of triazine herbicides in soil extracts has been carried out by SPE with alumina or Florisil and various mixtures of organic solvents have been used for eluting these compounds.

TABLE 8.6

Cleanup Techniques Used in the Multiresidue Analysis of Pesticides

Class	Technique	Solvent	References
H, I, F	LLP	Petroleum ether-diethyl ether (1:1, v/v)	[68]
I, F	LLP	Dichloromethane	[73]
H, I, F	SPE-C18	Acetone-hexane (20:80, v/v)	[78]
H, I, F, A	SPE-polymer	Dichloromethane–methanol (1:1, v/v)	[74]

H, herbicides; I, insecticides; F, fungicides; A, acaricides; LLP, liquid–liquid partition; SPE, solid-phase extraction.

In the analysis of multiclass herbicide mixtures, the cleanup of soil extracts has been carried out by SPE on Florisil or alumina, after LLP.

8.2.3.2   Insecticides and Fungicides

In general, extracts from soil samples have been cleaned up by means of chromatographic columns filled with alumina or Florisil as adsorbents and pesticides have been eluted with nonpolar or low polarity solvents (hexane, ethyl acetate). In some cases, more hydrophobic sorbents, such as carbon, have been used for low polarity insecticides. In addition, LLP of soil extracts between immiscible solvents is a method sometimes used. Moreover, solid-phase extraction with molecularly imprinted polymers (MISPE) is a novel selective method that has been used for the analysis of OPs in soil and proved to be a good tool for their selective extraction.

In the analysis of multiclass insecticide mixtures, good recoveries have been obtained using reversed-phase C18 cartridges and methanol as eluting solvent.

8.2.3.3   Multiresidue

Analysis of complex mixtures of pesticides in soil is a difficult problem because of the presence of a wide variety of compounds with different physical–chemical properties.

In modern analytical techniques, the classical methodology for the cleanup of extracts, based on LLP, has been replaced by miniaturized techniques for residue analysis that are less solvent consuming. SPE is a technique widely used to determine pesticide residues in soil after their extraction with water or aqueous mixtures of organic solvents. Octyl and octadecyl-bonded silica sorbents have been frequently used in the analysis of nonpolar and medium polarity pesticides in soil extracts.

8.2.4   Derivatization

The thermal instability and low volatility of some pesticides make analysis by gas chromatography (GC) difficult. Consequently, methods of analysis based on GC require, in some cases, the derivatization of pesticides to increase their volatility. In addition, pesticide derivatives are sometimes prepared to enhance the response of a pesticide to a specific detector in GC or high-performance liquid chromatography (HPLC) analyses.

8.2.4.1   Benzonitriles

The derivatization of the hydroxyl group usually involves perfluoroacylation with heptafluorobutyric anhydride to form perfluoroacylated derivatives, which are determined by GC [6].

8.2.4.2   Glyphosate

This compound is very polar and has a high solubility in water so direct determination by GC or HPLC is difficult. Derivatives for HPLC determination are prepared to improve the pesticide response and pre- or postcolumn reactions have been used with this aim. In postcolumn derivatization, the reaction is produced with *o*-phthalaldehyde (OPA) and mercaptoethanol and in precolumn derivatization 9-fluorenylmethyl chloroformate (FMOC-Cl) is used to form fluorescent derivatives with an improvement in the chromatographic determination [9].

8.2.4.3   Phenoxy Acid Herbicides

Because of their highly polar nature and low volatility, they cannot be directly determined by GC and have to be derivatized to their corresponding esters. Several derivatization procedures have been applied to make phenoxy acid herbicides amenable to GC analysis.

The carboxylic group is converted to the corresponding methyl ester by reacting with diazomethane [5,22] or by alternative less toxic methods such as esterification with methanol using an acid catalyst such as boron trifluoride [3] or with trimethylphenylammonium hydroxide [32]. The sensitivity towards electron-capture detection can be improved by using bromine–iodine to obtain the brominated methyl esters [5] or by reacting with pentafluorobenzyl bromine to obtain the halogenated aromatic esters [4,26].

8.2.4.4   Phenylureas

The analysis by direct GC of these compounds is difficult because of their thermal instability caused by the NH group. Phenylureas decompose in the sample inlet port and produce several peaks in the chromatogram (phenyl isocyanates).

Several analytical methods have been developed based on the possibility to obtain stable derivatives for GC determination, such as alkyl, acyl, and silyl derivatives. Other derivatization mode for phenylureas is the ethylation with ethyl iodide and hydrolysis to *N*-ethyl derivatives [14].

8.2.4.5   Sulfonylureas

Gas chromatographic analysis of sulfonylureas is difficult owing to their strongly polar nature. Pentafluorobenzyl derivatives, which have enhanced detection properties, have been used since the method is more sensitive than with ethyl or methyl derivatives [17].

8.2.4.6   Carbamates

Carbamates are thermally decomposed into the corresponding phenols and methyl isocianate. HPLC methods for carbamates are preferred over GC determination and they are based on postcolumn basic hydrolysis to release methylamine, which subsequently reacts with the OPA reagent to form isoindol derivatives, which are determined by fluorescence (FL) detection [49].

8.3 DETERMINATION OF PESTICIDE RESIDUES

Gas and liquid chromatography are the most widely used analytical techniques for the determination of pesticide residues in soil. Thermal stability and volatility are the main characteristics that a pesticide must possess in order to be suitable for gas chromatographic analysis. Initially, GC was performed with short glass or steel columns packed with a stationary phase; however, nowadays fused silica capillary columns are almost exclusively employed. The stationary phases used are usually polysiloxanes with different functional groups to increase the polarity.

Table 8.7 summarizes the GC methods used to determine pesticide residues in soil. Electron-capture detection (ECD) is adequate for halogenated compounds or

TABLE 8.7
GC Methods Used for the Determination of Pesticide Residues in Soil

Detector	Compound	LOD (μg/kg)	References
ECD	Organochlorines	0.1–12.9	[44,54,63]
	Pyrethroids	1–200	[60,61]
	Sulfonylureas	0.1 pg	[17]
	Multiresidue	0.05–20	[21,29,42,50,51,77,78]
NPD	Dinitroanilines	10	[18]
	Organophosphorus	12–34	[46]
	Phenylureas	10	[14]
	Pyridine	10	[19]
	Strobilurins	5	[39]
	Triazines	5–30	[28]
	Multiresidue	0.1–20	[20,29–31,37,51,73]
FPD	Organophosphorus	0.5–100 μg/L	[65]
MS			
EI	Benzonitriles	1	[6]
	Dinitroanilines	10	[18]
	Organochlorines	2–100 ng/L	[62]
	Phenoxy acids	5	[3]
	Pyrethroids	0.1–3.7	[61]
	Pyridine	10	[19]
	Triazines	2–100	[28,36]
	Multiresidue	0.01–137.1	[20,22,30,31,37,48,66–68, 72,73,75,76,81,82]
NCI	Pyrethroids	0.1–2	[60,61]
MS/MS			
EI	Organochlorines	0.02–3.6	[59]
	Pyrethroids	0.08–0.54	[61]
	Multiresidue	0.1–3.7	[33,79]
NCI	Pyrethroids	0.4–1.2	[61]

ECD, electron-capture detector; NPD, nitrogen–phosphorus detector; FPD, flame photometric detector; MS, mass spectrometry; EI, electron impact; NCI, negative chemical ionization; MS/MS, tandem mass spectrometry. LOD, limit of detection.

those that contain electronegative atoms such as oxygen or sulfur, pyrethroids and OC pesticides being typical examples. A chromatogram of a mixture of fungicides analyzed by GC–ECD is depicted in Figure 8.2. On the other hand, the determination of pesticides that contain nitrogen or phosphorus atoms, such as triazines and OP pesticides, has been carried out with nitrogen–phosphorus detection (NPD) or flame photometric detection (FPD). Atomic emission and flame ionization detectors have also been employed in the determination of pesticide residues in soil.

Although these selective detectors allow quantitating residues at trace levels, the confirmation of the identity is achieved by mass spectrometry (MS) coupled to GC. The ionization technique most commonly used in GC–MS analysis is electron impact (EI), which produces characteristic ion fragments of compounds that are

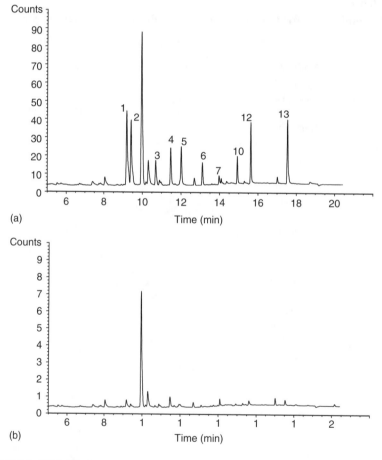

FIGURE 8.2 GC–ECD chromatograms. (a) A soil sample fortified at 0.05 μg/g and (b) a blank soil sample. Peak identification: 1 = Quintozene; 2 = chlorothalonil; 3 = tolclofos-methyl; 4 = dichlofluanid; 5 = triadimefon; 6 = procymidone; 7 = myclobutanil; 10 = ofurace; 12 = nuarimol; and 13 = fenarimol. (From Sánchez-Brunete, C. et al., *J. Chromatogr. A*, 976, 319, 2002. With permission.)

collected in spectral libraries. Full scan and selected ion monitoring (SIM) are the two working modes for EI-MS; SIM mode is more sensitive and selective than full scan. Most of the multiresidue methods developed in the last few years use MS as detection system as it offers the possibility of the simultaneous determination and identity confirmation of a large number of pesticides from different chemical classes in a single injection. Chemical ionization (CI) is a useful tool when molecular ions are not observed in EI mass spectra that can work with two different polarities, positive (PCI) and negative (NCI). Time of flight mass spectrometry (TOF-MS) is the result of the significant advances undergone by the analytical instrumentation that is beginning to be applied in the determination of pesticides since full mass-range spectrum and exact mass determination can be obtained for each pesticide without compromising sensitivity. Tandem mass spectrometry (MS/MS) coupled to GC has also been used to determine pesticides in soil with good selectivity and high sensibility.

HPLC is an analytical tool adequate for the determination of pesticides that are not thermally stable or not volatile. Reversed-phase HPLC has been widely used in the analysis of pesticides as most of these compounds present a low polarity. The HPLC methods developed for the determination of pesticides in soil are summarized in Table 8.8. Ultraviolet (UV) detection has been the most frequently used technique in liquid chromatography, although other selective detectors such as FL present higher selectivity and sensitivity. The drawback of FL detection is that it is limited

TABLE 8.8
HPLC Methods Used for the Determination of Pesticide Residues in Soil

Detector	Compound	LOD (µg/kg)	References
UV	Benzimidazoles	n.a.	[40,41]
	Carbamates	n.a.	[52]
	Phenoxy acids	3–50	[8,34]
		1–50 µg/L	[10]
	Organophosphorus	0.5–34	[38,45]
	Triazines	10–60	[28,35]
FL	Carbamates	1.6–3.7	[49]
MS			
APCI	Multiresidue	4.8–22[a]	[33]
ESI	Multiresidue	0.5–2.5	[74]
MS/MS			
APCI	Multiresidue	0.3–11[a]	[33]
ESI	Glyphosate	5	[9]
	Multiresidue	0.15–7.5[a]	[33]

UV, ultraviolet detector; FL, fluorescence detector; MS, mass spectrometry; APCI, atmospheric pressure chemical ionization; ESI, electrospray ionization; MS/MS, tandem mass spectrometry; n.a., not available.

[a] LOQ (limit of quantitation, µg/kg) instead of LOD (limit of detection).

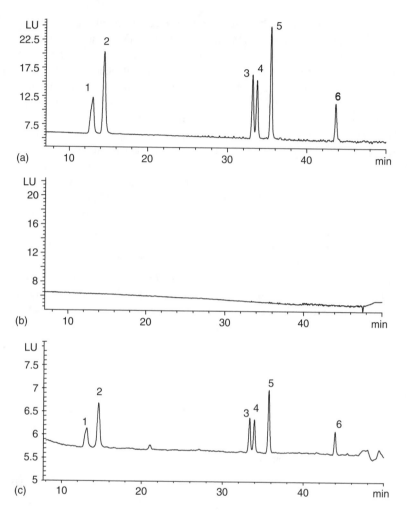

FIGURE 8.3 HPLC-Fl chromatograms. (a) A soil sample fortified at 0.1 μg/g, (b) a blank soil sample, and (c) a soil sample spiked at the LOQ level (0.01 μg/g). Peak identification: 1 = oxamyl, 2 = methomyl, 3 = propoxur, 4 = carbofuram, 5 = carbaryl, 6 = methiocarb. (From Sánchez-Brunete, C. et al., *J. Chromatogr. A*, 1007, 85, 2003. With permission.)

to compounds that fluoresce or else derivatization to obtain a fluorescent compound is required. Figure 8.3 shows a representative chromatogram of a mixture of carbamates that has gone through a postcolumn derivatization process.

The preparation of thermally stable derivatives for the subsequent gas chromatographic analysis is an alternative that nowadays is seldom applied because of the high sensitivity and selectivity achieved with liquid chromatography coupled with mass spectrometry (LC–MS). The implementation of robust ionization interfaces, such as electrospray ionization (ESI) and atmospheric pressure chemical ionization (APCI), is considered one of the main instrumental improvements. The selection of the ionization interface depends on the nature of the analyzed pesticide; APCI is adequate for moderately nonpolar pesticides such as triazines and phenylureas, whereas ESI is suitable for polar and ionic pesticides. Tandem mass spectrometry is also used to determine pesticides in soil with the advantage of achieving a better selectivity owing to the selection of daughter ions.

The analysis of pesticides has also been carried out with nonchromatographic methods. Capillary electrophoresis (CE) is an alternative analytical tool that has been applied in the determination of residues in soil samples [27,83,84]. CE presents different working modes, and micellar electrokinetic chromatography (MECK), capillary zone electrophoresis (CZE), and capillary electrochromatography (CEC) are the most frequently used. The application of sensors and biosensors in the determination of pesticides in environmental samples is also rapidly increasing. These portable analytical devices offer the possibility of in situ analysis [85]. Immunoassays, such as enzyme-linked immunoabsorbent assay (ELISA), have been also used to determine pesticides [86]. This technique, as well as the biosensors, is usually applied as screening tests rather than to quantitate residue levels, and the chromatographic methods are a more suitable alternative for this purpose.

8.4 APPLICATION TO REAL SAMPLES

In this section, principles of the main methods used in the determination of representative pesticide classes in soils are given.

8.4.1 BENZONITRILES

Bromoxynil and ioxynil are two hydroxybenzonitrile herbicides applied to soil as salts or esters, but they are decomposed rapidly by hydrolysis to their respective phenols. Derivatization of the hydroxyl group normally involves alkylation to form an ether or perfluoroacylation to form a butyryl derivative.

Soil samples (20–50 g) are extracted with 100 mL of methylene chloride and 15 mL of water acidified to pH 1 with 1 M HCl. The solution is decanted and the soil extracted again with methylene chloride. The extract is concentrated and transferred to a vial for derivatization with heptafluorobutyric anhydride (40 μL), hydrolyzed with 1 mL phosphate buffer (pH 6) and extracted with benzene (2×0.5 mL). Pesticide residues are determined by GC–MS–SIM. A nonpolar column is used with an oven temperature program from 70°C to 280°C and helium as carrier gas at a flow of 1 mL/min. The detection limit of the method is near 0.001 μg/g [6].

8.4.2 GLYPHOSATE

Glyphosate is a highly polar herbicide, very soluble in water, and insoluble in most organic solvents. GC analysis is normally carried out after obtaining acetyl derivatives and HPLC analysis after derivatization with FMOC.

Soil samples (5 g) are extracted by shaking with 10 mL of 0.6 M KOH. The extract is neutralized by adding some drops of HCl until pH 7 and derivatized with 120 μL of FMOC-Cl reagent. The derivative is acidified to pH 1.5 and analyzed by liquid chromatography coupled to electrospray tandem mass spectrometry with a limit of detection of 5 μg/kg. The method is rapid and selective for the determination of glyphosate at very low levels [9].

8.4.3 SULFONYLUREAS

GC of sulfonylurea herbicides is very difficult because of their thermally labile properties and strongly polar nature.

Soil samples (25 g) are extracted by shaking with 50 mL of methanol:glacial acetic acid (49 + 1) for 60 min. The extract is concentrated, transferred into a C8 SPE column, eluted with methanol (10 mL), and the solvent is evaporated. The residue is redissolved in acetone and ethyl piperidine (1 μL) and pentafluorobenzyl bromide (5 μL) are added. Derivatives are determined by GC–ECD. Amounts of herbicide residues as low as 0.1 pg can be detected [17].

8.4.4 CARBAMATES

Typical characteristics of carbamate insecticides are their high polarity and solubility in water and their thermal instability. Methods based on the derivatization of carbamates to thermally stable compounds have, in general, several limitations, which reduce their sensitivity.

Soil (5 g) is placed in a small column and extracted twice with 5 mL of methanol in an ultrasonic water bath. After extraction, the solvent is filtered. Residue levels in soil are determined by reversed-phase high-performance liquid chromatography (RP-HPLC) with FL detection after postcolumn derivatization by hydrolysis with NaOH solution to methylamine and reaction with OPA and thiofluor to form a highly fluorescent isoindol. The separation of carbamates is performed on a C8 column with water–methanol as mobile phase. The detection limits of carbamates range from 1.6 to 3.7 μg/kg. The emission and excitation spectra allow the confirmation of residues at levels around 0.1 μg/g. The method provides good response linearity and high precision [49].

8.4.5 ORGANOPHOSPHORUS

These insecticides have high polarity and solubility in water and are frequently analyzed by GC–NPD and HPLC. Soil (20 g) is extracted for 10 min by ultrasonic agitation with acetonitrile (20 mL). The acetonitrile is evaporated to dryness and the residue reconstituted in 0.4 mL of mobile phase (acetonitrile–water, 65:35, v/v). The determination of diazinon and fenitrothion is performed by HPLC with a reversed-phase C-18 column and UV photodiode detection at 245 and 267 nm, respectively. The quantification limits are 1 and 2 ng/g for fenitrothion and diazinon, respectively, with a good level of reproducibility and accuracy [45].

8.4.6 PYRETHROIDS

These compounds are retained in soil because of their low solubility in water. Chromatographic methods, GC as well as HPLC, are used for the determination of pyrethroids in soil.

Soil (2 g) is placed in a closed PTFE vessel for microwave-assisted extraction with 10 mL toluene and 1 mL water and irradiated during 9 min. Vessels are opened after cooling, the toluene extract is evaporated, and 2 mL of hexane is added. The hexane extract is passed through a 2 g Florisil column and pyrethroids eluted with 20 mL ethyl acetate:hexane (1:2, v/v). The determination of pyrethroid residues is carried out by GC with ion trap mass spectrometry (EI-MS-MS) and ECD. A nonpolar capillary column of 30 m is used with both detectors, with a temperature program from 60°C to 270°C. This method provides a high sensitivity and selectivity with LOD from 0.08 to 0.54 ng/g [61].

8.4.7 PYRIMETHANIL AND KRESOXIM-METHYL FUNGICIDES

Pyrimethanil (anilino-pyrimidine) and kresoxim-methyl (strobilurin) are two novel fungicides with broad-spectrum activity.

Soil (2 g) is placed in a vial with phosphate buffer solution (pH 7) and NaCl and immersed in a temperature-controlled oil bath at 100°C. The sample is agitated with a magnetic stirring bar during the head space SPME. The polyacrilate (PA) fiber is exposed to the headspace for 25 min, and then inserted in the injector of a GC, in which the fungicides are desorbed for 5 min. A low polarity capillary column of 30 m is used for the determination of fungicides with a temperature program from 100°C to 300°C and carrier gas at a flow rate of 2 mL/min. The detection limits are 0.001 and 0.004 μg/g for pyrimethanil and kresoxim-methyl, respectively [67].

8.4.8 MULTIRESIDUE

Because of the large number of pesticides used, multiresidue analytical methods require techniques that are able to determine the greatest possible number of these compounds in a single analysis.

Soil (5 g) is extracted twice in an ultrasonic water bath with 5 and 4 mL, respectively, of ethyl acetate for 15 min. The extracts are then evaporated to an appropriate volume (1 mL) and 2 μL injected in a GC for the chromatographic analysis. A capillary phenyl polysiloxane column (30 m $\times$ 0.25 mm $\times$ 0.25 μm) is employed. Pesticide residues are detected by GC–MS, and good precision and low LOD (0.02–1.6 μg/kg) are obtained [76]. Figure 8.4 shows the observed levels, in different agricultural fields, of various pesticides that were identified by the selected ions observed in their mass spectra.

8.5 FUTURE TRENDS

Determination of pesticides in soils usually involves conventional extraction methods that demand large volumes of hazardous organic solvents. Therefore, substantial efforts have been made to develop sample preparation techniques that could alleviate the drawbacks associated with the conventional methods. Various modern extraction techniques have been yielded good results, although they still require optimization for multiresidue analysis of pesticides in soil because of the disparity of chemical compounds involved. Automation of sample preparation and coupling with instrumental analysis are also important goals to reach.

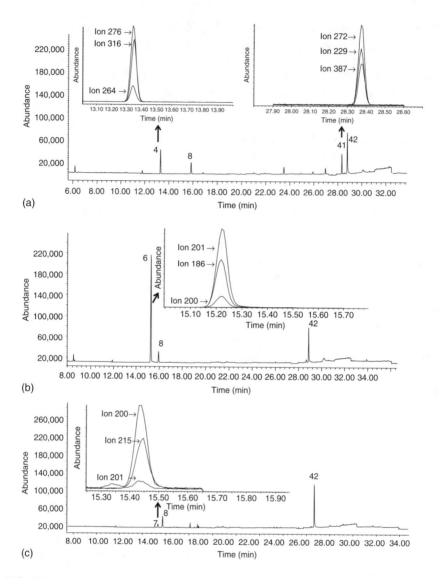

FIGURE 8.4 GC–MS–SIM chromatograms. (a) Soil sample collected from a tomato field, peak 4 = ethalfluralin (227 µg/kg) and peak 41 = endosulfan sulfate (70 µg/kg), (b) Soil sample collected from a forested field, peak 6 = simazine (446 µg/kg), and (c) Soil sample collected from a corn field, peak 7 = atrazine (11 µg/kg). Peaks 8 and 42 are internal standards. (From Sánchez-Brunete, C. et al., *J. Agric. Food Chem.*, 52, 1445, 2004. With permission.)

Analytical methodologies employed must be capable of residue measurement at very low levels and must also provide unambiguous evidence to confirm the identity of any residue detected. Gas chromatography–tandem mass spectrometry is a powerful tool to identify thermally stable pesticides in soils with high sensitivity and selectivity. However, the number of compounds that cannot be determined by GC because of their poor volatility and thermal instability has grown dramatically in the last few years. Thus, liquid chromatography coupled with mass spectrometry has become one important technique for the determination of pesticide residues. HPLC in combination with tandem MS is capable of discriminating more efficiently than HPLC–MS. Recently, several applications have described the use of MS–MS with both triple quadrupole and ion trap analyzers in multiresidue analysis of pesticides. Another analyzer employed is TOF–MS in negative and positive modes. This results in an improved mass spectrometric resolution, which is important in the detection of unknown compounds. Further optimization of sensitivity and quality is accomplished when mass spectrometers that have very fast MS–MS/MS switching and scanning capabilities are used. Most of the methods based on HPLC–MS–MS achieve satisfactory results even without making use of any cleanup step.

REFERENCES

1. Anonymous. Council Directive 91/414 EEC of 15 July 1991 concerning the placing of plant protection products on the market, *Off. J.*, L230, 32, 1991.
2. Anonymous. Federal insecticide, fungicide and rodenticide act, USA, 106, 2004.
3. Sánchez-Brunete, C., García-Valcárcel, A.I., and Tadeo, J.L. Determination of residues of phenoxy acid herbicides in soil and cereals by gas chromatography-ion trap detection, *J. Chromatogr. A*, 675, 213, 1994.
4. Lee, H.B. and Chau, A.S.Y. Analysis of pesticide residue by chemical derivatization. 6. Analysis of 10 acid herbicides, *J. Assoc. Off. Anal. Chem.*, 66, 1023, 1983.
5. Khan, S.U. Electron-capture gas liquid chromatographic method for simultaneous analysis of 2,4-D, dicamba, and mecoprop residues in soil wheat and barley, *J. Assoc. Off. Anal. Chem.*, 58, 1027, 1975.
6. Sánchez-Brunete, C., García-Valcárcel, A.I., and Tadeo, J.L. Determination of bromoxynil and ioxynil residues in cereals and soil by GC-ECD, *Chromatographia*, 38, 624, 1994.
7. West, S.D., Weston, J.H., and Day, E.W. Gas-chromatographic determination of residue levels of the herbicides trifluralin, benefin, ethalfluralin, and isopropalin in soil with confirmation by mass selective detection, *J. Assoc. Off. Anal. Chem.*, 71, 1082, 1988.
8. Moret, S., et al. The evaluation of different sorbents for the preconcentration of phenoxyacetic acid herbicides and their metabolites from soils, *J. Chromatogr. A*, 1099, 55, 2005.
9. Ibañez, M., et al. Residue determination of glyphosate, glufosinate and aminomethylphosphonic acid in water and soil samples by liquid chromatography coupled to electrospray tandem mass spectrometry, *J. Chromatogr. A*, 1081, 145, 2005.
10. Moret, S., Hidalgo, M., and Sánchez, J.M. Development of an ion-pairing liquid chromatography method for determination of phenoxyacetic herbicides and their main metabolites: application to the analysis of soil samples, *Chromatographia*, 63, 109, 2006.
11. Smith, A.E. and Briggs, G.G. Fate of herbicide chlortoluron and its possible degradation products in soils, *Weed Res.*, 18, 1, 1978.
12. Mudd, R.J., Hance, R.J., and Wright, S.J.L. The persistence and metabolism of isoproturon in soil, *Weed Res.*, 23, 239, 1983.
13. McKone, C.E. Determination of some substituted urea herbicides residues in soil by electron-capture, *J. Chromatogr.*, 44, 60, 1969.

14. Pérez, S., García-Baudín, J.M., and Tadeo, J.L. Determination of chlortoluron, isopro-
    turon and metoxuron in soil by GLC-NPD and confirmation using GLC-MS, *Fresenius'
    J. Anal. Chem.*, 339, 413, 1991.
15. Ramsteiner, K., Hörmann, W.D., and Eberle, D.O. Multiresidue method for the deter-
    mination of triazine herbicides in field-grown agricultural crops, water, and soils,
    J. Assoc. Off. Anal. Chem., 57, 192, 1974.
16. Lechon, Y., Sánchez-Brunete, C., and Tadeo, J.L. Influence of the laboratory incubation
    method on chlorotoluron and terbutryn degradation in soil, *J. Agric. Food Chem.*, 45,
    951, 1997.
17. Cotterill, E.G. Determination of the sulfonylurea herbicides chlorsulfuron and metsul-
    furon-methyl in soil, water and plant-material by gas chromatography of their pentafluoro-
    benzyl derivatives, *Pest. Sci.*, 34, 291, 1992.
18. García-Valcárcel, A.I., et al. Determination of dinitroaniline herbicides in environmental
    samples by gas chromatography, *J. Chromatogr. A*, 719, 113, 1996.
19. Pérez, R.A., Sanchez-Brunete, C., and Tadeo, J.L. Determination of thiazopyr in soil and
    plants by gas chromatography with nitrogen-phosphorus detection and confirmation by
    gas chromatography-mass spectrometry, *J. Chromatogr. A*, 778, 193, 1997.
20. Sánchez-Brunete, C., Martínez, L., and Tadeo, J.L. Determination of corn herbicides by
    GC-MS and GC-NPD in environmental samples, *J. Agric. Food Chem.*, 42, 2210, 1994.
21. Balinova, A.M. and Balinova, I. Determination of herbicide residues in soil in the presence
    of persistent organochlorine insecticides, *Fresenius' J. Anal. Chem.*, 339, 409, 1991.
22. Sánchez-Brunete, C. and Tadeo, J.L. Multiresidue analysis of herbicides in soil by
    GC-MS, *Quim. Anal.*, 15, 53, 1996.
23. Abian, J., Durand, G., and Barceló, D. Analysis of chlorotriazines and their degradation
    products in environmental samples by selecting various operating modes in thermospray
    HPLC/MS/MS, *J. Agric. Food Chem.*, 41, 1264, 1993.
24. Durand, G., et al. Comparison of gas chromatographic-mass spectrometric methods for
    screening of chlorotriazine pesticides in soil, *J. Chromatogr.*, 603, 175, 1992.
25. Crouch, R.V. and Pullin, E.M. Analytical method for residues of bromoxynil octanoate
    and bromoxynil in soil, *Pest. Sci.*, 5, 281, 1974.
26. Tsukioka, T. and Murakami, T. Capillary gas chromatographic-mass spectrometric
    determination of acidic herbicides in soils and sediments, *J. Chromatogr.*, 469, 351, 1989.
27. Hernandez-Borges, J., et al. Analysis of triazolopyrimidine herbicides in soils using field-
    enhanced sample injection-coelectroosmotic capillary electrophoresis combined with
    solid-phase extraction, *J. Chromatogr. A*, 1100, 236, 2005.
28. Stipicevic, S., et al. Comparison of gas and high performance liquid chromatography with
    selective detection for determination of triazine herbicides and their degradation products
    extracted ultrasonically from soil, *J. Sep. Sci.*, 26, 1237, 2003.
29. Guardia Rubio, M., et al. Determination of triazine herbicides and diuron in mud from
    olive washing devices and soils using gas chromatography with selective detectors, *Anal.
    Lett.*, 39, 835, 2006.
30. Pérez, R.A., et al. Analytical methods for the determination in soil of herbicides used in
    forestry by GC-NPD and GC-MS, *J. Agric. Food Chem.*, 46, 1864, 1998.
31. Sánchez-Brunete, C., et al. Multiresidue herbicide analysis in soil samples by means of
    extraction in small columns and gas chromatography with nitrogen-phosphorus and mass
    spectrometric detection, *J. Chromatogr. A*, 823, 17, 1998.
32. Kremer, E., Rompa, M., and Zygmunt, B. Extraction of acidic herbicides from soil by
    means of accelerated solvent extraction, *Chromatographia*, 60, S169, 2004.

33. Dagnac, T., et al. Determination of chloroacetanilides, triazines and phenylureas and some of their metabolites in soils by pressurised liquid extraction, GC-MS/MS, LC-MS and LC-MS/MS, *J. Chromatogr. A*, 1067, 225, 2005.

34. Patsias, J., Papadakis, E.N., and Papadopoulus-Mourkidou, E. Analysis of phenoxy-alkanoic acid herbicides and their phenolic conversion products in soil by microwave assisted solvent extraction and subsequent analysis of extracts by on-line solid-phase extraction-liquid chromatography, *J. Chromatogr. A*, 959, 153, 2006.

35. Papadakis, E.N. and Papadopoulus-Mourkidou, E. LC-UV determination of atrazine and its principal conversion products in soil, *Intern. J. Environ. Anal. Chem.*, 86, 573, 2006.

36. Shen, G. and Lee, H.K. Determination of triazines in soil by microwave-assisted extraction followed by solid-phase microextraction and gas chromatography-mass spectrometry, *J. Chromatogr. A*, 985, 167, 2003.

37. Vryzas, Z. and Papadopoulou-Mourkidou, E. Determination of triazine and chloroacetanilide herbicides in soils by microwave-assisted extraction (MAE) coupled to gas chromatographic analysis with either GC-NPD or GC-MS, *J. Agric. Food Chem.*, 50, 5026, 2002.

38. Guardino, X., et al. Determination of chlorpyrifos in air, leaves and soil from a greenhouse by gas chromatography with nitrogen-phosphorus detection, high performance liquid chromatography and capillary electrophoresis, *J. Chromatogr. A*, 823, 91, 1998.

39. Li, J.Z., Wu, X., and Hu, J.Y. Determination of fungicide kresoxim-methyl residues in cucumber and soil by capillary gas chromatography with nitrogen-phosphorus detection, *J. Environ. Sci. Health B*, 41, 427, 2006.

40. Lee, L.S., et al. Degradation of *N,N'*-dibutylurea (DBU) in soils treated with only DBU and DBU-fortified benlate fungicides, *J. Environ. Qual.*, 33, 1771, 2004.

41. Sassman, S.A., et al. Assessing *N,N'*-dibutylurea (DBU) formation in soils after application of *n*-butylisocyanate and benlate fungicides, *J. Agric. Food Chem.*, 52, 747, 2004.

42. Vig, K., et al. Insecticide residues in cotton crop soil, *J. Environ. Sci. Health B*, 36, 421, 2001.

43. Bladek, J., Rostkowski, A., and Miszczak, M. Application of instrumental thin-layer chromatography and solid-phase extraction to the analysis of pesticide residues in grossly contaminated samples of soil, *J. Chromatogr. A*, 754, 273, 1996.

44. Tor, A., Aydin, M.E., and Ozcan, S. Ultrasonic solvent extraction of organochlorine pesticides form soil, *Anal. Chim. Acta*, 559, 173, 2006.

45. Sanchez, M.E., et al. Determination of diazinon and fenitrothion in environmental water and soil samples by HPLC, *J. Liq. Chromatogr. Relat. Technol.*, 26, 483, 2003.

46. Zhu, X., et al. Selective solid-phase extraction using molecularly imprinted polymer for the analysis of polar organophosphorus pesticides in water and soil samples, *J. Chromatogr. A*, 1092, 161, 2005.

47. Ali, M.A. and Baugh, P.J. Pyrethroid soil extraction, properties of mixed solvents and time profiles using GC/MS-NICI analysis, *Intern. J. Environ. Anal. Chem.*, 83, 909, 2003.

48. Rial-Otero, R., et al. Parameters affecting extraction of selected fungicides from vineyard soils, *J. Agric. Food Chem.*, 52, 7227, 2004.

49. Sánchez-Brunete, C., Rodriguez, A., and Tadeo, J.L. Multiresidue analysis of carbamate pesticides in soil by sonication assisted extraction in small columns and liquid chromatography, *J. Chromatogr. A*, 1007, 85, 2003.

50. Castro, J., Sánchez-Brunete C., and Tadeo, J.L. Multiresidue analysis of insecticides in soil by gas chromatography with electron-capture detection and confirmation by gas chromatography-mass spectrometry, *J. Chromatogr. A*, 918, 371, 2001.

51. Sánchez-Brunete, C., Miguel, E., and Tadeo, J.L. Multiresidue analysis of fungicides in soil by sonication assisted extraction in small columns and gas chromatography, *J. Chromatogr. A*, 976, 319, 2002.

52. Sun, L. and Lee, H.S. Optimization of microwave-assisted extraction and supercritical fluid extraction of carbamate pesticides in soil by experimental design methodology, *J. Chromatogr. A*, 1014, 165, 2003.

53. O'Mahony, T., et al. Monitoring the supercritical fluid extraction of pyrethroid pesticides using capillary electrochromatography, *Intern. J. Environ. Anal. Chem.*, 83, 681, 2003.

54. Ling, Y.C. and Liao, J.H. Matrix effect on supercritical fluid extraction of organochlorine pesticides from sulfur-containing soils, *J. Chromatogr. A*, 754, 285, 1996.

55. Snyder, J.L., et al. The effect of instrumental parameters and soil matrix on the recovery of organochlorine and organophosphate pesticides from soils using supercritical fluid extraction, *J. Chromatogr. Sci.*, 31, 183, 1993.

56. Popp, P., et al. Application of accelerated solvent extraction followed by gas chromatography, high-performance liquid chromatography and gas chromatography-mass spectrometry for the determination of polycyclic aromatic hydrocarbons, chlorinated pesticides and polychlorinated dibenzo-*p*-dioxins and dibenzofurans in solid wastes, *J. Chromatogr. A*, 774, 203, 1997.

57. Concha-Graña, E., et al. Development of pressurized liquid extraction and cleanup procedures for determination of organochlorine pesticides in soils, *J. Chromatogr. A*, 1047, 147, 2004.

58. Hussen, A., et al. Development of a pressurized liquid extraction and cleanup procedure for the determination of α-endosulfan, β-endosulfan and endosulfan sulfate in aged contaminated Ethiopian soils, *J. Chromatogr. A*, 1103, 202, 2006.

59. Herbert, P., et al. Development and validation of a novel method for the analysis of chlorinated pesticides in soils using microwave-assisted extraction-headspace solid phase microextraction and gas chromatography-tandem mass spectrometry, *Anal. Bional. Chem.*, 384, 810, 2006.

60. Esteve-Turrillas, F.A., et al. Microwave-assisted extraction of pyrethroid insecticides from soil, *Anal. Chem. Acta*, 522, 73, 2004.

61. Esteve-Turrillas, F.A., Pastor, A., and de la Guardia, M. Comparison of different mass spectrometric detection techniques in the gas chromatographic analysis of pyrethroid insecticide residues in soil after microwave-assisted extraction, *Anal. Bional. Chem.*, 384, 801, 2006.

62. Zambonin, C.G., Aresta, A., and Nilsson, T. Analysis of organochlorine pesticides by solid-phase microextraction followed by gas chromatography-mass spectrometry, *Intern. J. Environ. Anal. Chem.*, 82, 651, 2002.

63. Zhao, R.S., et al. A novel headspace solid-phase microextraction method for the exact determination of organochlorine pesticides in environmental soil samples, *Anal. Bional. Chem.*, 384, 1584, 2006.

64. Magdic, S., et al. Analysis of organophosphorus insecticides from environmental samples using solid-phase microextraction, *J. Chromatogr. A*, 736, 219, 1996.

65. De Pasquale, C., et al. Use of SPME extraction to determine organophosphorus pesticides adsorption phenomena in water and soil matrices, *Intern. J. Environ. Anal. Chem.*, 85, 1101, 2005.

66. Lambropoulou, D. and Albanis, T.A. Determination of the fungicides vinclozolin and dicloran in soils using ultrasonic extraction coupled with solid-phase microextraction, *Anal. Chem. Acta*, 514, 125, 2004.

67. Navalon, A., et al. Determination of pyrimethanil and kresoxim-methyl in soils by headspace solid-phase microextraction and gas chromatography-mass spectrometry, *Anal. Bional. Chem.*, 379, 1100, 2004.

68. Papadopoulou-Mourkidou, E., Patsias, J., and Kotopoulou, A. Determination of pesticides in soils by gas chromatography-ion trap mass spectrometry, *J. AOAC Int.*, 80, 447, 1997.

69. Vinas, P., et al. Capillary gas chromatography with atomic emission detection for pesticide analysis in soil samples, *J. Agric. Food Chem.*, 51, 3704, 2003.

70. Snyder, J.L., et al. Comparison of supercritical fluid extraction with classical sonication and Soxhlet extraction for selected pesticides, *Anal. Chem.*, 64, 1940, 1992.

71. Babic, S., Petrovic, M., and Kastelan-Macan, M. Ultrasonic solvent extraction of pesticides from soil, *J. Chromatogr. A*, 823, 3, 1998.

72. Richter, P., et al. Screening and determination of pesticides in soil using continuous subcritical water extraction and gas chromatography-mass spectrometry, *J. Chromatogr. A*, 994, 169, 2003.

73. Fenoll, J., et al. Multiresidue analysis of pesticides in soil by gas chromatography with nitrogen-phosphorus detection and gas chromatography mass spectrometry, *J. Agric. Food Chem.*, 53, 7661, 2005.

74. Belmonte Vega, A., Garrido Frenich, A., and Martinez Vidal, J.L. Monitoring of pesticides in agricultural water and soil samples from Andalusia by liquid chromatography coupled to mass spectrometry, *Anal. Chem. Acta*, 538, 117, 2005.

75. Gonzalves, C. and Alpendurada, M.F. Assesment of pesticide contamination in soil samples from an intensive horticulture area, using ultrasonic extraction and gas chromatography-mass spectrometry, *Talanta*, 65, 1179, 2005.

76. Sánchez-Brunete, C., Albero, B., and Tadeo, J.L. Multiresidue determination of pesticides in soil by gas chromatography-mass spectrometry, *J. Agric. Food Chem.*, 526, 1445, 2004.

77. Tadeo, J.L., Castro, J., and Sánchez-Brunete, C. Multiresidue determination in soil of pesticides used in tomato crops by sonication-assisted extraction in small columns and gas chromatography, *Intern. J. Environ. Anal. Chem.*, 84, 29, 2004.

78. Rissato, S.R., et al. Multiresidue analysis of pesticides in soil by supercritical fluid extraction/gas chromatography with electron-capture detection and confirmation by gas chromatography-mass spectrometry, *J. Agric. Food Chem.*, 53, 624, 2005.

79. Goncalves, C., et al. Optimization of supercritical fluid extraction of pesticide residues in soil by means of central composite design and analysis by gas chromatography-tandem mass spectrometry, *J. Chromatogr. A*, 1110, 6, 2006.

80. Castro, J., et al. Analysis of pesticides volatilised from plants and soil by headspace solid-phase microextraction and gas chromatography, *Chromatographia*, 53, S-361, 2001.

81. Schmeck, T. and Wenclawiak, B.W. Sediment matrix induced response enhancement in the gas chromatographic-mass spectrometric quantification of insecticides in four different solvent extracts from ultrasonic and Soxhlet extraction, *Chromatographia*, 62, 159, 2005.

82. Hou, L. and Lee, H.K. Determination of pesticides in soil by liquid-phase microextraction and gas chromatography-mass spectrometry, *J. Chromatogr. A*, 1038, 37, 2004.

83. Sánchez, M.E., et al. Solid-phase extraction for the determination of dimethoate in environmental water and soil samples by micellar electrokinetic capillary chromatography (MECK), *J. Liq. Chromatogr. Relat. Technol.*, 26, 545, 2003.

84. Orejuela, E. and Silva, M. Rapid and sensitive determination of phosphorus-containing amino acid herbicides in soil samples by capillary zone electrophoresis with diode laser-induced fluorescence detection, *Electrophoresis*, 26, 4478, 2005.
85. Bäumner, A.J. and Schmid, R.D. Development of a new immunosensor for pesticide detection: a disposable system with liposome-enhancement and amperometric detection, *Biosens. Bioelectron.*, 13, 519, 1998.
86. Kramer, K., Lepschy, J., and Hock, B. Long-term monitoring of atrazine contamination in soil by ELISA, *J. AOAC Int.*, 84, 150, 2001.

9 Determination of Pesticides in Water

Jay Gan and Svetlana Bondarenko

CONTENTS

9.1   INTRODUCTION

Concerns over the contamination of water by pesticides generally arise from two scenarios, that is, concern over human health risks when water (e.g., groundwater) is used for drinking and concern over ecotoxicological effects when nontarget organisms (e.g., aquatic organisms and amphibians) are exposed to water in their habitats. Both the European Union (EU) and the United States have adopted stringent limits for pesticide presence in drinking water. For instance, EU regulations for drinking water quality set a limit of 0.5 μg/L for the sum of all pesticides and 0.1 μg/L for each compound. However, when acute or chronic toxicities or other ecological effects (e.g., bioaccumulation) are implied, water quality limits can be much lower than those for drinking water. For instance, in the total maximum daily loads (TMDL) established for diazinon and chlorpyrifos for a watershed in Orange County, California, the numerical targets for diazinon were set at 80 ng/L for acute toxicity and 50 ng/L for chronic toxicity, and those for chlorpyrifos at 20 ng/L for acute toxicity and 14 ng/L for chronic toxicity [1]. Regulatory requirements such as these have driven the development of increasingly more sensitive and rigorous methods for the analysis of pesticides in water.

9.1.1   METHOD CLASSIFICATION

A complete method for pesticide analysis in water, as in other matrices, always includes a sample preparation method and a pesticide detection method. The need for detecting pesticides at trace levels means that a water sample must be reduced many times in size so that a small aliquot of the final sample may provide adequate sensitivity for detection. The concentration magnification is achieved through phase transfer by using liquid–liquid extraction (LLE) or solid-phase extraction

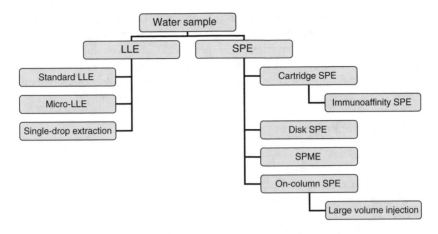

FIGURE 9.1 A general diagram describing preparation methods used for analysis of pesticides in water.

(SPE). Many other methods may be considered as variations of the traditional LLE and SPE methods (Figure 9.1). For instance, micro-LLE or single-drop extraction can be considered as a miniaturization of the standard LLE procedure. Variations of cartridge SPE include SPE disks and solid-phase microextraction (SPME). Methods can also be classified based on the mechanisms used for pesticide detection. However, as detection methods are usually common among different sample matrices and are not limited only to water, this chapter will mostly focus on sample preparation methods for water analysis, with exceptions made only for immunoassays and capillary electrophoresis (CE) because of their significant deviations from conventional chromatographic methods.

9.1.2 OBJECTIVES

Advancements and challenges in pesticide analysis in water are periodically updated in the form of journal review articles [2–6]. It must be noted that the number of publications on this topic is enormous, and that it is infeasible to thoroughly review all published studies. In this chapter, only a limited number of publications since 1990 are cited. The purpose is to evaluate and compare some of the most commonly used methods, and to provide the reader with condensed information on method principles, procedures, advantages, disadvantages, and trends. A few applications are further included in each method, which may lead the reader to more concrete details.

9.2 LIQUID–LIQUID EXTRACTION

9.2.1 STANDARD LLE

LLE is one of the earliest methods used for analyzing pesticides in water samples. Because of its simplicity and also its inclusion in EPA methods, LLE is still probably the most used method for pesticide analysis in water. Depending on the types of analytes, different solvents or other conditions may be used. In the United States, LLE procedures for different classes of pesticides are given in different EPA methods and are routinely used by commercial laboratories. The following method is a brief description of EPA method 8141, using separatory funnels for preparation of water samples containing organophosphate or carbamate residues.

9.2.1.1   General Procedures

- Measure out 1000 mL water sample using a 1 L graduated cylinder or by weighing in a container.
- Spike 100 μL of the surrogate spiking solution into each sample and mix well.
- For the sample in each batch selected for use as a matrix spike sample, add 100 μL of the matrix-spiking standard.
- Quantitatively transfer the sample to a 2 L glass separatory funnel, adding 50 g of sodium chloride. Use 100 mL of methylene chloride to rinse the sample container and transfer this rinse solvent to the separatory funnel.
- Seal and shake the separatory funnel vigorously for 1–2 min with periodic venting to release excess pressure.
- Allow the organic layer to separate from the water phase for a minimum of 10 min. If the emulsion interface between layers is more than one-third the size of the solvent layer, the analyst must employ mechanical techniques to complete the phase separation. The optimum technique depends upon the sample and may include stirring, filtration of the emulsion through glass wool, centrifugation, or other physical methods. Dry the extract by passing it through a drying funnel containing about 50 g of anhydrous sodium sulfate. Collect the solvent extract in a round bottom flask.
- Repeat the extraction two more times using fresh portions of solvent. Combine the three solvent extracts.
- Rinse the separation flask, which contained the solvent extract, with 20–30 mL of methylene chloride and add it to the drying column to complete the quantitative transfer.
- Perform the concentration, if necessary, using a vacuum evaporator. For further concentration, nitrogen blow down technique is used to adjust the extract to the final volume required.
- The extract may now be analyzed for the target analytes using the appropriate determinative technique(s).

9.2.1.2   Advantages

Standard LLE is a mature method that has been well used and tested. Its advantages include relatively minimal requirements for equipment and low demand on the analyst's skills, compatibility for a broad range of pesticides, and reliability. Variations in analyte recovery may be addressed by using a surrogate prior to the extraction. The surrogate can be either a similar compound or a stable-isotope labeled form of the target analyte, if detection is to be made by a selective detector such as mass spectrometry (MS).

9.2.1.3 Disadvantages

A number of drawbacks may be easily iterated regarding the standard LLE; Most notable is the consumption of large quantities of organic solvents, which makes LLE methods less environment-friendly. Analysis of a 1 L water sample typically needs about 300–500 mL solvent. The heavy use of solvents in LLE may pose a health concern to the analyst, and also produce large amounts of wastes. LLE is generally labor intensive, time consuming, and physically demanding. Extraction and preparation of 6–8 samples may easily take one day of the analyst's time. LLE is generally not suitable for analysis of polar pesticide compounds. LLE can also be less effective for water samples containing high levels of organic matter or suspended particles, such as runoff effluents and other surface water samples, because heavy emulsion often forms between the aqueous and solvent phases. This may prolong phase separation and make recovery variable.

9.2.2 MICRO-LLE

9.2.2.1 Principles and Procedures

Micro-LLE is a miniaturization of standard LLE in that only a very small amount of solvent is used for extraction. For instance, Zapf et al. [7] developed a micro-LLE method for the analysis of 82 various pesticides in tap water. Briefly, a 400 mL tap water sample in a 500 mL narrow-necked bottle was saturated with 150 g NaCl and buffered to a pH value of 6.5–7.0. The water sample was spiked with analyte mixtures in 100 μL methanol to achieve concentrations of 50, 100, and 500 ng/L. After addition of 500 μL toluene, the bottle was sealed and shaken for 20 min at 420 rpm. After phase separation, the solvent layer was brought up to the bottleneck by addition of a saturated NaCl solution using a Pasteur pipette connected to a separating funnel. About 150 μL of the toluene phase was transferred into 200 μL vials and 2 μL was injected into a gas chromatograph (GC) with electron capture detector (ECD) or nitrogen phosphorus detector (NPD) for detection. For 68 compounds, the recoveries were higher than 50%. The mean relative standard deviations (RSD) at spiking levels of 50, 100, and 500 ng/L were 7.9%, 6.6%, and 5.2%, respectively. In most cases, compounds were reproducibly detected at concentrations well below 0.1 μg/L.

de Jager and Andrews [8] have described a micro-LLE method, in which a single drop of water-immiscible solvent is attached to the tip of a syringe needle, for the analysis of organochlorine pesticides in water samples. This method is also called solvent microextraction (SME) or single-drop microextraction (SDME) [9]. In this method, a 2 μL drop of hexane containing 100 ng/mL of decachlorobiphenyl as internal standard was used as the extraction solvent and immersed in the stirred sample solution for a 5 min extraction time. The sample solution was stirred at a rate of 240 rpm, and a Hamilton 10 μL 701SN syringe fitted with a Chaney adapter (Hamilton, Reno, NV, USA) was used in all extractions and injections. By using the Chaney adapter, the maximum syringe volume was set to 2.2 μL and the delivery volume was set to 2.0 μL. For the extraction, 2.2 μL of hexane was drawn into the syringe and the plunger was depressed with the stop button engaged, causing 0.2 µL to be expelled. The microsyringe was then positioned in the extraction stand in such a way that the tip of the extraction needle protruded to a depth of about 8 mm below the surface of the aqueous solution. The syringe plunger was then completely depressed causing a 2 µL drop to form on the needle tip. The drop was suspended from the needle for 5 min at which time the plunger was withdrawn to 2.2 µL with the needle tip still submerged in the sample solution. The contents of the syringe were then injected into the GC for analysis. Total analysis time was less than 9 min, allowing 11 samples to be screened per hour. This method was therefore useful for quick screening of organochlorine compounds in water. Using a similar method, Liu et al. [9] was able to detect fungicides such as chlorothalonil, triadimefon, hexaconazole, and diniconazole in water at 0.006–0.01 µg/L with RSD < 8.6%.

9.2.2.2   Advantages

Micro-LLE is advantageous over the conventional LLE in that only a very small amount of organic solvent is used. As a significant fraction or all of the organic phase is used for detection, good sensitivity may be achieved. Micro-LLE is therefore far less time consuming and inexpensive.

9.2.2.3   Disadvantages

Micro-LLE operates at a phase ratio that does not favor pesticide enrichment into the organic phase. It is difficult to automate, and performance is likely dependent on the analyst's skills. The solvent chosen must be completely immiscible with water, and therefore micro-LLE is suitable only for nonpolar pesticides. Inconsistency in recovery may be overcome by using an internal standard at the extraction step. This method is more appropriate for rapid screening, rather than for routine analysis.

9.3   SOLID-PHASE EXTRACTION

9.3.1   STANDARD SPE

9.3.1.1   Principles

The trend in pesticide analysis in water has moved away from LLE to SPE. This is due to the better extraction efficiencies, ease of use, less use of solvents, potential for automation, and better selectivity of SPE. Compared with most other methods, SPE is a widely used and mature method. In SPE, the analyte is transferred from the aqueous phase onto a sorbent phase, which can then be recovered for analysis. Sorbents available in standard SPE include the common inorganic adsorbents used in liquid chromatography (LC), such as silica gel, as well as activated charcoal, bonded silica phases, and polymers [10]. The most popular phases are octadecyl (C18) and octyl-silica (C8), styrene-divinylbenzene copolymers, and graphitized carbon black.

Alkyl-bonded silica sorbents: The peak tailing and poor selectivity of silica gel led to the development of silica-based phases with an alkyl- or aryl-group substituted silanol. The functionality properties of the sorbent depend on the percentage of carbon loading, bonded-silica porosity, particle-size, and whether the phase is end-capped. Endcapping is used to reduce the residual silanols, but the maximum percentage of endcapping is 70%. The most popular sorbents from this group are C18 and C8.

Carbon sorbents: An important gain of graphitized carbon black (GCB) as the sorbent is that the recoveries do not decrease when environmental waters with dissolved organic carbon (DOC) are extracted. This is due to the fact that fulvic acids, which represent up to 80% of the DOC content in surface waters, are adsorbed on the anion-exchange sites of the GCB surface, and therefore they cannot compete with nonacidic pesticides for adsorption on the nonspecific sites of the sorbent. GCB has three main disadvantages: the collapsing of the sorbent, desorption problems during elution, and the possibility of reactions between the analytes and the sorbent surface, leading to incomplete sorption and desorption.

Polymeric resins: With these sorbents, the retention behavior of the analytes is governed by hydrophobic interactions similar to C18 silica, but, owing to the aromatic rings in the network of the polymer matrix, one can expect strong electro-donor interactions with aromatic rings of solutes.

Mixed phases: The advantages of each sorbent can be combined in the form of a mixture of sorbents used in the same SPE column.

9.3.1.2   General Procedures

A typical SPE sequence includes the activation of the sorbent bed (wetting), removal of the excess of activation solvent (conditioning), application of the sample, removal of interferences (cleanup) and water, elution of the sorbed analytes, and reconstitution of the extract [10]. Exact conditions are usually specified by the manufacturer, and may vary significantly in types of solvents used for conditioning and elution. A general procedure for using SPE cartridges is as follows [11]:

- Wash the cartridge with a small amount of relatively nonpolar solvent (e.g., ethyl acetate, acetone), followed by a relatively polar solvent (e.g., methanol), and finally water.
- Without letting the cartridge become dry, pass the water sample (e.g., 1 L) through the column under vacuum at a relatively fast rate (e.g., 15 mL/min).
- If the water sample contains an appreciable amount of suspended solids, filter the sample to remove suspended solids before loading.
- After the sample is loaded, wash the cartridge with a small amount of water and dry the cartridge by passing air for a short time.
- Elute the SPE cartridge with the same solvents used at the preparation step, except in a reversed order.
- The eluate is dried with a small amount of anhydrous sodium sulfate and further evaporated to dryness under a gentle stream of nitrogen.
- The residue is recovered in a small amount of solvent appropriate for GC or LC analysis.

9.3.1.3 Advantages

Compared with conventional LLE methods, SPE has several distinctive advantages. SPE generally needs a shorter analysis time, consumes much less organic solvents, and may be less costly than LLE [11]. SPE also offers the great advantage for easier transportation between laboratories or from the field to the laboratory, and for easier storage. For example, water samples can be processed at a remote site, and only the cartridges need to be transported back to the laboratory, which makes sampling at remote sites feasible. Automation or semiautomation may be potentially achieved for either off-line or on-line use of SPE, although manual, off-line is likely the dominant form that has been used.

9.3.1.4 Disadvantages

There are many different types of sorbents and configurations (e.g., mass of sorbent per tube), and each SPE is inherently best suited for a specific class of pesticide compounds. This, when combined with operational factors such as flow rate, conditioning, and elution, and the effect of sample matrix, can make the recovery of pesticides highly variable [11]. In addition, suspended solids and salts are known to cause blockage of SPE cartridges. Samples compatible with SPE must be relatively clean (e.g., groundwater). When surface water samples are analyzed, prefiltration is generally necessary to remove the suspended solids. This may not be desirable for hydrophobic compounds, because a significant fraction of the analyte is associated with the suspended solids.

Both low and enhanced recoveries have been observed when SPE is used for extracting pesticides from water samples. For instance, when using C18 SPE cartridges for the determination of 23 halogenated pesticides, Baez et al. [11] found that recoveries depended on the pesticides, and losses occurred with heptachlor, aldrin, and captan. Recoveries for vinclozolin and dieldrin from groundwater were lower than those obtained from nanopure water. In river water, losses of these compounds were higher. High losses were also observed for trifluralin, α-BHC, γ-BHC, triallate, and chlorpyrifos. In a follow-up study, Baez et al. [12] evaluated the use of C18 SPE columns for the determination of organophosphorus, triazine, and triazole-derived pesticides, napropamide, and amitraz. Under general extraction conditions, losses were found for amitraz, prometryn, prometon, dimethoate, penconazole, and propiconazole. At 100 ng/L, enhanced responses were observed for mevinphos, simazine, malathion, triadimefon, methidathion, and phosmet, which was attributed to matrix effects.

9.3.1.5 Trends

Current trends include the use of SPE on-line, coupling with selective or sensitive detectors, the use of stable isotopes to overcome the issue of variable recoveries, and automation. Bucheli et al. [13] reported a method for the simultaneous identification and quantification of neutral and acidic pesticides (triazines, acetamides, and phenoxy herbicides) at the low ng/L level. The method included the enrichment of the compounds by SPE on GCB, followed by the sequential elution of the neutral and acidic pesticides and derivatization of the latter fraction with diazomethane. Identification and quantification of the compounds was performed with GC–MS using atrazine-d5, [^{13}C6]-metolachlor, and [^{13}C6]-dichlorprop as internal standards. Absolute recoveries from nanopure water spiked with 4–50 ng/L were $85 \pm 10\%$, $84 \pm 15\%$, and $100 \pm 7\%$ for the triazines, the acetamides, and the phenoxy acids, respectively. Recoveries from rainwater and lake water spiked with 2–100 ng/L were $95 \pm 19\%$, $95 \pm 10\%$, and $92 \pm 14\%$ for the triazines, the acetamides, and the phenoxy acids, respectively. Average method precision determined with fortified rainwater (2–50 ng/L) was $6.0 \pm 7.5\%$ for the triazines, $8.6 \pm 7.5\%$ for the acetamides, and $7.3 \pm 3.2\%$ for the phenoxy acids. MDLs ranged from 0.1 to 4.4 ng/L. Crescenzi et al. [14] reported the coupling of SPE and LC/MS for determining 45 widely used pesticides having a broad range of polarity in water. This method involved passing 4, 2, and 1 L, respectively, of drinking water, groundwater, and river water through a 0.5 g GCB cartridge at 100 mL/min. In all cases, recoveries of the analytes were better than 80%, except for carbendazim (76%). For drinking water, MDLs ranged between 0.06 (malathion) and 1.5 (aldicarb sulfone) ng/L. Kampioti et al. [15] reported a fully automated method for the multianalyte determination of 20 pesticides belonging to different classes (triazines, phenylureas, organophosphates, anilines, acidic, propanil, and molinate) in natural and treated waters. The method, based on on-line SPE-LC-MS, was highly sensitive with MDLs between 0.004 and 2.8 ng/L, precise with RSDs between 2.0% and 12.1%, reliable, and rapid (45 min per sample).

9.3.1.6   Applications

Fernandez et al. [16] performed a comparative study between LLE and SPE with trifunctional bonding chemistry (tC18) for 22 organochlorine and 2 organophosphorus pesticides, 2 triazines, and 7 PCBs. Mean recovery yields were higher with the LLE method, although SPE for most of the 33 analytes surpassed 70%. The MDLs for both techniques were below 5 ng/L, except for parathion (7 ng/L), methoxychlor (8 ng/L), atrazine (35 ng/L), and simazine (95 ng/L). Patsias and Papadopoulou-Mourkidou [17] reported a rapid multiresidue method for the analysis of 96 target analytes in field water samples. Analytes were extracted from 1 L filtered water samples by off-line SPE on three tandem C18 cartridges. The sorbed analytes eluted with ethyl acetate were directly analyzed by GC-ion trap MS (GC–IT–MS). The mean recoveries, at the 0.5 µg/L level, for two-thirds of the analytes ranged from 75% to 120%; the recoveries for less than one-third of the analytes ranged from 50% to 75% and the recoveries for the 10 relatively most polar analytes ranged from 12% to 50%. The MDLs for 69 analytes were below 0.01 µg/L; the MDLs for 18 analytes were below 0.05 µg/L; for captan, carbofenothion, deltamethrin, demeton-S-methyl sulfone, fensulfothion, deisopropylatrazine, and metamitron, the MDL was 0.1 µg/L and for chloridazon and tetradifon, the MDL was 0.5 µg/L.

9.3.2 SPE DISKS

9.3.2.1 Principle and Procedures

In a special form of SPE, the sorbent is bonded to a solid support that is configured as a disk. During filtration, using SPE disks, the pesticides sorb to the stationary phase and then are eluted with a minimal amount of organic solvent. Empore disks (3 M, St. Paul, MN), bonded with a C18 or C8 solid phase, have been the most commonly used SPE disks [18]. The general procedure for using Empore disks is as follows, although details may vary for specific applications and for the types of SPE disks used [19].

- Before use, condition Empore disks by soaking in a solvent (e.g., acetone).
- Pass the water sample through the disk under vacuum on an extraction manifold. In some applications, a small amount of solvent modifier (e.g., methanol) is added to the water sample to improve pesticide recovery [20]. It is usually recommended that the disk should not be allowed to become dry during the extraction.
- After sample extraction, elute the disks with a small amount of solvent (e.g., dichloromethane–ethyl acetate mixture) or extract the disk by mixing the disk in an extracting solvent in a closed vessel.
- Evaporate the solvent extract to a small volume, and an aliquot of the final sample extract is injected into GC or LC for detection.

9.3.2.2 Advantages

Like SPE cartridges, the use of SPE disks also greatly reduces the volume of solvents, decreases sample preparation time and labor, and sometimes increases extract purity from water samples [21–23]. SPE disks can also be used for temporary pesticide storage [24,25], field extraction of pesticides [26], and shipping pesticides from one location to another [27,28].

Field extraction capability adds a new dimension to the sampling of natural water samples. When using the conventional approach, water samples are collected in glass containers and transported or shipped to a laboratory for extraction and analysis. With SPE disks, it is possible to extract pesticides from water in the field and transport only the disks to the laboratory for elution and analysis [26]. This eliminates the risk of glass breakage during collection, transport, and shipping, in addition to greatly reducing freight costs, and preserves some pesticides that are prone to hydrolysis. Numerous studies have shown that SPE disks can be used to extract pesticides from water and to preserve sample integrity until laboratory analysis [18,28–30]. Pesticide stability studies using Empore disks show that some pesticides have greater stability on C18 disks than in water at 4°C [25]. For instance, Aguilar et al. [27,31] stored SPE cartridges at room temperature, 4°C, and 20°C for 1 week or 3 months, and found minimal losses of pesticide for the lowest temperature at both time intervals. A multistate regional project showed that the pesticides atrazine, chlorpyrifos, and metolachlor could be retained on SPE disks and shipped to another laboratory for analysis with little pesticide losses [27].

9.3.2.3 Disadvantages

The main difficulties encountered with any kind of SPE configurations are caused by the presence of suspended particles in the sample. The particles of the alkyl bonded silica act as a mechanical filter that retains suspended soil or sediment particles, and the result is a loss of filtration due to clogging. This is very inconvenient when large volumes of sample are processed. To resolve this problem, acidification to a pH value of 2 is widely applied. Alternatively, the water sample is filtered prior to extraction. However, this treatment may not be desirable if the purpose of the analysis is to determine the total chemical concentration. In addition, although many studies have demonstrated the stability and good recovery of many pesticides from SPE disks, recoveries may vary with pesticide chemistry. It has also been shown that pesticide recovery from turbid water samples is less than that from deionized water samples [32]. Recoveries for compounds such as chlorpyrifos can be low and variable [29]. Therefore, field spikes, surrogates, and other quality assurance measures must be considered when using SPE disks for field samples.

9.3.2.4 Trends

A couple of problems may be encountered when using Empore SPE disks for pesticide extraction at one site followed by shipment to another site for elution and analysis. Once removed for shipping, it is impossible to perfectly realign disks onto another laboratory's extraction manifold so that the entire impregnated portion of the disk is exposed to the elution solvent. Realignment problems can result in reduced recovery from incomplete pesticide elution. This problem can be solved by combining the disks with the elution solvent in screw cap tubes, which are mixed on a shaker to extract pesticides from the disks [27]. In addition, surface water with high levels of particulates clogs disks and requires a filtration step prior to passing the water sample through the disk. Speedisks (J.T. Baker, Phillipsburg, NJ) offer an alternative to the use of traditional Empore SPE disks. Speedisks contain the extraction sorbent in a plastic housing, which is placed directly onto an extraction manifold, eliminating the realignment problems as noted earlier. The combination provides one-step filtration and extraction.

9.3.2.5 Applications

Numerous studies have reported the use of SPE disks for extracting or preserving pesticides from water samples. C18 Empore disks have been reported to extract some fungicides [33], carbamates and herbicides [34], or polar pesticides and herbicides [20] from waters. C8 Empore disks have been used to recover organochlorine pesticides, triazine herbicides, and other compounds from spiked water samples [35], and organochlorine, organophosphorus insecticides, triazine, and neutral herbicides from drinking water [23]. For instance, in Ref. [36], Empore C18 disks were used to extract a range of organophosphate compounds, including bromophos ethyl, bromophos methyl, dichlofenthion, ethion, fenamiphos, fenitrothion, fenthion, malathion, parathion ethyl, and parathion methyl. Using GC/MS or GC/FTD, MDLs were in the range of 0.01–0.07 µg/L and the recovery was from 60.7% to 104.1%.

9.3.3 SOLID-PHASE MICROEXTRACTION

9.3.3.1 Principles and Procedures

Although SPE methods use less amount of solvents, they are multiple-step procedures and are still somewhat time consuming. In 1990, an alternative extraction procedure employing SPME was introduced by Pawliszyn and coworkers [37,38]. In SPME, a thin fiber is coated with a sorbent and is exposed to the aqueous solution or the headspace of an aqueous sample to cause partitioning of some of the target analyte into the sorbent phase of the fiber. The fiber is then withdrawn, and introduced directly into a GC inlet to thermally desorb the enriched analyte into the GC column or eluted with the mobile phase in the mode of LC analysis. This technique fuses sample extraction and analysis into a single, continuous step, is compatible with GC and LC, and eliminates the use of any solvent for extraction. SPME is an equilibrium process that involves the partitioning of analytes between the sample and the extraction phase. Sampling conditions must therefore be systematically optimized to increase the partitioning of analytes in the coated fiber. Besides sampling conditions and analyte properties, the type of fiber coating is one of the most important aspects of optimization. Supelco (Bellefonte, PA, USA) is the main supplier of commercialized SPME fibers. Depending on the coating phase, the commercially available SPME fibers can be divided into absorbent- and adsorbent-type fibers. Absorbent-type fibers extract the analytes by partitioning of analytes into a "liquid-like" phase (e.g., polydimethylsiloxane or PDMS) whereas adsorbent-type fibers (e.g., activated carbon) extract the analytes by adsorption.

SPME consists of two extraction modes. One is the direct immersion mode, in which analytes are extracted from the liquid phase onto an SPME fiber, and the other is the headspace mode (HS–SPME), in which analytes are extracted from the headspace of a liquid sample onto the SPME fiber [39]. In general, direct SPME is more sensitive than HS–SPME for analytes present in a liquid sample, although HS–SPME gives lower background than direct SPME [40].

SPME can be coupled with either GC or LC. Coupling of SPME–GC is suitable for nonpolar and volatile or semivolatile pesticides. However, thermal desorption at high temperature creates practical problems such as degradation of the polymer, and furthermore, many nonvolatile compounds cannot be completely desorbed from the fiber. Solvent desorption is thus proposed as an alternative method through SPME–LC coupling. An organic solvent (static desorption mode) or the mobile phase (dynamic mode) is used to desorb the analytes from the SPME fiber.

9.3.3.2 Advantages

Several advantages can be pointed out in relation to SPME: it is solvent free, uses the whole sample for analysis, and requires only small sample amounts. The fibers are highly reusable (up to more than 100 injections). The success of SPME is based on its combining sampling, isolation, and concentration into a continuous step, and its compatibility with GC or LC.

9.3.3.3 Disadvantages

SPME suffers drawbacks such as sample carry-over, high cost, and a decline in performance with increased usage. The reluctance to adopting SPME in some cases can be also due to the steep learning curve expected for new users. To achieve good reproducibility, conditions such as fiber exposure time, solution stirring speed, fiber immersion depth, and fiber activation time and temperature must be precisely controlled, which may prove to be difficult if a manual assembly is used. In general, the use of manual SPME is tedious and gives low sample throughput. However, precise and easy handling of SPME can be realized using an automated SPME sampler such as the Combi-PAL autosampler made by Varian (Palo Alto, CA, USA).

9.3.3.4 Trends

In addition to the general purpose PDMS and polyacrylate (PA)-coated fibers, a large number of fiber coatings based on solid sorbents are available, namely the PDMS–divinylbenzene (PDMS–DVB), Carbowax–DVB (CW–DVB), CW–templated resin (CW–TR), Carboxen–PDMS, and DVB–Carboxen PDMS coated fibers [41]. SPME fibers with bipolar characteristics can be very useful for the simultaneous analysis of pesticides representing a wide range of polarities.

In-tube SPME is a new variation of SPME that has recently been developed using GC capillary columns as the SPME device instead of the SPME fiber. In-tube SPME is suitable for automation, and automated sample handling procedures not only shorten the total analysis time but also usually provide better accuracy and precision relative to manual SPME. In Ref. [42], an automated in-tube SPME method coupled with LC/ESI–MS was developed for the determination of chlorinated phenoxy acid herbicides. A capillary was placed between the injection loop and the injection needle of the autosampler. A metering pump was used to repeatedly draw and eject sample from the vial, allowing the analytes to partition from the sample matrix into the stationary phase. The extracted analytes were directly desorbed from the stationary phase by mobile phase, transported to the LC column, and then detected. The optimum extraction conditions were 25 draw/eject cycles of 30 ml of sample in 0.2% formic acid (pH = 2) at a flow rate of 200 ml/min using a DB-WAX capillary. The herbicides extracted by the capillary were easily desorbed by 10 μl acetonitrile. The calibration curves of herbicides were linear in the range 0.05–50 μg/L with correlation coefficients above 0.999. This method was successfully applied to the analysis of river water samples without interference peaks. The MDL was in the range of 0.005–0.03 μg/L. The repeatability and reproducibility were in the range of 2.5%–4.1% and 6.2%–9.1%, respectively.

9.3.3.5 Applications

Choudhury et al. [43] evaluated the use of SPME–GC analysis of 46 nitrogen- and phosphorus-containing pesticides defined in the EPA Method 507. Effects of pH, ionic strength, methanol content, and temperature on extraction were determined. Analytes were extracted into a PDMS fiber and then thermally desorbed in a GC injector and analyzed. When analyzed by SPME GC/NPD or by SPME GC/MS, 34 and 39 pesticides, respectively, were measured at levels lower than the EPA MDLs and precision requirements. This method was applied to the analysis of contaminated well water, watershed, and stream water and compared to U.S. EPA Method 507 findings. The results demonstrated that SPME was a valuable tool for the rapid screening of 39 EPA Method 507 nitrogen- and phosphorus-containing pesticides in water.

Jackson and Andrews [44] evaluated the use of SPME under nonequilibrium conditions for analysis of organochlorine pesticides. SPME is typically performed for a length of time that nears the equilibrium time of the analyte in the sample. However, equilibrium times for organochlorines fall in the range of 30–180 min. Studies show that linear responses having good precision are possible by using extraction times well short of equilibrium times [37,45]. With a 2 min extraction time and 100 μm PDMS fiber, analysis of a sample took less than 10 min, with MDLs in the order of 10 ng/L.

Chafer-Pericas et al. [46] compared the advantages and disadvantages of two different configurations for the extraction of triazines from water samples, on-fiber SPME coupled to LC, and in-tube SPME coupled to LC. In-tube SPME used a packed column or an open capillary column. In the on-fiber SPME configuration, the fiber coating was PDMS–DVB. The MDLs obtained with this approach were between 25 and 125 μg/L. The in-tube SPME approach with a C18 packed column (35 mm $\times$ 0.5 mm I.D., 5 μm particle size) connected to a switching microvalve provided the best sensitivity; under such configuration, the MDLs were between 0.025 and 0.5 μg/L. The in-tube SPME approach with an open capillary column coated with PDMS (30 cm $\times$ 0.25 mm I.D., 0.25 μm of thickness coating) connected to the injection valve provided MDLs between 0.1 and 0.5 μg/L.

9.4 CAPILLARY ELECTROPHORESIS

9.4.1 PRINCIPLES

CE is a relatively new analytical technique that is complementary to GC and LC. CE is a microvolume separation technique characterized by its relatively short analysis time, and nanoliter to picoliter sample volumes. In CE, a fused-silica capillary is filled with an electrolytic solution, known as the running buffer or background electrolyte. An electric field is applied to the capillary to cause migration of charged molecules in opposite directions. The mobility of the ion is governed by its charge-to-size ratio, and the size is dependent on the molecular weight, the three-dimensional structure, and the degree of solvation. The most common mode of CE is known as micellar electrokinetic chromatography (MEKC), which was introduced by Terabe et al. [47] in an effort to extend the use of CE to neutral molecules. In MEKC, surfactants are added to the background electrolyte to produce micelles, which will allow nonpolar compounds to interact with the micelles to cause separation. The most common surfactant is sodium dodecyl sulfate (SDS). Smith and coworkers [48–50] have introduced in situ charged micelles that are based on the complexation of borate or borate ions with the neutral surfactant that have polyolic polar head groups. The surface charge density of the micellar phases can be varied by either altering the borate or boronate concentration and pH of the running buffer, which further optimizes the resolution and peak capacity during MEKC analysis.

9.4.2 ADVANTAGES

CE offers several advantages over conventional chromatographic techniques, including use of little solvents, high resolution, small sample volume, and short run time. In addition, CE may be complementary to GC when the thermal lability of the analyte is a concern. CE is more efficient at separation when compared with common LC and GC techniques due to its maximum theoretical plate number. CE is especially applicable to the analysis of water samples as preconcentration treatments such as SPE or field amplification can be easily used to enhance the detection sensitivity.

9.4.3 DISADVANTAGES

The largest drawback to CE is its relatively low sensitivities, which is caused by the extreme degree of miniaturization involved in CE, and the limited availability of sensitive detection systems. For instance, when a UV detector is used, the sensitivity of CE is comparable to that of LC. This problem is slowly being resolved with the introduction of high flow cell capillaries and new detectors. For instance, coupling with MS enhances the sensitivity because of the decreased matrix interference. Preconcentration techniques such as sample stacking and SPE greatly enhance the sensitivity of CE, and this sample manipulation approach is suitable especially for water samples.

9.4.4 TRENDS

In general, CE application for routine analysis of pesticides is restricted by the relatively few developed methods, partly because of its short history. Sample pretreatment techniques such as SPE and on-column preconcentration methods such as sample stacking are used in CE to achieve better sensitivity for pesticide analysis in water. There are two types of stacking. The first stacking method involves the stacking of sample into a shorter zone during CE separation. The second method is known as field-amplified sample stacking, involving stacking with reversed polarity [51–53]. In addition, more sensitive and selective detectors have been coupled with CE to improve sensitivity. For instance, laser-induced fluorescence and photothermal systems offer enhanced sensitivities for certain compounds. MS with a proper interface has also been shown to provide much better sensitivities.

9.4.5 APPLICATIONS

Fung and Mak [54] used a two-step sample preconcentration (SPE and field-amplified sample stacking) and applied MEKC for the analysis of 14 pesticides (including aldicarb, carbofuran, isoproturon, chlorotoluron, metolachlor, mecoprop, dichlorprop, MCPA, 2,4-D, methoxychlor, TDE, DDT, dieldrin, and DDE) in drinking water. Good recoveries of pesticides were obtained using SPE with sample pH adjusted to 2–3. Field-amplified sample stacking was found to give additional enrichment up to 30-fold. The optimized background electrolyte consisted of 50 mM sodium dodecyl sulfate (SDS), 10 mM borate buffer, 15 mM β-cyclodextrin (β-CD), and 22% acetonitrile at pH 9.6, and running was performed under 25 kV with detection at 202 nm. Good linearity was obtained for all pesticides with detection limits down to 0.04–0.46 μg/L. Song and Budde [55] applied CE with electrospray negative ion MS (CE–ENI–MS) for the determination of chlorinated acid herbicides and several phenols in water. Sixteen acid herbicides were separated as anions in less than 40 min with a buffer consisting of 5 mM ammonium acetate in isopropanol–water (2:3, v/v) at pH 10. A sample stacking technique was used to achieve lower detection limits, along with selected ion monitoring (SIM) and internal standardization. Safarpour et al. [56] reported the use of CE–MS for the analysis of imazamox in water. Residues of imazamox were extracted from the water samples using reversed-phase SPE. Pesticide measurement was accomplished by CE–MS using electrospray ionization with SIM in the positive-ion mode. The MDL was 0.02 μg/L.

9.5 IMMUNOASSAYS

9.5.1 PRINCIPLES

Immunoassays (IAs) are based on the interaction of antibodies (Abs) with antigens (Ags). Antibodies are polymers containing hundreds of individual amino acids arranged in a highly ordered sequence. These polypeptides are produced by immune system cells (B lymphocytes) when exposed to antigenic substances or molecules. In Abs, there are recognition/binding sites for specific molecular structures of the Ags. According to the "key–lock" model, an Ab interacts in a highly specific way with its unique Ag. This feature constitutes the key to IA [57]. In most IAs, the antibodies (or the antigens) are immobilized on a solid support and a measurement of the binding sites by the antigens (or the antibodies) is made because the antibody occupancy reflects the concentration of analytes in the medium [58]. However, since the binding reaction does not produce a signal, a tracer must be added to allow estimation of the occupancy by measuring the tracer signal. The labels capable of detecting the immunological reaction with the purpose of quantitation can be fluorescent, chemiluminescent enzymes or radioisotopes.

Immunobased kits make possible the performance of analysis in different laboratories and also in field under standard conditions, owing to their reproducibility, ease of use, and good shelf life [59]. Approximately 90% of the developed IAs for pesticide residue analyses use the enzyme-linked immunosorbent assay (ELISA). In this technique, the analyte from the sample and a known amount of enzyme-tagged analyte compete for a limited number of antibody binding sites. Quantification is achieved by comparing the signal generated by an unknown sample with a standard curve. Immunoassay test kits include antibodies, reagents, standard, and substrates in field transportable units that are ready to use. Immunoassay kits are suitable for use under field conditions because they are fast, and many of the standard documentary and sample-handling procedures can be avoided. In general, no sample treatment is necessary. In some cases, only a filtration step is required.

9.5.2 ADVANTAGES

IAs are specific, sensitive, easy to perform, fast, relatively inexpensive, and highly portable [59]. Compared with chromatographic techniques, IAs are advantageous if large series of samples have to be analyzed. Also, no complex or sophisticated instrumentation is required and the use of organic solvents is minimal. IAs may be useful for polar pesticides that are difficult to analyze by standard techniques [60]. Immunoaffinity chromatography (IAC) is based on the use of antibodies, not on differences in polarity. Antibodies raised against a specific target pesticide can be immobilized on a solid phase and will selectively retain the pesticide, thus effecting both a preconcentration and a cleanup.

9.5.3 DISADVANTAGES

While numerous studies have shown excellent correlation between the results of IAs and the conventional chromatographic methods, there are also many instances suggesting that IAs can cause under- or overestimation of the true pesticide concentrations because of matrix interference and cross activity. Acceptance of IAs is dependent upon several factors, including the demonstration of quality and validity compared with more traditional methods. IAs are considered as a supplementary method for other more reliable methods or as a semiquantitative method for initial screening.

9.5.4 TRENDS

The earlier IAs dealt with single pesticides, but there is clearly a trend for developing class-specific assays [58]. Simultaneous detection of a plurality of analytes by IAs would answer many of the requirements of pesticide pollution monitoring. Further developments in ELISA include the automation of both plate and tube assays. Great efforts have been made to achieve the miniaturization and automation of immunoassay techniques. Also, an important premise in ELISA kits is their application in field assays, and a very interesting approach would be the introduction of stable immunoreagents. In this sense, there is a need for the development of new tracers, such as fluorescent labels, that avoid the drawbacks inherent to enzyme use (stability, cost, handling, and storage). Finally, new strategies for antibody production to increase sensitivity and selectivity through the use of recombinant antibodies and molecular-imprinted polymers are promising and attractive alternatives to conventional approaches for the development of antibodies.

One extension of IA is the development of immunosensors as prescreening techniques in environmental monitoring [61]. Immunosensors are based on the principles of solid-phase IA. They combine the power of antibodies as recognition agents and an appropriate physicochemical transduction mechanism to convert the recognition events to signals. The working principle of an immunosensor is based on the formation of an immunochemical interaction step that produces a physicochemical change in the system, which can be converted into a readable signal. To meet the requirement of continuous monitoring, the principles of flow injection analysis have been incorporated into the biosensor manufacturing, leading to the development of flow-through immunosensors. The bioactive surface of the biosensor can be regenerated to enable continual monitoring of the measured signal. Regeneration of the sensing surface is usually performed by displacement of the immunoreaction, using agents that are able to break the antibody–analyte association, such as organic solvents with acidic buffers, chaotropic agents, or digesting enzymes, or a combination of two or more of these methods.

Another application of IA is the development of immunoaffinity columns that can be coupled with MS for the determination of analytes. Zhang et al. [62] reported the use of IA–LC–MS–MS for analysis of diuron in water. This method used a sol–gel immunoaffinity column (20 mm × 4 mm I.D.) for on-line sample cleanup and enrichment, a monolithic analytical column (100 mm × 4.6 mm I.D.) for separation, and a triple quadrupole MS for quantitation. The optimized on-line protocol was emphasized by the observation that low MDL of 1.0 ng/L was achieved with only 2.5 mL sample. In addition, a satisfactory accuracy (about 90% of recovery) and precision (<6% RSD) at 50 ng/L concentration were also obtained.

9.5.5 APPLICATIONS

Bruun et al. [63] reported an IA method for the triazine metabolites hydroxypropazine, hydroxyatrazine, and hydroxysimazine. The assay was based on covalent immobilization of antigen in combination with an enzyme-labeled anti-hydroxy-s-triazine monoclonal antibody. This system enabled the development of an assay with variation coefficients below 3% and MDL below 0.01 μg/L hydroxyatrazine and hydroxypropazine. Analysis of hydroxyatrazine-spiked water of three different types yielded an average recovery of 102% at 0.1 μg/L hydroxyatrazine. Relative to hydroxyatrazine, assay cross-reactivity was 148% towards hydroxypropazine and 67% towards hydroxysimazine.

Schraer et al. [64] compared ELISA data from a surface water reconnaissance to GC data on cyanazine and metolachlor. A total of 535 surface water samples from locations in Mississippi, Louisiana, Arkansas, and Tennessee were collected. When ELISA analyses were duplicated, cyanazine and metolachlor detection was found to have highly reproducible results; adjusted R^2s were 0.97 and 0.94, respectively. When ELISA results for cyanazine were regressed against GC results, the models effectively predicted cyanazine concentrations from ELISA analyses (adjusted R^2s ranging from 0.76 to 0.81). The intercepts and slopes for these models were not different from 0 and 1, respectively. This indicates that cyanazine analysis by ELISA gave the same results as analysis by GC. However, regressing ELISA analyses for metolachlor against GC data provided more variable results (adjusted R^2s ranged from 0.67 to 0.94). Regression models for metolachlor analyses had two of three intercepts that were not different from 0. Slopes for all metolachlor regression models were significantly different from 1. This indicates that as metolachlor concentrations increased, ELISA over- or underestimated the concentration.

9.6 DETECTION METHODS

9.6.1 BACKGROUND

Following sample preparation, a wide range of methods can be used for analysis of pesticides in water samples. These methods include mainstream methods such as various GC and LC methods, and other methods such as immunoassays and CE analysis. Principles and applications of immunoassays and CE are described in the previous sections. In this section, discussion will be focused on the principles, advantages and disadvantages, and trends in GC and LC methods that are used for quantitative measurement of pesticides in water.

It is important to realize that pesticide types have changed through the years, going from persistent and more nonpolar pesticides (e.g., organochlorine insecticides) to more polar and sometimes thermal-labile compounds. In particular, the heavy use of herbicides such as sulfonylureas, imidazolinones, triazines, chlorophenoxy acids, and phenylureas has prompted the development of methods suitable for more thermolabile and low-volatility compounds. This trend has apparently driven the development of some new and more robust methods, including especially LC–MS based methods.

9.6.2 GC Detection Methods

From the early 1970s to the early 1990s, most routine pesticide residue analysis were conducted by GC in combination with ECD, NPD, flame ionization detector (FID) or flame photometric detection (FPD) [65]. The sensitivity of these conventional GC detectors is highly specific to the types of pesticides being analyzed. While ECD can be highly sensitive for halogenated pesticides, FID is generally less sensitive and NPD lies somewhere in between for N- or P-containing pesticides. These conventional GC detectors are not universal and are not desired for multiresidue analysis. More importantly, these detection methods do not provide any qualitative information on the structures being analyzed. Confirmation often requires the use of a second column of a different polarity. However, combination of GC with MS enables simultaneous determination and confirmation of pesticide residues with one instrument in one analytical run. In the scan mode, the sensitivity of MS approximates that of FID and poorer than ECD, while in the SIM or MS/MS mode, much better sensitivity can usually be obtained because of the greatly improved selectivity that suppresses the matrix background.

In GC–MS, ionization of pesticides can be achieved by electron impact ionization (EI) or positive or negative chemical ionization (PCI, NCI). Most of the published studies on residue analysis by GC–MS report on results obtained by single quadrupole instruments and EI [65]. Compared with EI, positive or negative CI–MS gives better selectivity for some pesticides. This is due to reduced matrix interference [66]. Figure 9.2 shows two chromatograms from the same surface water sample analyzed by GC–ECD or GC–MS (NCI) (with methane as the CI gas). Because of its much better selectivity, it is clear that GC–MS (NCI) was 1–2 orders of magnitude more sensitive at detecting pyrethroid pesticides in the same water sample than GC–ECD.

As in GC–MS (CI), a good suppression of matrix background can also be obtained by GC–MS/MS systems [67]. Some limitations in GC–MS/MS arise from the absence of a universal soft ionization mode that may be used for producing dominant molecular ions. CI generates high-intensity ions of only some pesticide classes, while the total ion current of EI is spread on many fragments, resulting in a low intensity of parent ions. In general, both GC–MS (CI) and GC–MS/MS are more advantageous than any of the conventional GC detection methods, because they provide both high sensitivity and the capability for structure confirmation.

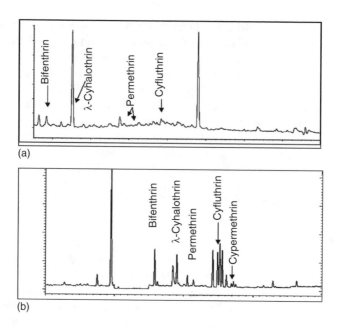

FIGURE 9.2 GC chromatograms from the urban runoff water sample subjected to two different analyses. (a) GC–ECD and (b) GC–MS (NCI).

9.6.3 LC DETECTION METHODS

Until a decade ago, methods based on LC were applied much less frequently than GC for pesticide residue analysis, because traditional UV and fluorescence detectors are less sensitive than the various GC detection methods [65]. However, in the last decade, the availability of atmospheric pressure ionization (API) made possible the coupling of LC with MS. Compared with traditional detectors, electrospray (ESI) or atmospheric pressure chemical ionization (APCI) in combination with MS has increased the sensitivity of LC detection by several orders of magnitude. Single quadrupole was the predominant configuration of LC–MS in the early 1990s. A disadvantage of single quadrupole LC–MS is the high intensity of background signals produced from sample matrix and LC solvent clusters. Because of this chemical noise in real samples, detection sensitivity was relatively poor, even though the instrumental sensitivity was good [68]. The chemical background can be reduced significantly if tandem MS in combination with selected reaction monitoring (SRM) is applied. Even if a coextracted matrix component has the molecular mass of a pesticide, usually both isobaric ions can be separated in SRM, because their fragmentation in the collision cell most often results in different product ions. Therefore, LC–MS/MS offers excellent sensitivity and unsurpassed selectivity. Moreover, when LC–MS/MS is operated in the SRM mode, it is unnecessary to perform LC column switching or extensive sample cleanup [68,69].

Time-of-flight (TOF)–MS in combination with LC is more often used in the high-resolution mode, which provides better discrimination of background [70].

The main advantage of this type of instrument is the identification of unknown peaks in a sample even if analytical standards are not available [71].

9.6.4 COMPARISON BETWEEN GC AND LC METHODS

Alder et al. [65] recently reviewed applications of GC–MS and LC–MS methods for pesticide residue analysis and made several important observations. LC–MS/MS was shown to be better suited for sulfonyl or benzoyl ureas, carbamates, and triazines than GC–MS. For the remainder pesticides, the application scope of LC–MS/MS was also wider than that of GC–MS. Both GC–MS and LC–MS based methods revealed a significant variation in sensitivity, covering at least a range of 3–4 orders of magnitude, depending on the pesticide analyzed. However, a comparison of the median of MDLs clearly showed much higher sensitivity if determinations were based on LC–MS/MS. The better performance of LC–MS/MS is probably determined by several reasons. Among them the larger injection volume used in LC–MS/MS (20 μL vs. 1 μL) and the lower amount of fragmentation during ionization (ESI vs. EI) may explain some of these differences.

Table 9.1 lists some of the most commonly used GC and LC methods for analysis of pesticides in water samples. These methods are also roughly ranked according to their sensitivity, selectivity, universal applicability, and ability for structure identification. An important factor to consider when selecting a detection method is whether the analysis is for screening of a wide range of pesticides or target analysis of a predefined set of compounds. For the screening of a wide range of

TABLE 9.1

Frequently Used GC and LC Methods for Analysis of Pesticides in Water and Their Relative Ranking in Detection Sensitivity, Universal Applicability, Matrix Background Suppression, and Ability for Structural Confirmation

Detection Method	Sensitivity	Universal Applicability	Matrix Suppression	Confirmation Capability
Gas Chromatography (GC)				
GC–FID	★	★★★	★	★
GC–NPD	★★	★	★	★
GC–ECD	★★★	★	★★	★★
GC–FPD	★★	★	★	★
GC–MS (SIM)	★	★★★	★★★	★★★
GC–MS (CI)	★★★★	★★★	★★★★	★★★★
GC–MS/MS	★★★	★★★	★★★★	★★★★
Liquid Chromatography (LC)				
LC–UV	★	★★	★	★
LC–fluoroscence	★★	★	★	★★
LC–MS	★★	★★★★	★★★	★★★
LC–MS/MS	★★★★	★★★★	★★★★	★★★★

analytes, as often required in monitoring studies, universal methods such as GC–MS, GC–MS/MS, or LC–MS/MS will be superior to the other more specific methods. On the other hand, a specific method may be selected for target analysis of a known group of pesticides. Last but not least, factors such as cost, availability of instruments, and skills of analysts can also contribute to the methods selected for pesticide analysis in water samples.

REFERENCES

1. U.S. Environmental Protection Agency, Diazinon and chlorpyrifos TMDL for San Diego Creek and Newport Bay watershed. http://www.waterboards.ca.gov/santaana/pdf/03-39.pdf 2003.
2. Hatrik, S., Tekel, J., Extraction methodology and chromatography for the determination of residual pesticides in water. *Journal of Chromatography A* 1996, 733(1–2), 217–233.
3. Liska, I., Slobodnik, J., Comparison of gas and liquid chromatography for analysing polar pesticides in water samples. *Journal of Chromatography A* 1996, 733(1–2), 235–258.
4. Wan, H.B., Wong, M.K., Minimization of solvent consumption in pesticide residue analysis. *Journal of Chromatography A* 1996, 754(1–2), 43–47.
5. van der Hoff, G.R., van Zoonen, P., Trace analysis of pesticides by gas chromatography. *Journal of Chromatography A* 1999, 843(1–2), 301–322.
6. Hogendoorn, E., van Zoonen, P., Recent and future developments of liquid chromatography in pesticide trace analysis. *Journal of Chromatography A* 2000, 892(1–2), 435–453.
7. Zapf, A., Heyer, R., Stan, H.J., Rapid micro liquid–liquid-extraction method for trace analysis of organic contaminants in drinking water. *Journal of Chromatography A* 1995, 694(2), 453–461.
8. de Jager, L.S., Andrews, A.R.J., Development of a rapid screening technique for organochlorine pesticides using solvent microextraction (SME) and fast gas chromatography (GC). *Analyst* 2000, 125(11), 1943–1948.
9. Liu, Y., Zhao, E.C., Zhou, Z.Q., Single-drop microextraction and gas chromatographic determination of fungicide in water and wine samples. *Analytical Letters* 2006, 39(11), 2333–2344.
10. Soriano, J.M., Jimenez, B., Font, G., Molto, J.C., Analysis of carbamate pesticides and their metabolites in water by solid phase extraction and liquid chromatography: A review. *Critical Reviews in Analytical Chemistry* 2001, 31(1), 19–52.
11. Baez, M.E., Lastra, O., Rodriguez, M., Solid phase extraction of halogenated pesticides from ground and surface waters and their determination by capillary gas chromatography. *HRC-Journal of High Resolution Chromatography* 1996, 19(10), 559–563.
12. Baez, M.E., Rodriguez, M., Lastra, O., Contreras, P., Solid phase extraction of organophosphorus, triazine, and triazole-derived pesticides from water samples. A critical study. *HRC-Journal of High Resolution Chromatography* 1997, 20(11), 591–596.
13. Bucheli, T.D., Gruebler, F.C., Muller, S.R., Schwarzenbach, R.P., Simultaneous determination of neutral and acidic pesticides in natural waters at the low nanogram per liter level. *Analytical Chemistry* 1997, 69(8), 1569–1576.
14. Crescenzi, C., DiCorcia, A., Guerriero, E., Samperi, R., Development of a multiresidue method for analyzing pesticide traces in water based on solid-phase extraction and electrospray liquid chromatography mass spectrometry. *Environmental Science & Technology* 1997, 31(2), 479–488.

15. Kampioti, A.A., da Cunha, A.C.B., de Alda, M.L., Barcelo, D., Fully automated multi-analyte determination of different classes of pesticides, at picogram per litre levels in water, by on-line solid-phase extraction-liquid chromatography-electrospray-tandem mass spectrometry. *Analytical and Bioanalytical Chemistry* 2005, 382(8), 1815–1825.

16. Fernandez, M.J., Garcia, C., GarciaVillanova, R.J., Gomez, J.A., Evaluation of liquid–solid extraction with a new sorbent and liquid–liquid extraction for multiresidue pesticides. Determination in raw and finished drinking waters. *Journal of Agricultural and Food Chemistry* 1996, 44(7), 1790–1795.

17. Patsias, J., Papadopoulou-Mourkidou, E., Rapid method for the analysis of a variety of chemical classes of pesticides in surface and ground waters by off-line solid-phase extraction and gas chromatography ion trap mass spectrometry. *Journal of Chromatography A* 1996, 740(1), 83–98.

18. Riley, M.B., Dumas, J.A., Gbur, E.E., Massey, J.H., Mattice, J.D., Mersie, W., Mueller, T.C., Potter, T., Senseman, S.A., Watson, E., Pesticide extraction efficiency of two solid phase disk types after shipping. *Journal of Agricultural and Food Chemistry* 2005, 53(13), 5079–5083.

19. Lambropoulou, D.A., Konstantinou, I.K., Albanis, T.A., Determination of fungicides in natural waters using solid-phase microextraction and gas chromatography coupled with electron-capture and mass spectrometric detection. *Journal of Chromatography A* 2000, 893(1), 143–156.

20. Brouwer, E.R., Lingeman, H., Brinkman, U.A.T., Use of membrane extraction disks for online trace enrichment of organic-compounds from aqueous samples. *Chromatographia* 1990, 29(9–10), 415–418.

21. Markell, C., Hagen, D.F., Bunnelle, V.A., New technologies in solid-phase extraction. *LC GC-Magazine of Separation Science* 1991, 9(5), 332–337.

22. Hagen, D.F., Markell, C.G., Schmitt, G.A., Blevins, D.D., Membrane approach to solid-phase extractions. *Analytica Chimica Acta* 1990, 236(1), 157–164.

23. Davi, L.M., Baldi, M., Penazzi, L., Liboni, M., Evaluation of the membrane approach to solid-phase extractions of pesticide-residues in drinking-water. *Pesticide Science* 1992, 35(1), 63–67.

24. Johnson, W.G., Lavy, T.L., Senseman, S.A., Stability of selected pesticides on solid-phase extraction disks. *Journal of Environmental Quality* 1994, 23(5), 1027–1031.

25. Senseman, S.A., Lavy, T.L., Mattice, J.D., Myers, B.M., Skulman, B.W., Stability of various pesticides on membranous solid-phase extraction media. *Environmental Science & Technology* 1993, 27(3), 516–519.

26. Mattice, J.D., Senseman, S.A., Walker, J.T., Gbur, E.E., Portable system for extracting water samples for organic analysis. *Bulletin of Environmental Contamination and Toxicology* 2002, 68(2), 161–167.

27. Mueller, T.C., Senseman, S.A., Wauchope, R.D., Clegg, C., Young, R.W., Southwick, L.M., Riley, M.B., Moye, H.A., Dumas, J.A., Mersie, W., Mattice, J.D., Leidy, R.B., Recovery of atrazine, bromacil, chlorpyrifos, and metolachlor from water samples after concentration on solid-phase extraction disks: Interlaboratory study. *Journal of AOAC International* 2000, 83(6), 1327–1333.

28. Mersie, W., Clegg, C., Wauchope, R.D., Dumas, J.A., Leidy, R.B., Riley, M.B., Young, R.W., Mattice, J.D., Mueller, T.C., Senseman, S.A., Interlaboratory comparison of pesticide recovery from water using solid-phase extraction disks and gas chromatography. *Journal of AOAC International* 2002, 85(6), 1324–1330.

29. Senseman, S.A., Mueller, T.C., Riley, M.B., Wauchope, R.D., Clegg, C., Young, R.W., Southwick, L.M., Moye, H.A., Dumas, J.A., Mersie, W., Mattice, J.D., Leidy, R.B.,

Interlaboratory comparison of extraction efficiency of pesticides from surface and laboratory water using solid-phase extraction disks. *Journal of Agricultural and Food Chemistry* 2003, 51(13), 3748–3752.

30. Cobb, J.M., Mattice, J.D., Senseman, S.A., Dumas, J.A., Mersie, W., Riley, M.B., Potter, T.L., Mueller, T.C., Watson, E.B., Stability of pesticides on solid-phase extraction disks after incubation at various temperatures and for various time intervals: Interlaboratory study. *Journal of AOAC International* 2006, 89(4), 903–912.

31. Aguilar, C., Ferrer, I., Borrull, F., Marce, R.M., Barcelo, D., Monitoring of pesticides in river water based on samples previously stored in polymeric cartridges followed by on-line solid-phase extraction liquid chromatography diode array detection and confirmation by atmospheric pressure chemical ionization mass spectrometry. *Analytica Chimica Acta* 1999, 386(3), 237–248.

32. Johnson, W.E., Fendinger, N.J., Plimmer, J.R., Solid-phase extraction of pesticides from water—Possible interferences from dissolved organic material. *Analytical Chemistry* 1991, 63(15), 1510–1513.

33. Salau, J.S., Alonso, R., Batllo, G., Barcelo, D., Application of solid-phase disk extraction followed by gas and liquid-chromatography for the simultaneous determination of the fungicides Captan, captafol, carbendazim, chlorothalonil, ethirimol, folpet, metalaxyl and vinclozolin in environmental waters. *Analytica Chimica Acta* 1994, 293(1–2), 109–117.

34. Chiron, S., Barcelo, D., Determination of pesticides in drinking-water by online solid-phase disk extraction followed by various liquid-chromatographic systems. *Journal of Chromatography* 1993, 645(1), 125–134.

35. Krautvass, A., Thoma, J., Performance of an extraction disk in synthetic organic-chemical analysis using gas-chromatography mass-spectrometry. *Journal of Chromatography* 1991, 538(2), 233–240.

36. Lambropoulou, D., Sakellarides, T., Albanis, T., Determination of organophosphorus insecticides in natural waters using SPE-disks and SPME followed by GG/FTD and GC/MS. *Fresenius Journal of Analytical Chemistry* 2000, 368(6), 616–623.

37. Arthur, C.L., Pawliszyn, J., Solid-phase microextraction with thermal-desorption using fused-silica optical fibers. *Analytical Chemistry* 1990, 62(19), 2145–2148.

38. Zhang, Z.Y., Yang, M.J., Pawliszyn, J., Solid-phase microextraction. *Analytical Chemistry* 1994, 66(17), A844–A853.

39. Sakamoto, M., Tsutsumi, T., Applicability of headspace solid-phase microextraction to the determination of multi-class pesticides in waters. *Journal of Chromatography A* 2004, 1028(1), 63–74.

40. Kataoka, H., Lord, H.L., Pawliszyn, J., Applications of solid-phase microextraction in food analysis. *Journal of Chromatography A* 2000, 880(1–2), 35–62.

41. Goncalves, C., Alpendurada, M.F., Comparison of three different poly(dimethylsiloxane)-divinylbenzene fibres for the analysis of pesticide multiresidues in water samples: Structure and efficiency. *Journal of Chromatography A* 2002, 963(1–2), 19–26.

42. Takino, M., Daishima, S., Nakahara, T., Automated on-line in-tube solid-phase microextraction followed by liquid chromatography/electrospray ionization-mass spectrometry for the determination of chlorinated phenoxy acid herbicides in environmental waters. *Analyst* 2001, 126(5), 602–608.

43. Choudhury, T.K., Gerhardt, K.O., Mawhinney, T.P., Solid-phase microextraction of nitrogen and phosphorus-containing pesticides from water and gas chromatographic analysis. *Environmental Science & Technology* 1996, 30(11), 3259–3265.

44. Jackson, G.P., Andrews, A.R.J., New fast screening method for organochlorine pesticides in water by using solid-phase microextraction with fast gas chromatography and a pulsed-discharge electron capture detector. *Analyst* 1998, 123(5), 1085–1090.

45. Page, B.D., Lacroix, G., Application of solid-phase microextraction to the headspace gas chromatographic analysis of semi-volatile organochlorine contaminants in aqueous matrices. *Journal of Chromatography A* 1997, 757(1–2), 173–182.

46. Chafer-Pericas, C., Herraez-Hernandez, R., Campins-Falco, P., On-fibre solid-phase microextraction coupled to conventional liquid chromatography versus in-tube solid-phase microextraction coupled to capillary liquid chromatography for the screening analysis of triazines in water samples. *Journal of Chromatography A* 2006, 1125(2), 159–171.

47. Terabe, S., Otsuka, K., Ichikawa, K., Tsuchiya, A., Ando, T., Electrokinetic separations with micellar solutions and open-tubular capillaries. *Analytical Chemistry* 1984, 56(1), 111–113.

48. Smith, J.T., Nashabeh, W., Elrassi, Z., Micellar electrokinetic capillary chromatography with in-situ charged micelles.1. Evaluation of N-D-gluco-N-methylalkanamide surfactants as anionic borate complexes. *Analytical Chemistry* 1994, 66(7), 1119–1133.

49. Smith, J.T., Elrassi, Z., Micellar electrokinetic capillary chromatography with in situ charged micelles. 3. Evaluation and comparison of octylmaltoside and octylsucrose surfactants as anionic borate complexes in the separation of herbicides. *Journal of Microcolumn Separations* 1994, 6(2), 127–138.

50. Smith, J.T., Elrassi, Z., Micellar electrokinetic capillary chromatography with in-situ charged micelles. 4. Influence of the nature of the alkylglycoside surfactant. *Journal of Chromatography A* 1994, 685(1), 131–143.

51. Albin, M., Grossman, P.D., Moring, S.E., Sensitivity enhancement for capillary electrophoresis. *Analytical Chemistry* 1993, 65(10), A489–A497.

52. Chien, R.L., Burgi, D.S., On-column sample concentration using field amplification in CZE. *Analytical Chemistry* 1992, 64(8), A489–A496.

53. Aebersold, R., Morrison, H.D., Analysis of dilute peptide samples by capillary zone electrophoresis. *Journal of Chromatography* 1990, 516(1), 79–88.

54. Fung, Y.S., Mak, J.L.L., Determination of pesticides in drinking water by micellar electrokinetic capillary chromatography. *Electrophoresis* 2001, 22(11), 2260–2269.

55. Song, X.B., Budde, W.L., Determination of chlorinated acid herbicides and related compounds in water by capillary electrophoresis-electrospray negative ion mass spectrometry. *Journal of Chromatography A* 1998, 829(1–2), 327–340.

56. Safarpour, H., Asiaie, R., Katz, S., Quantitative analysis of imazamox herbicide in environmental water samples by capillary electrophoresis electrospray ionization mass spectrometry. *Journal of Chromatography A* 2004, 1036(2), 217–222.

57. Marco, M.P., Gee, S., Hammock, B.D., Immunochemical techniques for environmental analysis. 1. Immunosensors. *Trac-Trends in Analytical Chemistry* 1995, 14(7), 341–350.

58. Hennion, M.C., Barcelo, D., Strengths and limitations of immunoassays for effective and efficient use for pesticide analysis in water samples: A review. *Analytica Chimica Acta* 1998, 362(1), 3–34.

59. Gabaldon, J.A., Maquieira, A., Puchades, R., Current trends in immunoassay-based kits for pesticide analysis. *Critical Reviews in Food Science and Nutrition* 1999, 39(6), 519–538.

60. Houben, A., Meulenberg, E., Noij, T., Gronert, C., Stoks, P., Immune affinity extraction of pesticides from surface water. *Analytica Chimica Acta* 1999, 399(1–2), 69–74.

61. Mallat, E., Barcelo, D., Barzen, C., Gauglitz, G., Abuknesha, R., Immunosensors for pesticide determination in natural waters. *TRAC-Trends in Analytical Chemistry* 2001, 20(3), 124–132.

62. Zhang, X.L., Martens, D., Kramer, P.M., Kettrup, A.A., Liang, X.M., On-line immuno-affinity column-liquid chromatography-tandem mass spectrometry method for trace

analysis of diuron in wastewater treatment plant effluent sample. *Journal of Chromatography A* 2006, 1133(1–2), 112–118.

63. Bruun, L., Koch, C., Jakobsen, M.H., Aamand, J., New monoclonal antibody for the sensitive detection of hydroxy-*s*-triazines in water by enzyme-linked immunosorbent assay. *Analytica Chimica Acta* 2000, 423(2), 205–213.

64. Schraer, S.M., Shaw, D.R., Boyette, M., Coupe, R.H., Thurman, E.M., Comparison of enzyme-linked immunosorbent assay and gas chromatography procedures for the detection of cyanazine and metolachlor in surface water samples. *Journal of Agricultural and Food Chemistry* 2000, 48(12), 5881–5886.

65. Alder, L., Greulich, K., Kempe, G., Vieth, B., Residue analysis of 500 high priority pesticides: Better by GC-MS or LC-MS/MS? *Mass Spectrometry Reviews* 2006, 25(6), 838–865.

66. Hernando, M.D., Aguera, A., Fernandez-Alba, A.R., Piedra, L., Contreras, M., Gas chromatographic determination of pesticides in vegetable samples by sequential positive and negative chemical ionization and tandem mass spectrometric fragmentation using an ion trap analyser. *Analyst* 2001, 126(1), 46–51.

67. Goncalves, C., Alpendurada, M.F., Solid-phase micro-extraction-gas chromatography-(tandem) mass spectrometry as a tool for pesticide residue analysis in water samples at high sensitivity and selectivity with confirmation capabilities. *Journal of Chromatography A* 2004, 1026(1–2), 239–250.

68. Hernandez, F., Sancho, J.V., Pozo, O.J., Critical review of the application of liquid chromatography/mass spectrometry to the determination of pesticide residues in biological samples. *Analytical and Bioanalytical Chemistry* 2005, 382(4), 934–946.

69. Seitz, W., Schulz, W., Weber, W.H., Novel applications of highly sensitive liquid chromatography/mass spectrometry/mass spectrometry for the direct detection of ultra-trace levels of contaminants in water. *Rapid Communications in Mass Spectrometry* 2006, 20(15), 2281–2285.

70. Ferrer, I., Garcia-Reyes, J.F., Mezcua, M., Thurman, E.M., Fernandez-Alba, A.R., Multi-residue pesticide analysis in fruits and vegetables by liquid chromatography-time-of-flight mass spectrometry. *Journal of Chromatography A* 2005, 1082(1), 81–90.

71. Garcia-Reyes, J.F., Ferrer, I., Thurman, E.M., Molina-Diaz, A., Fernandez-Alba, A.R., Searching for non-target chlorinated pesticides in food by liquid chromatography/time-of-flight mass spectrometry. *Rapid Communications in Mass Spectrometry* 2005, 19(19), 2780–2788.

10 Sampling and Analysis of Pesticides in the Atmosphere

Maurice Millet

CONTENTS

10.1   INTRODUCTION

The intensive use of pesticide leads to the contamination of all compartments of the environment. The atmosphere is known to be a good pathway for the worldwide dissemination of pesticides. Pesticides can enter into the atmosphere by "spray drift" during application, postapplication volatilization from soils and leaves, and by wind erosion when pesticides are sorbed to soil particles and entrained into the atmosphere on windblown particles.[1] There are few data on the significance of this pathway, and on the quantitative effects of soil and environmental factors that influence this process.[2] This process is most important for herbicides as they are applied either at pre-emergence or postemergence at an early growth stage of the crops (e.g., summer cereals, maize) when there is low soil coverage.[3]

Spray drift phenomenon can be defined as the proportion of the output from an agricultural crop sprayer that is deflected out of the target area by the action of wind. Drift losses can occur either as vapor or as droplets.[4] These particles are so small that they do not reach the target area and cannot be effectively captured by drift collectors. The proportion of a pesticide spray application that exists in the gas phase and as aerosol is therefore a loss, and should be considered in addition to drift. Vapor drift could be a problem with volatile active substances, with applications at high temperatures and strong wind conditions to nontarget aquatic and terrestrial organisms. Other factors such as spray droplet size, the height of spraying, the direction of the wind, and development of the vegetation can influence strongly the drift of pesticides to nontarget areas during application. In general the drift reduces when the development of the vegetation is high. Some authors state that losses of pesticides through spray drift can vary between 1% and 30% of the quantities applied.[5] Drift can be calculated using drift tables.[6]

Volatilization is defined as the physicochemical process by which a compound is transferred to the gas phase. It can result from evaporation from a liquid phase, sublimation from a solid phase, evaporation from an aqueous solution, or desorption from the soil matrix. Volatilization of pesticides from soil is governed by a combination of several factors[2] such as the physicochemical properties of the compounds (vapor pressure, solubility, adsorption coefficient, molecular mass, chemical nature, and reactivity), the soil properties (water content, soil temperature, soil density, organic matter content, clay content/texture, pH), the meteorological conditions (air temperature, solar radiation, rain/dew, air humidity, wind/turbulences), and agricultural practices (application rate, application date, ploughing/incorporation, type of formulation). Most of these parameters are closely linked and interact with each other. Their combined effects on the volatilization process are therefore far from linear.[7]

Pesticide volatilization from plant surfaces may occur very quickly after treatment. Volatilization of more than 90% of the application dose was observed. Even though the rate of volatilization from plants seems to be higher than that from soil, little data are available, as pointed out by many authors.[7] Volatilization from plant volatilization is up to three times as high as soil volatilization under similar meteorological conditions.

Vapor pressure is a key factor driving volatilization and is therefore a good trigger for screening compounds in a tiered risk assessment scheme. Another important factor is Henry's law coefficient (H), mostly given as the result of $(V_p \times M)/S$ where V_p is the vapor pressure, M is the molecular weight, and S is the water solubility. Under liquid conditions, H may also be used as a trigger and is therefore only effective directly after spraying, when the spraying solution has not yet dried.

The FOCUS Air group[8] has defined that substances that are applied to plants and have a vapor pressure less than 10^{-5} Pa (at 20°C), or are applied to soil and have a vapor pressure less than 10^{-4} Pa (at 20°C), need not be considered in the short-range risk assessment scheme. Substances that exceed these triggers require evaluation at the second tier, which is done by modelling.

When in the atmosphere pesticides can be distributed between the gas and particle phases depending on their physical and chemical properties (vapor pressure, Henry's law constant, etc.) and of environmental and climatic conditions (concentration of particles, temperature, air humidity, etc.). The knowledge of the gas/particle partitioning of pesticides is important since this process affects the potential removal of pesticides by wet and dry deposition and by photolysis. It can also, together with photolysis, play a role in the atmospheric transport of pesticides to short or long distances.

Compounds adsorbed to particulate matter are mostly found in wet deposition.[9] Compounds mostly in the vapor phase are likely to be more evenly divided between wet and dry deposition. Pesticides in the gas phase generally have longer atmospheric residence time. In this case, the rate of removal is strongly influenced by Henry's law constant (H). Compounds with a low H value will be more selectively washed out by rain.

On the other hand, the gaseous organic compounds with high H values will demonstrate long atmospheric residence time since they will not be removed neither by precipitation nor by particle deposition.[10]

The capacity for pesticides to be transported over long distances is also a function of their atmospheric lifetime, which is the result of emission and removal processes. In fact long-range transport of pesticides will occur when compounds have a significant lifetime.[11] Photooxidative processes (indirect photolysis) and light-induced reactions (direct photolysis) are the main transformation pathways for pesticides in the atmosphere. According to Finlayson-Pitts and Pitts,[12] four processes can be considered (the first three being photooxidative processes and the fourth being direct photolysis): reactions with OH-radicals, which are considered to be the major sink for most air pollutants, including pesticides,[13,14] due to the reaction with double bonds, the H abstractive power of hydroxyl, and its high electrophilicity,[15–17] reactions with O_3 (ozone), which are only efficient with molecules with multiple bonds,[13] reactions with NO_3-radicals, which are potentially important for compounds containing double bonds,[11] and direct photolysis, which acts only with molecules absorbing at $\lambda > \sim290$ nm which corresponds to the cutoff region of sunlight UV radiation.

Pesticides are present in the atmosphere in the gas phase (from volatilization processes) and in the particle phase (including aerosols). For pesticides in the gas phase, removal by chemical transformation processes involves photolysis, reactions with OH radicals, NO_3 radicals, O_3, and possibly with HNO_3 in polluted urban areas. In the particle phase, reactions with OH-radicals, O_3, and photolytic reactions are assumed to be the major chemical transformation processes based on information from the gas phase.[11]

"Deposition" is defined as the entry path for transport of airborne substances from the air as an environmental compartment to the earth's surface, i.e., to an aquatic or terrestrial compartment. It is also a loss pathway for substances from the air. Dry and wet deposition should be considered separately because they are subject to different atmospheric physical processes. In essence, wet deposition is the removal of pesticides in precipitation, while dry deposition of particulates is due to a settling out effect (often referred to as the deposition velocity). Indeed, the removal rate of pesticides from the atmosphere by dry and wet deposition depends partly on the Henry's law coefficient, to some extent on their diffusivity in air, and on meteorological conditions (wind speed, atmospheric stability, precipitation) and on the conditions of the surface (for dry deposition only).

The presence of modern pesticides, such as 2,4-D, in rainwater was first reported, in the mid 1960s, by Cohen and Pinkerton[18] but until the late 1980s, no special attention was given to this problem. Van Dijk and Guicherit[18] and Dubus et al.[19] published, in the beginning of the 2000s, reviews on monitoring data of current-used pesticides in rainwater for European countries. Some other measurements were also performed in the United States[20,21] and in Japan[22] and more recently in France,[23] Germany,[24,25] Poland,[26] Belgium,[27] and Denmark.[28]

Pesticides are generally present in precipitation from few ng L^{-1} to several μg L^{-1} [18] and the highest concentrations were detected during application of pesticides to crops.

Generally, local contamination of rainwater by pesticides was observed, but some data show contamination of rainwater by pesticides in regions where the pesticides are not used.[18] These data suggest the potentiality of transport and consequently the potentiality of the contamination of ecosystems far from the site of the pesticide application.

The actual concentration of a pesticide in rainwater or wet deposition of a pesticide does not only depend on its properties and the meteorological conditions at the observational site, but also on the geographical distribution of the amount of pesticide applied, the type of surface onto which it is applied, and the meteorological conditions in the area of which the emissions contribute to the concentration at the measuring site.

From studies preformed on the monitoring of the contamination of the atmosphere by pesticides, it appears that atmospheric concentrations were function of applied quantities, physical–chemical properties of pesticides, climatic and soil conditions, and site localization.

In general all of the year, residues of pesticides in the atmosphere were very low in comparison with volatile organic compounds (VOCs) or PAHs in atmospheric concentrations. Some very punctual peaks of pollution have been observed with levels sometimes higher than other pollutants during application periods. However, this strong contamination remains very short in terms of duration. These assumptions are in accordance with EPCA report,[29] which concludes that extremely low levels of Crop Protection Products can be detected in rain and fog, redeposition rates are about 1000 times lower than normal application rates less than 1 g per year, levels detected in precipitation and air pose no risk to man and any environmental impact, particularly to aquatic organisms, is extremely unlikely.

Pesticides can also contaminate indoor air as a result of indoor as well as outdoor applications (residential and occupational uses). It has been demonstrated that pesticide residues may translocate from their original points of application as vapors, bound to particles, or through physical transport processes. The principal factors that influence their movement are the compounds' physicochemical properties, the substrates contacted, and the physical activities of humans and their pet animals.[30]

Bouvier et al.[31] state that domestic pesticide uses include pet treatments, extermination of household pests, removal of lice, and garden and lawn treatments while professional uses include crop, greenhouse, cattle and pet treatments, but also pest control operations in buildings.

Barro et al.[32] used pyrethroids because they are widely applied as insecticides in households and greenhouses, as well as for the protection of crops. Releases into the air represent the most important emission pathway for these insecticides. Because of that, inhalation is an important route of exposure for humans, especially just after spraying application in domestic indoors or agricultural close areas. The Occupational Safety and Health Administration (OSHA) has established the occupational exposure limit for an 8 h workday, 40 workweek, at 5 mg of pyrethrins and pyrethroids per cubic meter of workplace air (5 mg m^{-3}).

Bouvier et al.[31] summarized the exposure studies of the general population, conducted in different countries, including residential and personal measurements. The results from these studies suggest that people were exposed at home to various insecticides, such as organochlorines, organophosphates, and pyrethroids and also to wood preservatives, some herbicides and fungicides.

10.2 MONITORING OF PESTICIDES IN THE ATMOSPHERE

Pesticides are present in the atmosphere at very low concentrations, except when measurements are performed directly near the field where treatments are performed. Because of the low concentrations, high volumes of air, rain, or fog are needed to assess the atmospheric levels together with concentration and purification steps before analysis.

10.2.1 SAMPLING AND EXTRACTION OF PESTICIDES IN AMBIENT AIR

Methods used for the sampling and extraction of pesticides in the atmosphere are not diverse. Generally, the sampling is carried out by pumping the air onto traps and extraction of pesticides on traps are performed by solid–liquid extraction.

10.2.1.1 Sampling of Pesticides in Ambient Air

Pesticides in ambient air are sampled by conventional high-volume samplers on glass fiber or quartz filters followed by solid adsorbents, mainly polyurethane foam (PUF) or polymeric resin (XAD-2 or XAD-4), for the collection of particle and gas phases, respectively.

Depending on the high-volume sampler used, length or diameter of filters varied generally between 200 × 250 mm (Andersen sampler), 102 mm diameter (PS-1 Tisch

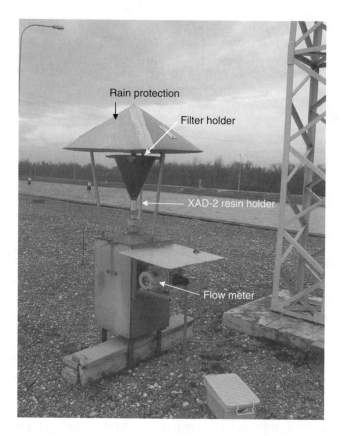

FIGURE 10.1 High-volume sampler developed in the LPCA. (From Scheyer, A., PhD thesis, University of Strasbourg, 2004.)

Environmental, Inc., Village of Cleves, OH) to 300 mm (LPCA collector, home made) diameter (Figure 10.1). Generally 10–20 g of XAD-2 resin, a styrene–divinylbenzene sorbent that retains all but the most volatile organic compounds, is employed to trap the gaseous phase and is used alone or sandwiched between PUF plugs (75 mm × 37 mm). White et al.[33] used 100 g of XAD-2 resin between 2 PUF plugs.

XAD has been previously used to collect a variety of pesticides including diazinon, chlorpyrifos, disulfoton, fonofos, mevinphos, phorate, terbufos, cyanazine, alachlor, metolachlor, simazine, atrazine, deethyl atrazine, deisopropyl atrazine, molinate, hexachlorobenzene, trifluralin, methyl parathion, dichlorvos, and isofenphos.[34]

In a recent study, the efficiency of trapping gaseous current-used pesticides on different traps, including PUF, XAD-2 resin, XAD-4 resin, and PUF/XAD-2/PUF and PUF/XAD-4/PUF sandwich, was determined.[35] From this study, it appears that XAD-2 and PUF/XAD-2/PUF are the better adsorbent for current-used pesticides (27 pesticides tested) and the sandwich form is slightly more efficient than XAD-2 alone while PUF plugs is the less efficient.

The duration of sampling depends mainly on the purpose of the sampling and on the detection limits of the analytical method used. Generally, sampling varied between 24 h and 1 week and the total air pumped varied between 250 m^3,[36,37] 525–1081 m^3,[33] and 2500 m^3 of air.[38] A sampling time of about 24 h is generally sufficient to reach the detection limit of pesticides in middle latitude atmosphere and avoid clogging-up the filters.[39–41]

10.2.1.2 Extraction of Pesticides in Ambient Air

After sampling, traps are separately extracted by using Soxhlet extraction with different solvents used alone, such as acetone,[38] or as a mixture, such as 36% ethyl-acetate in n-hexane,[42] (85:15) n-hexane/CH$_2$Cl$_2$,[40,43] 25% CH$_2$Cl$_2$ in n-hexane,[44] (50:50) n-hexane/acetone,[34] or (50:50) n-hexane/methylene chloride[36,37] for 12–24 h. In some studies, the ASTM D4861–91 method was followed.[33]

After Soxhlet extraction, extracts were dried with sodium sulfate and reduced to 0.5 mL using a Kuderna Danish concentrator followed by nitrogen gas evaporation[42] or were simply concentrated to about 1 mL by using a conventional rotary evaporator.[36,37,41]

Depending on the authors and on the analytical method used, a cleanup procedure can be performed after concentration. Foreman et al.[42] passed extracts through a Pasteur pipet column containing 0.75 g of fully activated Florisil overlain with 1 cm of powdered sodium sulfate. Pesticides were eluted using 4 mL of ethyl acetate into a test tube containing 0.1 mL of a perdeuterated polycyclic aromatic hydrocarbon used as internal standard. The extract was evaporated to 150 mL using nitrogen gas, transferred to autosampler vial inserts using a 100 mL toluene rinse. Sauret et al.[41] and Scheyer et al.[36,37] used GC–MS–MS for the analysis of airborne pesticides and they do not perform a cleanup procedure.

Badawy,[44] who used GC–ECD for the analysis of pesticides in particulate samples, concentrated Soxhlet extracts to 5 mL and firstly removed elemental sulphur by reaction with mercury. After that, extracts were quantitatively transferred to a column chromatography for separation into two fractions using 3 g of 5% deactived alumina. Fraction one (FI), which contains chlorobiphenyls, chlorobenzenes, and hexachlorocyclohexane, was eluted with 16 mL of n-hexane. Second fraction (FII), includes permethrin, cypermethrin, deltamethrin, and chloropyrophos (rosfin), was eluted with 6 mL of 20% ether in hexane.

In the 1990s, a method using fractionation by HPLC on a silica column was used for the cleanup of atmospheric extracts.[45,46] After extraction, samples were fractionated on a silica column using an n-hexane/MTBE gradient for isolating nonpolar, medium-polar, and polar pesticides, which were analyzed by specific methods including GC–ECD and HPLC–UV. In the method developed by Millet et al.,[46] three fractions were obtained; the first one contains pp′DDT, pp′DDD, pp′DDE, aldrin, dieldrin, HCB, fenpropathrin, and mecoprop, the second one contains methyl-parathion, and the third one contains aldicarb, atrazine, and isoproturon. This step was necessary since fractions 2 and 3 were analyzed by HPLC–UV, a nonspecific method.

10.2.1.3 Cleaning of Traps for the Sampling of Pesticides in Ambient Air

Traps (XAD and PUF foam) were precleaned before use by Soxhlet successive cleaning steps or by one cleaning step depending on authors. Scheyer et al.[36,37] precleaned the filters and the XAD-2 resin by 24 h Soxhlet (50:50) with n-hexane/ CH_2Cl_2 and stored them in clean bags before use, while Peck and Hornbuckle[34] precleaned the XAD-2 resin with successive 24 h Soxhlet extractions with methanol, acetone, dichloromethane, hexane, and 50/50 hexane/acetone prior to sampling.

Some authors (i.e., Coupe et al.[21]) used a heater to clean filters (backing at 450°C for example). In all cases, a blank analysis is required to check the efficiency of the cleaning and storage before use.

The ultrasonic bath is poorly used for the extraction of filters and resins after sampling. Haraguchi et al.[39] used this technique for their study of pesticides in the atmosphere in Japan.

10.2.2 Sampling and Extraction of Pesticides in Rainwater Samples

10.2.2.1 Sampling of Rainwater

Rainwater samples are collected using different systems depending on studies and authors. Asman et al.[47] and Epple et al.[24] used for their study on pesticides in rainwater in Denmark and Germany, respectively, a cooled wet-only collector of the type NSA 181/KE made by G.K. Walter Eigenbrodt Environmental Measurements Systems (Konigsmoor, Germany). It consists of a glass 2(Duran) funnel of ~500 cm diameter connected to a glass bottle that is kept in a dark refrigerator below the funnel at a constant temperature of 4°C–8°C. A conductivity sensor is activated when it starts to rain and then the lid on top of the funnel is removed. At the end of the rain period the lid is again moved back onto the funnel. With this system, no dry deposit to the funnel during dry periods is collected. Millet et al.[48] and Scheyer et al.[49,50] used also a wet-only rainwater sampler built by Précis Mécanique (France). This collector is agreed by the French Meteorological Society (Figure 10.2). It consists of a PVC funnel of 250 mm diameter connected to a glass bottle kept in the dark. No freezing of the bottle was installed and the stability of the sample was checked for one week in warm months. This collector is equipped with a moisture sensor which promotes the opening of the lid when rain occurs.

Quaghebeur et al.[27] used for their study in Belgium, a bulk collector made in stainless steel by the FEA (Flemish Environmental Agency, Ghent, Belgium). The sampler consists of a funnel ($D \sim 0.5$ m) the sides of which meet at an angle of 120°. The outlet of the funnel is equipped with a perforated plate ($D \sim 0.05$ m). The holes have a diameter of 0.002 m. The funnel is connected with a collecting flask.

Haraguchi et al.[22] and Grynkiewicz et al.[26] used a very simple bulk sampler which consists of a stainless steel funnel (40 cm or 0.5 m^2 diameter, respectively) inserted in a glass bottle for their study of pesticides in rainwater in Japan and Poland, respectively.

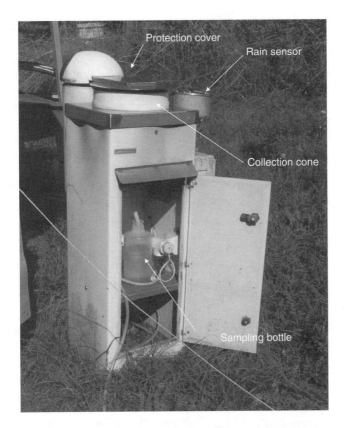

FIGURE 10.2 Wet-only rainwater collector. (From Scheyer, A., PhD thesis, University of Strasbourg, 2004.)

10.2.2.2 Extraction of Pesticides from Rainwater

Extraction of pesticides was made using the conventional method used for water; liquid–liquid extraction (LLE), solid-phase extraction (SPE), and solid-phase microextraction (SPME).

10.2.2.2.1 Liquid–liquid extraction

This method was used by many authors. Chevreuil et al.[51] extracted pesticides from rainwater by LLE three times with a mixture of 85% n-hexane/15% methylene chloride. Recoveries obtained were higher than 95% except for atrazine degradation metabolites (>75%). Depending on the chemical nature of the pesticide, Quaghebeur et al.[27] used different LLE extraction methods. Organochlorine pesticides, polychlorinated biphenyls, and trifluralin were extracted from the rainwater sample using petroleum ether (extraction yield > 80%) while organophosphorous and organonitrogen compounds (i.e., atrazine) were extracted with dichloromethane (extraction yield > 80%).

Kumari et al.[52] for their study of pesticides in rainwater in India used the following procedure to extract pesticides from rainwater. Representative (500 mL) sample of water was taken in 1 L separatory funnel and 15–20 g of sodium chloride was added. Liquid–liquid extraction (LLE) with 3 × 50 mL of 15% dichloromethane in hexane was performed. The combined organic phases were filtered through anhydrous sodium sulphate and this filtered extract was concentrated to near dryness on rotary vacuum evaporator. Complete removal of dichloromethane traces was ensured by adding 5 mL fractions of hexane twice and concentrating on gas manifold evaporator since electron capture detection (ECD) was used for the analysis of some pesticides.

All these authors do not use a cleanup procedure after LLE of rainwater samples mainly since they used very specific methods such as GC–ECD, GC–NPD, and GC–MS.

10.2.2.2.2   Solid-phase extraction

Solid-phase extraction (SPE) was used by Haraguchi et al.,[22] Millet et al.,[46] Coupe et al.,[21] Grynkiewicz et al.,[26] Bossi et al.,[53] and Asman et al.[28]

These authors used XAD-2 resin or C_{18} cartridges and they follow the classical procedure of SPE extraction consisting of conditioning of the cartridge, loading of the sample, and elution of pesticides by different solvents. Haraguchi et al.[22] used dichloromethane for the elution of pesticides trapped on XAD-2 cartridge while Asman et al.[28] used 5 mL of ethylacetate/hexane mixture (99:1 v/v) for the elution of pesticides from Oasis HLB 1000 mg cartridges (Waters) before GC–MS analysis. A 200 μL volume of isooctane was added to the extract as a keeper to avoid losses of more volatile compounds during evaporation. For LC–MS–MS analysis, these authors used Oasis HLB 200 mg cartridges (Waters) and pesticides were eluted with 8 mL methanol. The extracts were evaporated to dryness and then redissolved in 1 mL of a Millipore water/methanol mixture (90:10 v/v) before LC–MS–MS in ESI mode analysis.

Grynkiewicz et al.[26] used Lichrolut EN 200 mg cartridges (Merck) for the extraction of pesticides in rainwater. Pesticides were eluted with 6 mL of a mixture of methanol and acetonitrile (1:1). After it, a gentle evaporation to dryness under nitrogen was performed before analysis by GC–ECD (organochlorine pesticides) and GC–NPD (organophosphorous and organonitrogen).

Epple et al.[24] have compared two kinds of SPE cartridges for the extraction of pesticides in rainwater samples and their analysis by GC–NPD: Bakerbond C_{18} solid-phase extraction cartridges (Baker, Phillipsburg, NJ, USA) and Chromabond HR-P SDB (styrene–divinyl–benzene copolymer) cartridges 200 mg (Macherey-Nagel, Duren, Germany). The latter one is more efficient for polar compounds, such as the triazine metabolites. Prior to SPE extraction, rainwater samples were filtered by a glass fiber prefilter followed by a nylon membrane filter 0.45 nm. After that, filtered rainwater was filled with 5% of tetrahydrofuran (THF).

Elution was carried out with 5 mL of THF, the solvent evaporated, and the residue dried with a gentle stream of nitrogen and then dissolved in 750 μL of ethyl acetate. The sample was then cleaned by small silica-gel columns to remove polar components from precipitation samples. For this, 3 mL silica-gel columns

(5 × 0.9 cm boro silicate glass) with Teflon frits were used. The silica-gel type (60, 70–230 mesh, Merck) was dried overnight at 130°C, mixed with 5% by weight of water, and transferred into glass tubes as a mixture with ethyl acetate, so that each column contained 0.8 g of silica gel. The sample (750 μL) was transferred to the column and eluted with 4 mL of ethyl acetate before GC–NPD analysis.

Recoveries of the method for all the pesticides studied are summarized in Table 10.1.

TABLE 10.1
Relative Standard Deviations, RSD, Recoveries, Rec., and Determination Limits, DL, ($n = 10$, $P = 95\%$) for Determination of Pesticides in Wet-deposition Samples

Pesticide	Bakerbond C_{18}			Chromabond HR-P SDB		
	RSD (%)	Rec. (%)	DL (ng L^{-1})	RSD (%)	Rec. (%)	DL (ng L^{-1})
Desethyl atrazine 2	1.64	31	15	1.39	102	13
Desethyl terbuthylazine 2	1.91	95	19	1.20	102	12
Simazine 2	1.09	98	10	1.15	98	11
Atrazine 2	1.10	99	10	1.19	99	11
Propazine 2	1.38	101	13	1.56	98	15
Terbuthylazine 2	1.84	97	18	1.48	97	15
Diazinon 1	2.58	89	5	3.62	87	6
Triallate 3	2.64	106	110	3.36	85	130
Sebuthylazine 2	1.23	96	11	—	—	—
Metribuzin 3	4.21	81	95	3.35	79	75
Parathion-methyl 1	3.14	105	6	2.21	83	4.5
Metalaxyl 4	1.04	99	75	1.82	93	120
Prosulfocarb 3	2.48	100	60	3.81	94	90
Metolachlor 4	1.35	104	105	1.03	98	80
Parathion 1	3.11	103	6	3.16	82	6
Metazachlor 3	1.36	98	30	1.03	103	25
Pendimethalin 3	2.17	90	50	52.8	41	1300
Triadimenol 3	1.45	100	60	1.82	87	75
Triadimenol 3	1.42	101	60	2.44	88	100
Napropamide 3	1.13	101	35	—	—	—
Flusilazol 3	1.70	97	40	2.52	81	60
Propiconazol 3	1.90	94	75	2.74	77	110
Propiconazol 3	1.25	98	50	1.52	92	60
Tebuconazole 3	2.26	93	50	3.46	74	75
Bifenox 4	0.95	83	210	—	—	—
Pyrazophos 1	6.10	103	25	3.31	95	15
Prochloraz 4	3.32	86	300	—	—	—

Source: From Epple, J. et al., *Geoderma*, 105, 327, 2002. With permission.
Concentration ranges: (1) 5–50 ng L^{-1}; (2) 20–200 ng L^{-1}; (3) 100–1000 ng L^{-1}; (4) 250–2500 ng L^{-1}.
Enantiomeric pairs numbered in the order of their elution times.

Millet et al.[48] used also SPE extraction on Sep-Pak C_{18} cartridges (Waters) and elution with methanol for the analysis of pesticides in rainwater. Before analysis, they performed a HPLC fractionation as described earlier.[46]

10.2.2.2.3   Solid-phase microextraction

Among studies on pesticides in precipitation, extraction of pesticides was performed using classical developed methods for surface water. No special development was specifically done for atmospheric water. More recently, Scheyer et al.[49,50] used SPME for the analysis of pesticides in rainwater by GC–MS–MS. They used direct extraction for stable pesticides and a derivatization step coupled to SPME extraction for highly polar pesticides or thermo labile pesticides. These developments were derived from studies in water. SPME is a very interesting method for a fast and inexpensive determination of organic pollutants in water, including rainwater. The main advantage of SPME techniques is that it integrates sampling, extraction, and concentration in one step. This method is actually poorly used for the extraction of organic pollutants in atmospheric water probably because of low levels commonly found in precipitation.

For the evaluation of the spatial and temporal variations of pesticides' concentrations in rainwater between urban (Strasbourg, East of France) and rural (Erstein, East of France) areas, Scheyer et al.[49] have developed a method using SPME and ion trap GC–MS–MS for the analysis of 20 pesticides (alachlor, atrazine, azinphos-ethyl, azinphos-methyl, captan, chlorfenvinphos, dichlorvos, diflufenican, α and β-endosulfan, iprodione, lindane, metolachlor, mevinphos, parathion-methyl, phosalone, phosmet, tebuconazole, triadimefon, and trifluralin) easily analyzable by gas chromatography (GC). For some seven other pesticides (bromoxynil, chlorotoluron, diuron, isoproturon, 2,4-MCPA, MCPP, and 2,4-D), Scheyer et al.[50] used SPME and GC–MS–MS but they add, prior to GC analysis, a derivatization step. SPME was chosen because it permits with accuracy a rapid extraction and analysis of a great number of samples and MS–MS enables the analysis of pesticides at trace level in the presence of interfering compounds without losing identification capability because of a drastic reduction of the background noise.

The first step in developing a method for SPME is the choice of the type of fiber. To do that, all other parameters are fixed (temperature, pH, ionic strength, etc.). The fiber depth in the injector was set at 3.4 cm and the time of the thermal desorption in the split–splitless injector was 5 min at 250°C, as recommended by Supelco and confirmed by Scheyer et al.[49] Deeper fiber in the injector gave rise to carryover effects and less deeper fiber caused loss of response. The liner purge was closed during the desorption of the analytes from the SPME fiber in the split–splitless injector (2 min delay time). A blank must be carried out with the same fiber to confirm that all the compounds were desorbed within 5 min of thermal desorption.

In the method of Scheyer et al.,[49] extractions were performed by immersion of the fiber in 3 mL of sample, with permanent stirring and temperature control at 40°C, during 30 min. Indeed, a headspace coating of the fiber is possible but, in the case of pesticides, this method cannot be used with efficiency because of the general low volatility of pesticides from water (Figure 10.3). However, for some volatile pesticides

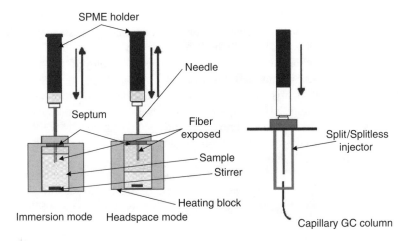

FIGURE 10.3 Principle of SPME extraction. (From Scheyer, A., PhD thesis, University of Strasbourg, 2004.)

such as some organophosphorous pesticides, headspace coating of the fiber can be developed.

Since the SPME technique depends on an equilibrium process that involves the adsorption of analytes from a liquid sample into the polymeric phase according to their partition coefficient, the determination of the time (duration of extraction) required to reach this equilibrium for each compound is required.

The equilibration rate is limited by the mass transfer rate of the analyte through a thin static aqueous layer at the fiber–solution interface, the distribution constant of the analyte, and the thickness and the kind of fiber coating[54] Moreover analytes with high molecular masses are expected to need longer equilibrium times because of their lower diffusion coefficient since the equilibrium time is inversely proportional to the diffusion coefficient.[55]

The temperature and the duration of extraction are associated since when increasing the temperature, it is possible to reach the equilibrium faster. Temperature can also modify the partition coefficient of the fiber and consequently decrease the amount of extracted compound.[54] A compromise has to be determined between the temperature and the duration of the extraction in order to obtain a sensitive method for the analysis of pesticides in rainwater.

To increase the extraction efficiencies, it is possible to add some salts which have for effect to modify the ionic strength and to decrease the solubility of the molecules in the water.

SPME of Pesticides in Rainwater with a Derivatisation Step. The SPME technique, firstly developed for GC analysis, integrates sampling, extraction, and concentration in one step followed by GC analysis, even the use of HPLC is possible.

However, many pesticides such as phenyl ureas (PUHs), phenoxy acids, or carbamates cannot be analyzed directly by GC because of their low volatility or thermal instability. GC analysis of these molecules requires a derivatization step to stabilize or increase their volatilities.

The use of SPME with derivatization is not commonly used for pesticides, especially in the simultaneous determination of many class of pesticides such as phenyl ureas, phenoxy acids, phenolic herbicides, etc.

Derivatization (sylilation, alkylation, acylation) is employed for molecules where properties cannot permit their direct analysis by GC.[56,57]

Alkylation with PFBBr is a very common reaction and permits the derivatization of molecules containing NH groups (chlorotoluron, diuron, and isoproturon), –OH groups on aromatic ring (bromoxynil) and –COOH groups (MCPP, 2,4-D, 2,4-MCPA). The mechanism of reaction on a molecule containing a hydrogen acid is a bimolecular nucleophile substitution (SN_2).[58]

After extraction, samples present in organic solvents are derivatized by addition of a small amount of derivatizing agent. In the case of SPME, no solvent is present and some approaches have been tested for combining derivatization and SPME.[54]

Derivatization directly in the aqueous phase followed by SPME extraction (direct technique).

Derivatization on the fiber. This method consists of headspace coating of PFBBr for 10′ of the fiber followed by SPME extraction. In this case, extraction and derivatization are made simultaneously.

Extraction of the analytes present in water followed by derivatization on the fiber or onto the GC injector.

For the direct technique, it is necessary to adjust the pH of the water below of the pKa of the molecules to be derivatized (i.e., <2.73, which is the lowest pKa value for 2,4-D) since in this case they are protonated and consequently derivatization becomes possible.

Scheyer et al.[50] clearly showed that the exposure of the fiber to the derivatization reagent followed by extraction gave the better results and this method was used for the analysis of the seven pesticides, which required derivatization before analysis by GC, in rainwater.

10.2.3   Evaluation of Soil/Air Transfer of Pesticides (Spray Drift and Volatilization)

As shown in the precedent paragraph, pesticides in ambient air are commonly sampled by high-volume samplers on filters and adsorbents (PUFs, XAD-2). After sampling, compounds trapped on the adsorbent must be released before determination. For this, a solvent for desorption with Soxhlet or ultrasonic extraction, followed by a concentration step, is commonly used. It is generally time consuming and the different steps (extraction, cleanup, concentration, etc.) induced many losses and subsequently increased detection limits.

Even if the association of high-volume sampling and solvent extraction is accurate for the measurement of ambient level of trace contaminants, this method cannot be applied to assess spray drift and volatilization processes. Indeed, this kind of study required a short sampling periodicity to be close to the variation of atmospheric dissipation processes.

As quoted by Majewski,[59] estimation of volatilization rate in the field is classically carried out using the aerodynamic profile. It gives an estimate of this mass transfer under actual field conditions and its variation with time. This method, based on the measurements of vertical profiles of pesticide concentrations in the atmosphere, needs a good precision for the estimation of these concentrations. Also, the determination of concentration gradients requires the measurements of concentrations at four heights at least and consequently greatly increases the number of samples to analyze.

Thermal desorption can present a novel approach since it substantially simplifies analyses (no concentration step is needed) and increases sensitivity (a large part of the preconcentrated material may be recovered for determination), and detection limits and background noise are lower because of the disappearance of solvent components. Moreover, this technique is easily automatable. Because of these aspects, it seems to be an interesting alternative to solvent extraction to assess atmospheric transfer of pesticides during and after application. Thermal desorption has often been used for the analysis of VOCs in indoor and outdoor atmospheres. Thermal desorption for the analysis of pesticides has already been described for the volatile and stable pesticides trifluralin and triallate[60,61] in field measurements and atrazine in laboratory volatilization experiments.[62]

Thermal desorption was extended to six pesticides in order to evaluate atmospheric transfer of pesticides following application (spray drift and volatilization).[63] To the best of my knowledge, this was the first time that a thermal desorption unit–GC was interfaced with a mass selective detector to provide both pesticide quantification and confirmation. From the first results obtained in this study, it appears that thermal desorption followed by GC–mass spectroscopy (MS) analysis is accurate and sensitive but presents some limitations, especially as a result of the physicochemical properties of pesticides such as thermal stability and low volatility.

The principle of thermal desorption is detailed in Figure 10.4. It consists of two steps: (1) primary desorption, which consists of desorption of pesticides adsorbed on

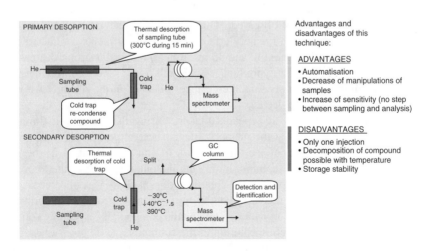

FIGURE 10.4 Principle of thermal desorption.

the resin of the sampling tube and accumulation on a trap maintained at $-30°C$ by peltier effect and (2) secondary desorption, which consists of the rapid heating of the trap before introduction on the GC column maintained at 50°C.

Application of thermal desorption for pesticides presents some difficulties mainly because of the very low volatility of some of them. Briand et al.[64] have extended the method developed by Clément et al.[63] to deethylatrazine (DEA), deisopropylatrazine (DIA), carbofuran, cyprodinil, epoxyconazole, iprodione, 3,5-dichloroaniline, lindane, α-HCH, metolachlor, terbuconazole, and trifluralin.

The main problem dealing with the extraction of pesticides is the memory effects on the thermal desorption system. This problem was located on the cold trap containing glass wool.

To visualize an eventual memory effect for the 10 pesticides and metabolites under study, first experiments were performed as follows: A 400 ng amount of each compound was deposited at the end of a tube which was placed in the thermal desorption unit followed by four empty tubes (tubes without adsorbent) which were analyzed following the spiked tube.

To check the influence of the amount of pesticides accumulated on the cold trap, two parameters can be modified in the ATD system, the inlet-split flow rate (initially at 0 mL min^{-1}), situated between the tube and the trap and the outlet-split flow rate (initially at 20 mL min^{-1}), located after the cold trap just before injection into the analytical column. This last flow rate imposed the gas velocity in the cold trap.

To evaluate the memory effect, two kinds of experiments were performed: one modifying inlet-split flow rate of 10 mL min^{-1} (outlet-split flow rate 20 mL min^{-1}) and second modifying outlet-split flow rate (inlet-split flow rate 0 mL min^{-1}). From the first experiment (inlet-split flow rate of 10 mL min^{-1}), a strong decrease of the memory effect in all empty tubes, analyzed after sample tube, was observed since it remained only for cyprodinil (0.93%) and tebuconazole (1.70%). Thus, the amount of pesticides reaching the cold trap seems to be the reason for the observed memory effect. However, a strong loss of sensitivity (20%–60%) especially for the most volatile compounds (DIA, DEA, α-HCH, trifluralin, carbofuran, lindane, atrazine, and alachlor) was observed. Thus, increasing the inlet-split flow rate cannot be used to resolve the memory effect problem. Experiments conducted with increasing outlet-split flow rate (30 and 35 mL min^{-1}) induced a strong decrease of the memory effect: 0.90% for cyprodynil with 30 mL min^{-1}, 1% for iprodione with 35 mL min^{-1} in the first empty tube. Percentages obtained in the second tube were not significant and can be neglected.

From these experiments, it appeared that increasing the outlet-split flow rate from 20 to 30 mL min^{-1} limit the memory effect. These outlet splits correspond to 5% and 3.3%, respectively, of the total amount of spiked compound in the tube actually injected into the GC-column. Increasing the outlet-split flow rate to 35 mL min^{-1} will not be accurate, since a too great loss of sensitivity was observed.

Loss of sensitivity when increasing outlet-split can be compared to the principle of GC split/splitless injector, in which more volatile compounds (especially solvent) are preferentially removed before entering in the column.

Experiments were conducted without glass wool in the cold trap to remove the memory effect completely. These experiments showed that the memory effect was very low (maximum 0.10% for epoxyconazole) and disappeared completely in the second empty tube. The resolution and sensitivity of each pesticide and metabolite under study were not affected by this removal since no significant decrease of areas was observed.

An experiment was performed by changing desorption rate of the cold trap from >40 to 5°C s^{-1}. This change greatly improved the peak resolution. From the different tests performed, it appeared that the memory effect was located in the cold trap, and that it could be partially removed by using an empty trap. Following these observations, complementary tests were performed with decreased outlet-split flow rates (25 and 20 mL min^{-1}) to increase the method sensitivity. Tests performed with spiked tube at 400 ng showed recurrence of the memory effect with an outlet-split flow rate under 25 mL min^{-1}; therefore, no more tests were conducted. Decreasing the outlet-split flow rate could be envisaged for very low amounts of pesticides to improve method performances.

ATD optimal conditions for the quantitative desorption of the 10 pesticides and metabolites under study are presented in Table 10.2.

This study used an ATD 400 from Perkin-Elmer Corp. (Norwalk, CT, USA) where some temperature ranges are limited (transfer line, valve). With Turbomatrix new systems, temperature can be increased and can improve the efficiency of thermal desorption for pesticides analysis.

10.2.3.1 Method Performances

10.2.3.1.1 ATD–GC/MS repeatability and calibration range

For repeatability experiments, five assays were conducted successively with conditions defined in Table 10.2. From this experiment, it appears that repeatability (determined by five replications) was good for each compound, with a relative standard deviation of 9%–12% (deviation due to the manual tube spiking step is included in this result).

TABLE 10.2
ATD Conditions

Parameter	Initial Conditions	Optimal Conditions
Oven temperature for tube	350°C	350°C
Desorb flow and time for tube	60 mL min^{-1}; 15 min	60 mL min^{-1}
Inlet-split	0 mL min^{-1}	0 mL min^{-1}
Temperature of cold trap	−30°C	−30°C
Temperature of desorption for the trap	390°C	390°C
Desorb time for the trap	15 min	15 min
Trap fast (−30°C to 390°C)	Yes (>40°C s^{-1})	No
Outlet-split	20 mL min^{-1}	25 mL min^{-1}
Temperature of the transfer valve	250°C	250°C
Temperature of the transfer line	225°C	225°C

Source: From Briand, O. et al., *Anal. Bioanal. Chem.*, 374, 848, 2002.

A calibration range was performed between 1 and 100 ng deposited on tubes. Linear range was observed as listed below:

- 1–100 ng for carbofuran and epoxyconazole
- 2–100 ng for alachlor and cyprodinyl and −HCH and trifluralin
- 4–100 ng for atrazine, iprodione, metolachlor, and tebuconazole
- 10–100 ng for desethylatrazine, disopropylatrazine, and 3,5-dichloroaniline

Detection limits were determined as two times lower than values of the quantification limit. No memory effect was observed in these range of concentrations.

10.2.3.1.2　Pesticides recoveries from Tenax

The optimal temperature for sampling tube desorption was 350°C. No trace of compounds had been observed during the second desorption of the tube. Recovery efficiencies obtained from Equation 10.1 equal 100%.

$$\text{R.Ei } (\%) = \frac{A_{i,1} - A_{Bi}}{(A_{i,1} + A_{i,2}) - A_{Bi}} \times 100\%, \tag{10.1}$$

where
　$R.E_i$ is the recovery efficiency for the analyte i
　$A_{i,1}$ is the peak area of analyte i for the first desorption of the spiked tube
　$A_{i,2}$ is the peak area of analyte i for the second desorption
　A_{Bi} is the count of analyte i from the adsorbent blank (if any)

No additional peak was observed in GC/MS, which seems to indicate that no thermal degradation occurs during tube desorption. Recovery efficiencies obtained at the other temperatures were lower than those at 350°C and were directly correlated to desorption temperature. Recovery efficiencies ranged from 17%, 22%, and 35% at 225°C for low volatile pesticides (iprodione, epoxyconazole, and tebuconazole, respectively) to more than 90% at 300°C. The other compounds gave recovery efficiencies of 60%–95% at 225°C–300°C.

10.2.3.1.3　Resin efficiency

Performance of Tenax TA to retain pesticides under study was tested by an experiment with three tubes in series and a GC oven. This technique offers some advantages such as simplicity and low cost, or the possibility to investigate two parameters at the same time, to evaluate adsorbent performances or reliabilities, retention efficiency, and breakthrough percentage.

For this, three tubes were connected in series. A heating system was combined with a stream of gas to sweep volatile pesticides from solid (125 mg of Tenax® enclosed in tube 1) into the vapor phase. Pesticides were then adsorbed on sample tubes (tubes 2 and 3), also packed with 125 mg of Tenax®.

Tube number 1, located in GC oven, spiked with a known amount of pesticides, was connected with Teflon tubes to a pump at one extremity and to two preconditioned Tenax® tubes (kept at room temperature) on the other. The tube 1 was then heated in GC oven at the same temperature than the first step of the ATD (350°C). After 15 min, the temperature of GC oven was brought down to and maintained at 60°C for 2 h 45 min. During all experiments, a stream of clean air was continuously passed through the first tube to carry volatiles in subsequent tubes 2 and 3. In total, 300 L of air were passed through tubes for 3 h to simulate field conditions. Tubes 2 and 3 were maintained at ambient temperature (20°C–25°C) with a stream of compressed air on their surface.

At the end of the experiment, tubes were separated and analyzed by ATD–GC/MS. For each compound, peak areas were then compared to a reference value (achieved by direct injection on the top of Tenax® tube just before analysis).

With this experiment, it was possible to calculate the actual quantity of pesticides which was volatilized (Equation 10.2, Tenax® retention efficiency Equation 10.3) and to collect nonretained pesticides with the third tube in order to estimate break-through percentage (Equation 10.4).

$$\text{V·E}_i \ (\%) = \frac{A_{i,\text{ref}} - A_{i,\text{T1}}}{A_{i,\text{ref}}} \times 100\%, \tag{10.2}$$

where

V·E_i is the volatilization efficiency for the analyte i

$A_{i,\text{ref}}$ is the peak area of analyte i for the reference desorption (20 ng injected)

$A_{i,\text{T1}}$ is the peak area of analyte i for the tube 1 analyzed

$$\text{T}_R\text{·E}_i \ (\%) = \frac{A_{i,\text{T2}}}{A_{i,\text{ref}} \times \text{V·E}_i} \times 100\%, \tag{10.3}$$

where

$\text{T}_R\text{·E}_i$ is the Tenax® retention efficiency for the analyte i

$A_{i,\text{T2}}$ is the peak area of analyte i for the tube 2

$A_{i,\text{ref}} \times \text{V·E}_i$ represents the actual volatilized quantity

$$\text{B·P}_i \ (\%) = \frac{A_{i,\text{T3}}}{A_{i,\text{ref}} \times \text{V·E}_i} \times 100\%, \tag{10.4}$$

where

B·P_i is the breakthrough percentage for the analyte i

$A_{i,\text{T3}}$ is the peak area of analyte i for the tube 3

10.2.3.1.4 Tenax® TA retention efficiency

Testing the capacity of an adsorbent to quantitatively retain all molecules present in the air during the sampling duration is fundamental in terms of accuracy and precision of the method.

Determining the maximum quantity of air passed through the adsorbent with 100% retention of molecules or breakthrough volume is required when air sampling is performed. Generally, breakthrough is determined by using two sampling tubes in series. Molecules going to the second indicate the limit of the sampling method.

To test the Tenax® TA retention efficiency, the same device as the one used for the resin efficiency was used. This experiment refers to the physical interaction between a molecule of gas, coming from the tube 1, and a solid surface, the porous polymer of the adsorbent. The sorption capacity was determined by passing a known amount M_i of analyte i through the sorbent bed and then analyzing the tube and measuring the amount of retained pesticides.

Vapor pesticide mixture comes from tube 1, where compounds that were first in adsorbed form were volatilized by heating action, and transferred to tube 2.

In lack of suitable standard gaseous mixtures, this test was an alternative from a direct liquid injection on the cartridge, and must be more representative of field experiments where pesticides are in vapor phase or coming as an aerosol. Values of efficiency obtained ranged between 68.4% and 99.1%. Two phenomenon could explain this variability: a competitive adsorption (the molecules with the highest affinity for Tenax® displace those of lowest affinity previously adsorbed and produce a migration in the sorbent bed) or kinetics of capture (which are different for each compound).

Presence of pesticides in the third tube indicated that some of them had penetrated through the front section. Thus, in the first tube, adsorption capacity was exceeded so that some layers of the sorbent bed must be partially or completely saturated and breakthrough occurs.

However, breakthrough percentage gives an indicative value of nonretained pesticides for a known volume of gas passed through the tube but cannot replace breakthrough volume or breakthrough time measurements using stable standard atmosphere and a continuous effluent monitoring with an appropriate detector. These conditions are rarely obtained for pesticides studies.

Breakthrough percentage was never more than 0.75%, whatever the compound, for about 300 L passed through the tubes. This appeared to be very low and have a direct application on field experiment since this volume covers greatly all field-sampling volumes.

Nevertheless, an increase to 10% of the breakthrough in relation with increasing ambient temperature from 20°C to 60°C was observed.

10.2.3.1.5   Recoveries and method detection limits

From the previous results described (resin retention efficiency and recoveries from Tenax®), no corrections of the atmospheric concentrations were needed.

According to the type of studies, determination of spray drift or characterization of postapplication transfers, or determination of volatilization fluxes, sampling periods can be very different and conduct variable detection limits of the method.

For spray drift, sampling periods are short, about a few minutes. Detection limits ranged from 50 to 500 ng m^{-3} (carbofuran, epoxyconazole, and metabolites, respectively) based on a 20 L air volume sampled.

In postapplication, on account of night–day cycles, sampling periods are longer; generally a few hours. For this study, they were fixed at 4 h so that detection limits ranged from 2 to 20 ng m^{-3}, based on a 500 L air volume sampled.

These results illustrate the effectiveness of this present method to assess atmospheric pesticide concentrations. Performances could be compared to conventional method (liquid extraction). For example, Demel et al.[65] have obtained detection limits between 1 and 9 μg m^{-3} based on 1 m^3 air volume sampled (trapping on Tenax® of propiconazole, deltamethrine, etc.). These differences confirm the interest of thermodesorption to analyze atmospheric pesticides in exposed area.

10.2.4 INDOOR AIR

To evaluate the population exposure to pesticides from indoor air, some monitoring studies are needed. These studies used mainly the same techniques that were used for outdoor air but adapted to confined atmosphere.

10.2.4.1 Sampling of Pesticides for Indoor Air Studies

Stout and Mason[30] for their study on the distribution of chlorpyrifos following a crack and crevice type application in the framework of the US EPA Indoor Air Quality Research House program used commercially available sampling tubes. These tubes consist of polyurethane foam open faced tube with no particle cutoff inlet (76 × 20 mm PUF plug in glass filter housing) and of OSHA versatile sampler or OVS tubes (SKC Inc., Eighty-Four, PA). The OVS tube consisted of a 74 × 13 mm glass housing containing a quartz filter and two 140 and 270 mg beds of XAD-2 sandwiched between PUF partitions.

The two types of tubes were suspended at 100 cm above the floor in the living room and sampling was done for 24 h at a flow rate of 3.8 and 1.0 L min^{-1} for PUF and OVS, respectively, by using an SKC Universal XR sample pump.

The sample inlets were directed towards the floor. Samples were collected prior to the application and at 1, 3, 7, 14, and 21 days postapplication. Following sample collection, the PUF and OVS tubes were capped with aluminium foil and individually sealed in plastic bags. The tubes were put in ice chests at reduced temperatures for transport.

Bouvier et al.[31] for their comparison of the pesticide exposure of nonoccupationally exposed subjects and some occupational exposure used a MiniPartisol air sampler 2100 (Rupprecht and Patashnik, East Greenbush, NY, USA) and a glass cartridge containing a polyurethane foam (PUF), (SKC, Blandford Forum, UK) for the collection of aerosols and a QM-A 1851 quartz fiber filter (Whatman, Maidstone, UK) for the collection of particulate matters. The MiniPartisol was placed on a table or working furniture at a height of approximately 1.60 m, in the main room of the workplace or in the living-room of the residences. Flow rate of the pumps was checked before and after each sampling with a mini-Buck Calibrator debitmeter (A.P. Buck, Orlando, FL, USA). The sampling lasted for 24 h without interruption at a flow rate of 5 L min^{-1}; the mean volume sampled was 7.1 m^3/24 h. Sampling was performed during a working day and this sampling method was based upon the ASTM D 4861–00 standard.

Barro et al.[32] for the analysis of pyrethroids as well as other components of frequently used domestic insecticide preparations in indoor air used a vacuum pump working at 100 L min^{-1} (Telstar model S-8, Tarrasa, Spain). A known volume of air was pumped through a glass tube containing 25 mg of an adsorbent; Tenax TA of mesh size 60–80 (Supelco) and Florisil (activated overnight at 105°C) of 60–100 mesh size (Aldrich, Steinheim, Germany).

Teflon (PTFE) tubing was used for all connections. This method of sampling was previously used for polychlorobiphenyls.[32]

Yoshida et al.[66] used quartz fiber filter disk (type 2500 QAT-UP, 47 mm diameter, ~0.64 mm thickness, Tokyo Dylec, Tokyo, Japan) and an Empore disk (type C18FF, 47 mm diameter, ~0.50 mm thickness, 3 M, MN, USA) connected to a small suction pump for 24 h at a flow rate of 5.0 L min^{-1} resulting in a total volume of air passed through the adsorbents of 7.2 m^3.

Prior to the sampling of airborne pesticides and other semivolatile organic compounds (SVOCs), the Empore disks were subjected to ultrasonic cleaning five times for 10 min each by 10 mL of acetone per disk, and the quartz fiber filter disks were heated at 400°C for 4 h.

The sampler was fixed using a tripod in the center of the target room at a height of 1.2–1.5 m from the floor.

SPME was also used for the analysis of pesticides in confined atmosphere, in particular on greenhouses. Dichlorvos, an organophosphate pesticide, was first studied[67] followed by an extension to 10 pesticides (bioallethrin, chlorpyriphos-methyl, folpet, malathion, procymidone, quintozene, chlorothalonil, fonofos, penconazole, and tri-methacarb).[68] Sampling was performed by using polydimethylsiloxane–solid-phase microextraction (PDMS–SPME) fibers immersed in a 250 mL sampling flask through which air samples were dynamically pumped from the analyzed atmosphere (Figure 10.5). After a 40 min sampling duration, samples were analyzed by GC/MS.

Calibration was performed from a vapor-saturated air sample (Figure 10.6). The linearity of the observed signal versus pesticide concentration in the vapor phase was proved from spiked liquid samples whose headspace concentrations were measured by the proposed method (Figure 10.7).

The same all experiments, as those used for extraction of pesticides by SPME from water samples, were performed.

10.2.4.2   Extraction of Pesticides for Indoor Air Studies

Classical methods using Soxhlet or an ultrasonic bath were used. Bouvier et al.[31] used Soxhlet extraction with 150 mL of dichloromethane Pestanal for 16 h for the

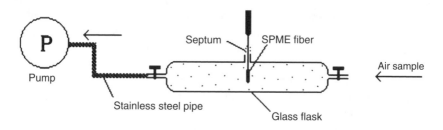

FIGURE 10.5 Assembly used in greenhouse for SPME samplings. (From Ferrari, F. et al., *Anal. Bioanal. Chem.*, 379, 476, 2004.)

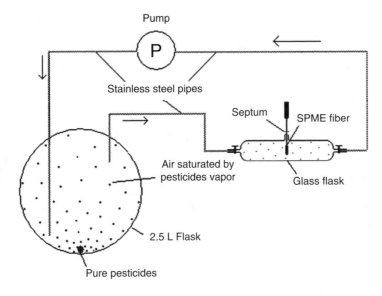

FIGURE 10.6 Laboratory assembly used for calibration of SPME samplings of pesticides vapors. (From Ferrari, F. et al., *Anal. Bioanal. Chem.*, 379, 476, 2004.)

extraction of pesticides from PUF plugs and quartz fiber filters. Barro et al.[32] for their study on pyrethrenoids pesticides in indoor atmosphere used an ultrasonic bath for a few minutes. They put traps on a 22 mL glass vial and analytes were extracted into an appropriate volume of organic solvent (*n*-hexane or ethyl acetate).

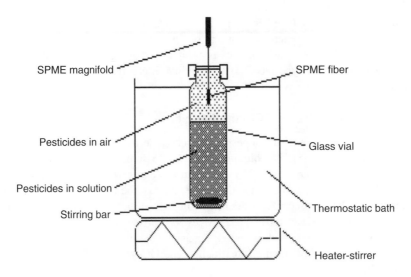

FIGURE 10.7 Assembly used for studying the linearity of HS–SPME samplings. (From Ferrari, F. et al., *Anal. Bioanal. Chem.*, 379, 476, 2004.)

Yoshida et al.[66] used also 15 min ultrasonic bath for the extraction of pesticides from glass fiber filters and Empore disks followed by shaking for 10 min. For this, they add 8.0 mL of acetone to a 10 mL centrifuge tube containing the two absorbents.

As mentioned in the paragraph of sampling of pesticides for indoor studies, the extraction of pesticides after sampling is made by direct exposure of the fiber in the split–splitless injector of the gas chromatograph.

10.3 ANALYSIS OF PESTICIDES IN THE ATMOSPHERE

Pesticides are analyzed after extraction by conventional GC or high performance liquid chromatography (HPLC). The detectors used in GC are electron capture detectors (ECD) for the analysis of pesticides containing halogens (organochlorines, pyrethrenoids, alachlor, etc.), nitrogen phosphorous detectors (organophosphates, triazines, etc.), and mass detection in the single ion monitoring mode (SIM). For HPLC, detectors are diode array detectors, fluorescence detectors for carbamates after postcolumn derivatization and MS.

10.3.1 ANALYSIS BY GAS CHROMATOGRAPHY

ECD and NPD are very sensible and selective detectors, but few recent studies used these detectors. They have been used by various authors such as Millet et al.,[48] Sanusi et al.,[69] Epple et al.,[24] Quaghebeur et al.,[27] Bouvier et al.,[31] and Kumari et al.[52] for the analysis of current-used pesticides in air and rainwater. The detection and analysis by ECD was commonly used for the analysis of organochlorine pesticides and GC–NIMS (negative ionization MS) tend to replace this detector especially because of the uncertainty on the identification with ECD.

The use of GC–MS is more developed since it provides sensitivity, specificity, and selectivity. Indeed, with mass spectroscopic detection, the identification of the compound can be done together with the identification of coeluted compounds.

Columns used are generally nonpolar of semipolar columns (30 m × 0.25 mm, 0.25 μm film thickness) and helium is used as carrier gas. A 5% phenyl/95% polydimethylsiloxane (type DB-5, HP-5, Optima-5, etc. depending on manufacturers) was used in many cases.[26,36,46,49,50,68,69]

10.3.1.1   Analysis by GC–ECD and GC–NPD

For the separation and analysis of pesticides (see Table 10.1 for the list) by GC–NPD, Epple et al.[24] used a SE-54 column (30 m × 0.25 mm, 0.25 μm film thickness; J&W Scientific, Folsom, CA, USA) and helium as carrier gas. The injection (1 μL) was made in the splitless mode and the temperature of the injector and detector was maintained at 250°C. Because of the fluctuating sensitivity of the detector, quantification of pesticides extracted by C_{18} cartridges was carried out by the internal standard method by using 2,3-diethyl-5-methylpyrazine and quinazoline. Detection limits and uncertainty of all the methods (extraction and GC–NPD analysis) are presented in Table 10.1.

Authors, because of uncertainty of the identification by GC–NPD, for most of the GC–NPD analysis, a verification by GC–MS using a GC HP 5890 II Plus, a MS

5989 B Engine, a column HP 5 MS (30 m × 0.25 mm, 0.25 μm film thickness), and crosslinked (Hewlett-Packard, Palo Alto, CA, USA) have been performed. Identification was performed by comparing the retention time and mass–peak relations with the standard substance.

Millet et al.[46] and Sanusi et al.[69] used GC–ECD for the analysis of organochlorine pesticides in atmospheric samples (air, fog, and rainwater) after fractionation of the samples by HPLC.

Detection limits obtained by Millet et al.[46] varied between 0.01 and 0.8 mg L^{-1} corresponding to 33 and 333 pg m^{-3} for a 24 h sampling at 12.5 m^3 h^{-1}.

10.3.1.2 Analysis by GC–MS

GC–MS is employed for the analysis of pesticides in atmospheric samples for its capacity to deliver results with high sensitivity and guaranty on the identification. In many cases, quadripole GC–MS in the SIM is employed and quantification is performed by the internal standard method by using various deuterated compounds, including pesticides.[22,33,63,64,70]

Ion trap was also employed in the SIM mode by Ferrari et al.[68] for the analysis of 11 pesticides in confined atmosphere after atmospheric sampling using SPME. Ion trap was also used in the MS–MS mode by Sauret et al.[41] and Scheyer et al.[36] for air samples and by Scheyer et al.[49,50] for the analysis of pesticides in rainwater after SPME extraction. The use of MS–MS permits a better sensitivity, a higher specificity, and a more important structural information on molecules in comparison to single MS and is also better for the quantification. To improve the specificity of the detection, in MS–MS only the daughter ions characteristic of the studied pesticides were used for quantification. The parent ion was systematically excluded from the quantitative analysis, since this parent ion could be obtained from several molecules and consequently have a low specificity. Indeed, the presence of the parent ion on the MS/MS spectrum meant that a fraction of this ion had not been fragmented by the Collision Induced Dissociation (CID) phenomenon, necessary to produce daughter ions.

10.3.2 DERIVATIZATION

Some pesticides cannot be analyzed directly by GC. This is the case for some phenoxy acids (2,4 D, MCPA, and MCPP) or ureas herbicides (chlorotoluron, diuron, and isoproturon). Prior to their analysis by GC, a derivatization step is required. Scheyer et al.[36] have used pentafluorobenzyl bromide (PFBBr) for the derivatization of seven herbicides before their analysis by GC–MS–MS.

Phenyl ureas (PUHs), phenoxy acids, and bromoxynil show very different physical–chemical properties and molecular structures. It was necessary to find a derivatization agent which can react simultaneously and easily with all the pesticides studied. An alkylation reaction with pentafluorobenzyl bromide (PFBBr) seems to be a good compromise. The mechanism implies a nucleophilic substitution with a bimolecular mechanism, without formation of carbocations (Figure 10.8).

The mechanism for the reaction of PFBBr with a molecule that has an acidic hydrogen atom is a bimolecular nucleophilic substitution SN_2. The functional groups

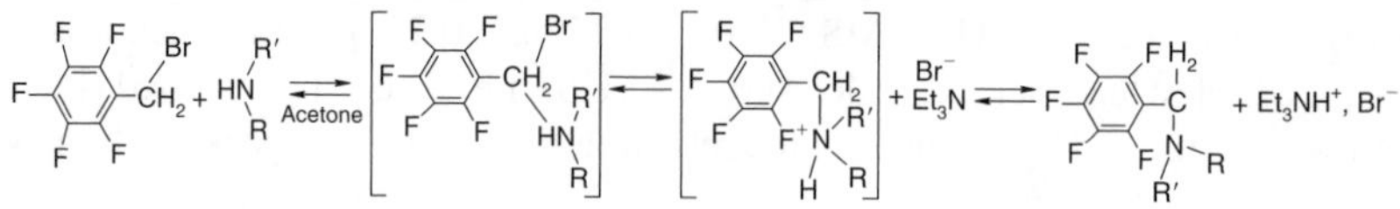

FIGURE 10.8 Mechanism of reaction with pentafluorobenzyl bromide (PFBBr) for amine group as an example. (From Scheyer, A. et al., *Anal. Bioanal. Chem.*, 381, 1226, 2005.)

present on pesticides which can react with PFBBr are —**NH** in the α position with respect to a carbonyl group (chlorotoluron, diuron, and isoproturon), —**OH** on an aromatic ring (i.e., bromoxynil) and —**COOH** (2,4 D, MCPA, and MCPP). The better solvent for this reaction must be aprotic and polar. This is why acetone was used. The reaction can also be performed in the presence of a base, such as triethylamine, which plays the role of proton acceptor.

PFBBr seems to be a good derivatization agent for the phenoxy acid herbicides but the method is less efficient for the PUHs.

This method was employed for the analysis of these herbicides in rainwater after SPME as mentioned in the section of extraction by SPME.[50]

10.3.3 ANALYSIS BY HIGH PERFORMANCE LIQUID CHROMATOGRAPHY

Liquid chromatography was used especially for the analysis of polar or acidic compounds. Detectors used were UV–Visible or Diode Array detectors or MS. Columns used are mainly C_{18} phases.

10.3.3.1 Analysis by LC–UV or LC–DAD

HPLC–UV or HPLC–DAD was used especially for triazines, ureas herbicides, or carbamates. Generally, a fractionation step is performed since the detection used is not specific.

HPLC–UV was used by Millet et al.[46] and Sanusi et al.[66] for the analysis of triazines herbicides (i.e., atrazine), urea herbicides (i.e., isoproturon), and carbamates (aldicarb). The HPLC quantification was done by the internal standard method after fractionation by normal phase HPLC.

Quaghebeur et al.[27] used HPLC–DAD for the analysis of phenylureas herbicides and their aniline degradation products in rainwater samples after solid-phase extraction. Quantification of the results is obtained using an internal standard.

Bouvier et al.[31] used also HPLC–DAD for the analysis of some pyrethroids, urea herbicides, and some fungicides. A postcolumn derivatization followed by fluorescence detection was also used for the specific analysis of carbamates pesticides.

10.3.3.2 Analysis by LC–MS

Bossi et al.[53] have developed and validated a LC–MS–MS method for the analysis of 53 pesticides, degradation products, and selected nitrophenols in rainwater. After extraction of rainwater by solid-phase extraction on Oasis HLB columns, extracts were analyzed by LC–MS–MS with electrospray ionization. All samples were analyzed in negative and in positive ionization mode, respectively, for acidic and neutral compounds.

Indeed, most of the modern pesticides and their degradation products are characterized by medium polarity and thermal lability. For these reasons liquid chromatography (LC) is the most appropriate analytical method. To quantify and identify the target analytes at trace levels, MS in the SIM mode has to be employed.

REFERENCES

1. Glotfelty, D.E. et al. Volatilisation and wind erosion of soil surface applied atrazine, simazine, alachlor, and toxaphene. *J. Agric. Food Chem.*, 37, 546, 1984.
2. Van den Berg, F. et al. Emission of pesticides in the air. Fate of pesticides in the atmosphere: implications for environmental risk assessment. *Water Air Soil Pollut.*, 115, 195, 1999.
3. Fritz, R. Pflanzenschutzmittel in der Atmosphäre. *Pflanzenschutznachrichten Bayer*, 46, 229, 1993.
4. Miller, P. The measurement of spray drift. *Pesticides Outlook*, October, 205, 2003.
5. Van der Werf, H.M.G. Assessing the impact of pesticides in the environment. *Agric. Ecosys. Environ.*, 60, 81, 1996.
6. Ganzelmeier, H., Rautmann, D., Spangenberg, R., Streloke, M., Herrmann, M., Wenzelburger, H.J., and Walter, H.F. Studies on the spray drift of plant protection products. *Mitteilungen aus der Biologischen Bundesanstalt für Land- und Forstwirtschaft*, Nr. 305, 111 S, Berlin, 1995.
7. Bedos, C. et al. Mass transfer of pesticides into the atmosphere by volatilisation from soils and plants: overview. *Agronomy*, 22, 21, 2002.
8. Kubiak, R. et al. Pesticides in air: considerations for exposure asssessment, FOCUS air group, SANCO, 2007, in press.
9. Unsworth, J.B. et al. Significance of long range transport of pesticides in the atmosphere. *Pure Appl. Chem.*, 71, 1359, 1999.
10. Sanusi, A. et al. Gas-particles partitioning of pesticides in atmospheric samples. *Atmos. Environ.*, 33, 4941, 1999.
11. Atkinson, R. et al. Transformation of pesticides in the atmosphere: a state of the art. *Water Air Soil Pollut.*, 115, 219, 1999.
12. Finlayson-Pitts, B.J. and Pitts, J.N. Jr. *Atmospheric Chemistry*. Wiley Ed., New York, 1986.
13. Klöpffer, W. et al. Testing of the abiotic degradation of chemicals in the atmosphere: the smog chamber approach. *Ecotox. Environ. Safety*, 15, 298, 1988.
14. Klöpffer, W., Kaufmann, G., and Frank, R. Phototransformation of air pollutants: rapid test for the determination of k_{OH}. *Z. Naturforsch.*, 40A, 686, 1985.
15. Atkinson, R. Kinetics and mechanisms of the gas phase reactions of the hydroxyl radical with organic compounds under atmospheric conditions. *Chem. Rev.*, 86, 69, 1986.
16. Becker, K.H. et al. Methods for ecotoxicological evaluation of chemicals. Photochemical degradation in the gas phase, Vol. 6, Report 1980–1983, Kernforschunganlage Jülich GmbH, Projektträgerschaft Umveltchemikalien, Jül-Spez-279, 1984.
17. Atkinson, R. et al. Kinetics and mechanisms of the reaction of the hydroxyl radical with organic compounds in the gas phase. In: *Advances in Photochemistry*, Vol. 11, 375–488, Wiley Ed., New York, 1979.

18. Van Dijk, H.F.G. and Guicherit, R. Atmospheric dispersion of current-use pesticides: a review of the evidence from monitoring studies. *Water Air Soil Pollut.*, 115, 21, 1999.

19. Dubus, I.G., Hollis, J.M., and Brown, C.D. Pesticides in rainfall in Europe. *Environ. Pollut.*, 110, 331, 2000.

20. McConnell, L.L. et al. Chlorpyrifos in the air and surface water of Chesapeake Bay: prediction of atmospheric deposition fluxes. *Environ. Sci. Technol.*, 31, 1390, 1997.

21. Coupe, R.H. et al. Occurence of pesticides in rain and air in urban and agricultural areas of Mississippi. *Sci. Total Environ.*, 248, 227, 1998.

22. Haraguchi, K. et al. Simultaneous determination of trace pesticides in urban precipitation. *Atmos. Environ.*, 29, 247, 1995.

23. Briand, O. et al. Influence de la pluviométrie sur la contamination de l'atmosphère et des eaux de pluie par les pesticides. *Rev. Sci. Eau.*, 15, 767, 2002.

24. Epple, J. et al. Input of pesticides by atmospheric deposition. *Geoderma*, 105, 327, 2002.

25. De Rossi, C., Bierl, R., and Riefstahl, J. Organic pollutants in precipitation: monitoring of pesticides and polycyclic aromatic hydrocarbons in the region of Trier (Germany). *Phys. Chem. Earth*, 28, 307, 2003.

26. Grynkiewicz, M. et al. Pesticides in precipitation from an urban region in Poland (Gdańsk-Sopot-Gdynia tricity) between 1998 and 2000. *Water Air Soil Pollut.*, 149, 3, 2003.

27. Quaghebeur, D. et al. Pesticides in rainwater in Flanders, Belgium: results from the monitoring program 1997–2001. *J. Environ. Monit.*, 6, 182, 2004.

28. Asman, W. et al. Wet deposition of pesticides and nitrophenols at two sites in Denmark: measurements and contributions from regional sources. *Chemosphere*, 59, 1023, 2005.

29. EPCA. Residues of crop protection products in precipitation and air. 1998. D/98/GRG/3255.

30. Stout, D.M. and Mason, M.A. The distribution of chlorpyrifos following a crack and crevice type application in the US EPA Indoor Air Quality Research House. *Atmos. Environ.*, 37, 5539, 2003.

31. Bouvier, G. et al. Pesticide exposure of non-occupationally exposed subjects compared to some occupational exposure: a French pilot study. *Sci. Total Environ.*, 366, 74, 2006.

32. Barro, R. et al. Rapid and sensitive determination of pyrethroids indoors using active sampling followed by ultrasound-assisted solvent extraction and gas chromatography. *J. Chromatogr. A*, 1111, 1, 2006.

33. White, L.M. et al. Ambient air concentrations of pesticides used in potato cultivation in Prince Edward Island, Canada. *Pest Manag. Sci.*, 62, 126, 2006.

34. Peck, M. and Hornbuckle, K.C. Gas-phase concentrations of current-use pesticides in Iowa. *Environ. Sci. Technol.*, 39, 2952, 2005.

35. Dobson, R. et al. Comparison of the efficiency of trapping of current-used pesticides in the gaseous phase by different types of adsorbents using the technique of high-volume sampling. *Anal. Bioanal. Chem.*, 386, 1781, 2006.

36. Scheyer, A. et al. A multiresidue method using ion trap GC-MS/MS by direct injection or after derivatization with PFBBr for the analysis of pesticides in the atmosphere. *Anal. Bioanal. Chem.*, 381, 1226, 2005.

37. Scheyer, A. et al. Variability of atmospheric pesticide concentrations between urban and rural areas during intensive pesticide application. *Atmos. Environ.*, 41, 3604, 2007.

38. Yao, Y. et al. Spatial and temporal distribution of pesticide air concentrations in Canadian agricultural regions. *Atmos. Environ.*, 40, 4339, 2006.

39. Haraguchi, K. et al. Simultaneous determination of trace pesticides in urban air. *Atmos. Environ.*, 28, 1319, 1994.

40. Sanusi, A. et al. A multiresidue method for determination of trace levels of pesticides in atmosphere. *Analusis*, 25, 302, 1997.

41. Sauret, N. et al. Analytical method using gas chromatography and ion trap tandem mass spectrometry for the determination of S-triazines and their metabolites in the atmosphere. *Environ. Pollut.*, 110, 243, 2000.

42. Foreman, W.T. et al. Pesticides in the atmosphere of the Mississipi River Valley, part II-air. *Sci. Total Environ.*, 248, 213, 2000.

43. Albanis, T.A., Pomonis, P.J., and Sdoukos, A.Th. Seasonal fluctuations of organochlorine and triazines pesticides in the aquatic system of Ionnina basin. *Sci. Total Environ.*, 58, 243, 1986.

44. Badawy, M.I. Organic insecticides in airborne suspended particulates. *Bull. Environ. Contam. Toxicol.*, 60, 693, 1998.

45. Seiber, J.N. et al. A multiresidue method by high performance liquid chromatography-based fractionation and gas chromatographic determination of trace levels pesticides in the air and water. *Arch. Environ. Contam. Toxicol.*, 19, 583, 1990.

46. Millet, M. et al. A multiresidue method for determination of trace levels of pesticides in air and water. *Arch. Environ. Contam. Toxicol.*, 31, 543, 1996.

47. Asman, W.A.H. et al. Wet deposition of pesticides and nitrophenols at two sites in Denmark: measurements and contributions from regional sources. *Chemosphere*, 59 1023, 2005.

48. Millet, M. et al. Atmospheric contamination by pesticides: determination in the liquid, gaseous and particulate phases. *Environ. Sci. Pollut. Res.*, 4, 172, 1997.

49. Scheyer, A. et al. Analysis of trace levels of pesticides in rainwater using SPME and GC–tandem mass spectrometry. *Anal. Bioanal. Chem.*, 384, 475, 2006.

50. Scheyer, A. et al. Analysis of trace levels of pesticides in rainwater by SPME and GC–tandem mass spectrometry after derivatisation with PFFBr. *Anal. Bioanal. Chem.*, 387, 359, 2007.

51. Chevreuil, M. et al. Occurrence of organochlorines (PCBs, pesticides) and herbicides (triazines, phenylureas) in the atmosphere and in the fallout from urban and rural stations of the Paris area. *Sci. Total Environ.*, 182, 25, 1996.

52. Kumari, B., Madan, V.K., and Kathpal, T.S. Pesticide residues in rain water from Hisar, India. *Environ. Monit. Assess.*, 2007, in press.

53. Bossi, R. et al. Analysis of polar pesticides in rainwater in Denmark by liquid chromatography–tandem mass spectrometry. *Chemosphere*, 59, 1023, 2002.

54. Lord, H. and Pawliszyn, J. Evolution of solid-phase microextraction technology. *J. Chromatogr. A*, 885, 153, 2000.

55. Louch, D., Motlagu, S., and Pawliszyn, J. Dynamics of organic compound extraction from water using liquid-coated fused silica fibers. *Anal. Chem.*, 64, 1187, 1992.

56. Cserhati, T. and Forgacs, E. Phenoxyacetic acids: separation and quantitative determination. *J. Chromatogr. B*, 717, 157, 1998.

57. Boucharat, C., Desauzier, V., and Le Cloirec, P. Experimental design for the study of two derivatization procedures for simultaneous GC analysis of acidic herbicides and water chlorination by-product. *Talanta*, 47, 311, 1998.

58. Majewski, M.S. Micrometeorological methods for measuring the post-application volatilisation of pesticides. *Water Air Soil Pollut.,* 115, 83, 1999.

59. Cessna, A.J. and Kerr, L.A. Use of an automated thermal desorption system for gas chromatographic analysis of the herbicides trifluralin and triallate in air samples. *J. Chromatogr. A*, 642, 417, 1993.

60. Pattey, E. et al. Herbicides volatilization measured by the relaxed eddy-accumulation technique using two trapping media. *Agric. Forest Meteorol.*, 76, 201, 1995.

61. Foster, W., Ferrari, C., and Turloni, S. Environmental behaviour of herbicides. Atrazine volatilisation study. *Fresenius Environ. Bull.*, 4, 256, 1995.
62. Briand, O. et al. Assessing transfer of pesticide to the atmosphere during and after application. Development of a multiresidue method using adsorption on Tenax/thermal desorption-GC/MS. *Anal. Bioanal. Chem.*, 374, 848, 2002.
63. Clément, et al. Adsorption/thermal desorption-GC/MS for the analysis of pesticides in the atmosphere. *Chemosphere*, 40, 49, 2000.
64. Briand, O. et al. Atmospheric concentrations and volatilisation fluxes of two herbicides applied on maize. *Fresenius Environ. Bull.*, 12, 675, 2003.
65. Demel, J., Buchberger, W., and Malissa, H. Jr. Multiclass/multiresidue method for monitoring widely applied plant protecting agents in air during field dispersion work. *J. Chromatogr. A*, 931, 107, 2001.
66. Yoshida, et al. Simultaneous determination of semivolatile organic compounds in indoor air by gas chromatography–mass spectrometry after solid-phase extraction. *J. Chromatogr. A*, 1023, 255, 2004.
67. Sanusi, A. et al. Pesticides vapours in confined atmospheres. Determination of dichlorvos by SPME/GC/MS at the μg m-3 level. *J. Environ. Monit.*, 5, 574, 2003.
68. Ferrari, F. et al. Multi-residue method using SPME for the determination of various pesticides with different volatility in confined atmosphere. *Anal. Bioanal. Chem.*, 379, 476, 2004.
69. Sanusi, A. et al. Airborne pesticides concentrations and α/γ HCH ratio in Alsace (East France) air. *Sci. Total Environ.*, 263, 263, 2000.
70. Scheyer, A. PhD thesis, University of Strasbourg I, 2004.

11 Levels of Pesticides in Food and Food Safety Aspects

Kit Granby, Annette Petersen,
Susan S. Herrmann, and
Mette Erecius Poulsen

CONTENTS

11.1 INTRODUCTION

Monitoring programs for pesticide residues in food are performed in many countries around the world to ensure that consumers are not exposed to unacceptable levels of pesticides and that only pesticides approved by the authority are used and for the right applications with respect to crop, application dose, time, and intervals. The food products are permitted as long as they comply with the maximum residue levels (MRLs) set by the authorities. Another purpose with the pesticide residue monitoring in food may be to assess the food safety risk due to the dietary exposure of the population to pesticides.

The present chapter deals with monitoring programs for pesticide residues in food in general. It also covers monitoring results in fruits, vegetables, cereals, food of animal origin, and processed food like drink, infant and baby food. In addition, risk assessments of consumer exposure based on dietary intake estimates are described and examples of exposure assessments from studies worldwide are shown.

11.2 MONITORING PROGRAMS; RESIDUE LEVELS IN FOOD

11.2.1 LEGISLATION

In many countries, there is national legislation regulation on which pesticides are authorized. Many countries also have national legislation on the maximum amounts of pesticide residues in different food commodities. Such upper limits are also referred to as MRLs or tolerances (in the United States). In countries with no national legislation, the MRLs set by the Codex system are often used. MRLs are normally set for raw agricultural commodities (RAC), for example, banana with peel, lettuce, and apples.

The Codex Alimentarius Commission (CAC) is an international body that aims to protect the health of consumers, ensure fair trade practices in the food trade, and promote coordination of all food standards work undertaken by international governmental and nongovernmental organizations. CAC also set MRLs, which are indicative and not statutory. The Codex MRLs are to be used as guidance on acceptable levels when there is no other legislation in place; for example, in countries without their own national MRLs or they can be used if national MRLs have not been set for a particular compound.

MRLs set by Codex are evaluated and negotiated through a stepwise procedure. Initially, the Joint FAO/WHO Meeting on Pesticide Residues (JMPR)[1] considers recognized use patterns of good agricultural practice (GAP) and evaluates the fate of residues, animal and plant metabolism data, and analytical methodology as well as residue data from supervised trials conducted according to GAP. Based on these data, MRLs are proposed for individual pesticides. Toxicologists evaluate the toxicological data related to the pesticides and propose acceptable daily intakes (ADI) and acute reference doses (ARfD). The toxicological data originate from animal studies and include both studies on the short-term and long-term effects. The ADI is a measure of the amount of specific substance (in this case, a pesticide) in foods and drinks that can be consumed over a lifetime without any appreciable health risk. ADIs are expressed as milligram/kilogram body weight/day. The ARfD of a substance (here pesticide) is an estimate of the amount a substance in food or drinks, normally expressed on a body weight basis that can be ingested in a period of 24 h or less without appreciable health risks to the consumer on the basis of all known facts at the time of the evaluation. ARfD apply only to pesticides that cause acute effects, for example, phosphorus pesticides that are cholinesterase inhibitors.

The Codex Committee on Pesticide Residues (CCPR) considers at their annual meetings the MRLs proposed by the JMPR. CCPR is an intergovernmental meeting with the prime objective to reach agreement on proposed MRLs. The MRLs are discussed in an eight-step procedure and after the final step the CCPR recommends MRLs to CAC, for adoption as Codex MRLs. To protect the health of the consumers, the intake calculated using the proposed MRLs is compared with the ADI or the ARfD and if the calculated intake exceeds one of these two values the MRL cannot be accepted.

Often when national MRLs are set, an evaluation is performed on a national level, that in many ways are similar to the evaluation performed by JMPR. Some countries also set their own ADIs or ARfDs. As part of the evaluation of pesticides within the European Union (EU) ADIs and ARfDs are set on the EU level which then apply in all Member States. These values can differ from the values set by Codex.

The Member States within the EU, which includes 27 countries, set harmonized EU MRLs for pesticides. All harmonized legislation can be found on the Web site of the EU Commission.[2] At the moment not all pesticides have harmonized MRLs and for these pesticides nationally MRLs can be set. In April 2005, new legislation (Regulation 396/2005)[2] entered into force in which only harmonized EU MRLs can be set and all national legislation are turned into EU legislation. The new regulation does, however, not apply at the moment, as all the annexes to the regulation are not yet established.

Some countries publish their MRLs on the Internet, for example, United States,[3] Canada,[4] Australia,[5] New Zealand,[6] India,[7] Japan,[8] South Africa,[9] Thailand,[10] and Korea.[11] In Australia,[5] New Zealand,[12] and the United States (USDA[13]), authorities have compiled information about legislation and MRLs worldwide. Other countries do not have their own legislation and MRLs published on Web sites but the information can be gathered by contacting the relevant authorities. For countries that have published MRLs on Web sites be aware that addresses changes and the most recent legislation is not yet published.

11.2.2   Monitoring Programs; General Aspects

There is a growing interest in pesticide residues in food from all aspects of the food chain from "the farm to the fork." It is the national governments that are responsible for regular monitoring of pesticide residues in food. Besides the national governments, monitoring activities or surveillance are also performed by nongovernmental organizations or by scientists studying the occurrence and fate of pesticides in relation to environment, agriculture, food, or human health. Food companies may also monitor pesticide residues in their products to secure and demonstrate good food safety quality of their products and/or prevent economical losses.

The monitoring sampling may be surveillance sampling where there is no prior knowledge or evidence that a specific food shipment contain samples exceeding the MRLs. The surveillance sampling may also include more frequent sampling of food groups with samples frequently exceeding the MRLs. Compliance sampling is defined as a direct follow-up enforcement sampling, where the samples are taken in case of suspicion for previously found violations. The follow-up enforcement may be directed to a specific grower/producer or to a specific consignment. To cover both the control aspect and the food safety aspect regarding exposure assessments, the design of a monitoring program may be a mixture of a program where the different food types are weighted relative to the consumption or sale and one where the food groups with samples exceeding the MRLs are weighted higher. In order to be able to have more samples of the same type for comparisons, all sample types may not be monitored annually as the selection of some (minor) sample types may change from year to year.

The monitoring programs do often include imported as well as domestically produced foods. Domestic samples may be collected as close to the point of production as possible, for food crops the sampling may be at the farm or at wholesalers or retailers. Imported samples may be collected by the customs authorities or at the import firms or retailers. The samples are often raw food, for example, fruits, vegetables, cereals, or food of animal origin. In addition, different kinds of processed foods are monitored, for example, dried, extracted, fermented, heated, milled, peeled, pressed, washed, or otherwise prepared products. The different kinds of processing, in most cases, lead to a decrease in levels of pesticides compared with the contents in the raw food.

An important parameter for a monitoring program is the choice and the number of pesticides investigated. To cover as many pesticides as possible, both multi-methods and single residue methods may have to be included in the monitoring program. In 2003, the U.S. Food and Drug Administration (US FDA) was able to monitor roughly half of the 400 pesticides for which U.S. Environmental Protection Agency (EPA) had set tolerances.[14] The same year all states participating in the EU monitoring program together analyzed for 519 different pesticides. However, most of the individual countries analyzed for a much smaller number of pesticides, for example, about 100–200.[15] In addition to the selection and number of pesticides analyzed for, the detection limits of the pesticides in the different foods are determining for how frequent findings of pesticide residues are.

On a worldwide scale, two major monitoring programs including many states exists: the EU monitoring programs and the US FDA program, both programs publishing their annual results at their respective Web sites. As an example, the "Monitoring of pesticide Residues in Products of Plant Origin in the European Union, Norway, Iceland and Liechtenstein 2004" included a total of 60,450 samples of which 92% of the samples were fresh fruits, vegetables, and cereals and 8% were processed foods.[16] The US FDA program included 7234 samples of fruits, vegetables, cereals, and food of animal origin.[14]

11.2.3 RESULTS FROM MONITORING PROGRAMS

The results of pesticide residues in different foods were found in internationally published surveys and monitoring programs on pesticide residues. The results are attempted to reflect the pesticide residue results in food worldwide. However, many countries either do not have monitoring results for pesticide residues or do not publish them so they are not available internationally. The European Commission compiles monitoring data from the 25 member states and Norway, Iceland, and Liechtenstein in annual reports[15–17] and the US FDA as well publish annual reports on their monitoring and surveillance program for pesticides in food.[14]

11.2.3.1   Fruits and Vegetables

In general, fresh fruits and vegetables account for the largest proportion of samples analyzed within pesticide monitoring programs. In 2003, the US FDA monitoring program of vegetables included 1132 domestic samples and 2494 imported samples, the major part of the total samples imported from Mexico, China, the Netherlands, and Chile.[14] Pesticide residues ≤MRL were detected in 30% of the domestic and 21% of the imported vegetable samples, whereas violations were detected in 1.9% of the domestic and 6.7% of the imported vegetable samples. The frequency of fruit samples with detected pesticide residues ≤MRL is somewhat higher: 49% of 813 domestic samples and 31% of 1537 imported fruit samples. The violations comprised 2.2% of the domestic and 5.3% of the imported fruit samples. Pesticide residues were detected in approximately half of the apple and pears and 60%–70% of the citrus fruits.

The "EU Monitoring of pesticides in Products of Plant Origin 2004" included 50,428 fruit and vegetable samples for surveillance monitoring of which 42% contained residues ≤MRL and in 5% of the samples the residue concentrations exceeded the MRL. In addition to the surveillance sampling, in 2004, 4% of all the samples were follow-up enforcement samples. The more targeted nature of the follow-up sampling resulted in a higher percentage of the samples exceeding the MRL, that is, 10.2% of the 2211 fruit and vegetable samples.

The overall trend in the presence of pesticide residues was followed from 1996 to 2004 for fruits, vegetables, and cereals (of which cereals comprise only ~5%). The percentage of samples with residues below or at the MRL (national or EC-MRL) has increased from 32% in 1999 to 42% in 2004. The percentage of samples with residues above the MRL varied from 3% in 1996 to 5.5% in 2002/2003. In 2004, the 5% violations were slightly lower than the last 2 years. In addition, the frequency of multiple residues in samples has increased from 14% in 1998 to 23% in 2004. Different factors may have contributed to the trend in the results. During that period, the average number of pesticides detected for increased from ~126 to 169, which may result in more findings. The legislative situation has also changed in recent years and will continue to change in direction of more MRLs set at the limit of detection (LOD).

The most frequently found pesticides in the monitoring of fruits and vegetables (in descending order) were dithiocarbamates,* chlorpyriphos,[†] imazalil,* procymidone,* benomyl group,* iprodione,* thiabendazole,* chlormequat,[‡] bromide,[§] and orthophenylphenol.* Approximately half of the 677 compounds detected for were actually detected.

Within the EU monitoring program, the Commission has designed a coordinated program, where eight alternating commodities were analyzed for a certain number of

---

* Fungicide.
[†] Insecticide.
[‡] Growth regulator.
[§] Indicator of bromofumigants.

TABLE 11.1

Examples of Frequencies of Pesticide Residues Found in the Fruit and Vegetable Commodities of the EU-Coordinated Monitoring Program

Commodity	Year	No. of Samples Analyzed	% Samples with Residues ≤ MRL	% Samples with Residues > MRL
Apples	2004	3133	59	1.8
Apples	2001	2641	47	1.1
Bananas	2002	883	56	1
Grapes	2003	2163	57	5
Grapes	2001	1721	60	1.8
Oranges	2002	2144	78	4
Pears	2002	1330	21	2
Strawberries	2004	2668	63	2.8
Strawberries	2001	1652	51	3.3
Cucumber	2003	1150	24	3
Cucumber	2000	1176	16	1.4
Head cabbage	2004	918	23	2.3
Leek	2004	769	16	1.3
Lettuce	2004	2301	48	3.3
Lettuce	2001	1838	49	3.9
Peas	2003	519	19	2
Peas	2000	730	20	3.0
Peppers	2003	1754	34	6
Tomatoes	2004	2665	36	0.9
Tomatoes	2001	2016	33	1.5

Source: From http://ec.europa.eu/food/fvo/specialreports/pesticides_index_en.htm

pesticides. In 2004, the program included 47 pesticides and the most frequent detections of particular pesticide/commodity combinations were cyprodinil,* fenhexamid,* tolyl-fluanid,* and azoxystrobin,* each found in 13%–34% of the strawberries; iprodione* and dithiocarbamates* in 22%–23% of the lettuce; benomyl group,* chlorpyriphos,[†] diphenylamine,* and captan (-folpet)* in 15%–20% of the apples. Examples of results from the EU-coordinated program 2000–2004 are shown in Table 11.1.

Two studies show the pesticide residues in Egyptian fruits and vegetables. In 1997, 2318 samples of different fruits and vegetables were collected from eight Egyptian markets throughout the country.[18] The samples were analyzed for 54 pesticides. The samples of 19% contained detectable pesticide residues and 1.9% exceeded the MRLs. Root and leafy vegetables showed low contamination frequencies, whereas 29% of the fruit samples contained residues—among them 2.3% violating the MRLs. Dicofol and dimethoate were the most frequently found pesticides. In another study, 78 vegetable samples and 44 fruit samples were

* Fungicide.
† Insecticide.

collected in Alexandria 1997–1998.[19] Cypermethrin, dimethoate, profenofos insecticides, and dithiocarbamate fungicides were analyzed in samples of tomato, eggplant, cucumber, potato, apple, grape, and orange. The most frequent findings were dithiocarbamates in 73% of the tomatoes, 80% of the eggplants and cucumbers, and 50% of the apples and grapes. The concentrations ranged from 0.002 to 0.29 mg/kg. The potatoes of 50% contained fenitrothion at a mean of 0.03 mg/kg. Profenofos was detected in 70% of the grapes in the range 0.005–0.025 mg/kg.

In Brazil, the dithiocarbamates were found in 61% of 520 food samples with the highest levels (up to 3.8 mg/kg) in strawberry, papaya, and banana.[20]

Pesticide residue monitoring has been performed in different Asian countries. In Taiwan, 1997–2003, pesticide residues were detected in 14% of 9955 samples (analyzed for 79 pesticide residues) and 1.2% were violating the MRLs.[21] In India, 60 vegetables were analyzed for organochlorine, pyrethroid, carbamate, and organophosphorus pesticides during 1996–1997.[22] Among the samples (okra, smooth gourd, bitter gourd, cucumber, tomato, and brinjal), 92% contained organochlorine pesticides (OCPs), 80% organophosphorus pesticides, 41% pyrethroides, and 30% carbamates. p,p'-DDT was the most dominant DDT compound, indicating recent use of the DDT in the fields. ΣDDT was, for example, 0.28 ± 0.41 mg/kg in okra. Chlorpyriphos in two brinjal samples exceeded the MRL of 0.2 mg/kg and eight samples of brinjal with triazophos also exceeded the MRL. In 2000–2002, in Karachi, Pakistan, 206 samples of different vegetables were analyzed for 24 pesticides.[23] 63% of the samples contained residues and 46% of them were violating the MRLs. However, the violations showed a downward trend with 62%, 56%, 37%, and 31% during the period 2000–2003. The pesticides that contributed to the violations were methamidophos, cypermethrin, cyhalothrin, carbofuran, and dimethoate. Of the 27 different vegetables analyzed, the violations were found in, for example, 4 of 7 carrot samples, 4 of 6 garlic samples, and 5 of 10 spinach samples.

The growth regulator chlormequat is an example of a pesticide that has been regulated during the period 2001–2006 and due to the systemic effect the residues remained in the pear trees from one year to another, causing residues in the pears even in harvest seasons without application of chlormequat. Chlormequat was studied in UK foods.[24] In 2001, the EU MRL of 3 mg/kg for pears was changed to a temporary MRL of 0.5 mg/kg, which was reduced in two steps to end in 2006 at the LOD level of 0.05 mg/kg. Surveys in 1997 and 1998 showed chlormequat contents of 0.05–16 mg/kg ($n = 54$) and 0.05–11 mg/kg ($n = 48$), respectively. In 1999, the half of 97 pear samples contained chlormequat and 10% exceeded the MRL of 3 mg/kg. In 2000, 79% of 136 samples contained chlormequat, but none of the samples exceeded the MRL of 3 mg/kg. A small survey in 2002 showed that only 42% of 75 samples contained chlormequat all below the MRL of 0.5 mg/kg.

11.2.3.2 Processed Fruits and Vegetables Including Processing Studies

The MRL is established for residues in the whole commodity. Hence for control purposes in the monitoring program, the pesticide residues are mostly determined in raw commodities. However, many foods are eaten after different kinds of processing. The processing of the food is defined as any operation performed on a food or food product from the point of harvest through consumption. The processing may take place when preparing the food at home or be commercial food processing. Typical home processing includes washing, peeling, heating, or juicing, whereas the commercial food processing additionally may include drying, canning, fermenting, oil extraction, refining, preserving, jamming, mixing with other ingredients, and so on.

The processing may affect the pesticide residue levels in the food products mainly by reducing the levels. The extent to which a pesticide is removed during processing depends on a variety of factors such as chemical properties of the pesticide, the nature of the food commodity, the processing step, and time of processing.[25] The reductions may be predicted by the solubility, sensitivity toward hydrolysis, octanol–water proportioning, and the volatility; for example, lipophilic pesticides tend to concentrate in tissues rich in lipids. Examples to the contrary are increased pesticide levels after drying or refining. The effects of processing on pesticide residues in food are compiled in a review by Holland et al.[25] and several studies on the effect of processing on the pesticide levels are made related to commercial or home processing. Information on processing may also be found in the annual pesticide evaluations reported by JMPR[26] and in the EFSA conclusions[27] performed as part of the overall evaluation in the EU. During these joint meetings on pesticide residues (JMPR), selected pesticides have been reviewed including the effects of processing but the company data presented here may be in a compiled form without detailed information.

A majority of the pesticides applied directly to crops are mainly found on the surface of the crops,[25] as the crops cuticular wax serves as a transport barrier for pesticides. Hence, the majorities of the pesticide residues may be found in the peel and when the peel is not an edible part, this will reduce the pesticide levels taken in through the diet. This is often the case for citrus fruits, where an investigation showed that >90% of the pesticide residues were found in the peel.[28]

In a study on apple processing, juicing and peeling significantly reduced the levels of 14 pesticides investigated compared with the unprocessed apple.[29] However, none of the pesticide residues were significantly reduced when the apples were subjected to simple washing or coring. The effect of processing was compared for two different apple varieties, Discovery and Jonagold, and the pesticides selected for field application were the most commonly used in the Danish apple orchards or those most often detected in the national monitoring program. The concentrations of chlorpyriphos in unprocessed and processed apples (Figure 11.1) show, for example, that peeling reduced the chlorpyriphos concentration by 93% and juicing reduced the chlorpyriphos concentrations by 95% compared with the unprocessed samples.

In a study on commercial processing, samples of tomatoes, peppers, asparagus, spinach, and peaches were exposed to three insecticides and four dithiocarbamates.[30] In most cases, canning operations gradually decreased the residue levels in the finished product, particularly through washing, blanching, peeling, and cooking processes. Washing and blanching led to >50% loss in pesticide residues except for peaches. The total amount of pesticide reduced by all the combined canning operations ranged from 90% to 100% in most products.

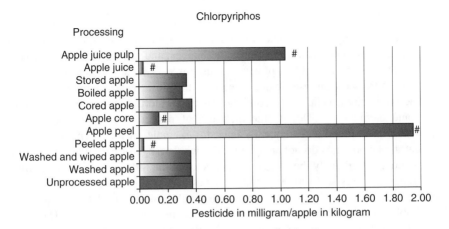

FIGURE 11.1 Concentrations (mg/kg) of chlorpyriphos in apples of the variety Discovery before and after different kinds of processing. (After Rasmussen, R.R., Poulsen, M.E., and Hansen, H.C.B., *Food Addit. Contam.*, 20, 1044, 2003.) #: Significant changes at the 95% confidence level ($n = 5$).

The reduction of pesticides are not necessarily beneficial; the pesticide may be degraded to a metabolite more hazardous like the ethylenethiourea (ETU) formed during degradation of dithiocarbamates. The formation of ETU was studied for different food processing steps, for example, 80% of ethylenebisdithio-carbamate was metabolized to ETU in the drinkable beer.[31] The persistence of the ETU varies in different matrices and it may be stable for up to 200 days in canned tomato puree.[32]

Both the reduction due to processing of wine and the pesticide residues in wine were compiled in an Italian study.[33] The different pesticides behaved differently according to their physicochemical properties and some of the pesticides disappeared totally or partly during the wine-making either due to degradation in the acidic environment, degradation during the fermentation process, or adsorption by the lees and the cake. Only a few pesticides passed from the grape to the wine without showing appreciable reduction among them: dimethoate, omethoate, metalaxyl, and pyrimethanil. In 1998–1999, 449 wine samples were analyzed for ~120 pesticides. Only very few pesticides were found in wine and at low levels.

Pesticide residues in processed food are monitored, for example, within EU. In 2004, 6% of the samples or 3678 samples were processed products.[16] The percentages of monitoring samples with residues were significantly lower in processed food than in fresh products. Residues ≤MRL were found in 24% of the samples, and residues exceeding the MRL were found in 1.2% of the samples. The processed food comprise of many kinds of food including vegetable oil, canned products, olives, cereal products, beverages, juices, and wine. In the EU-coordinated monitoring program, 704 samples of orange juice were taken. Residues below or at MRL were detected in 23% of these samples and in 2.3% of the juice samples, the pesticide concentrations detected exceeded the MRLs.[16]

11.2.3.3   Cereals

Cereals cover a range of crops like wheat, rye, barley, rice, maize, and millet. Cereals are sprayed with insecticides, fungicides, herbicides, and growth regulators through the whole growing period. To protect against insects, the stored cereals are often postharvest-treated with insecticides. Therefore, the most frequently found pesticides are the insecticides malathion, pyrimiphos-methyl, chlorpyrifos-methyl, deltamethrine, and dichlorvos.[34] Despite the high use of pesticides in cereal production, residues can be found less frequently than, for example, in fruits. The reason may be that the laboratories do not analyze for the whole range of pesticides used in the production. Additionally the samples, if collected at the mills, can be mixtures from different producers with different usage of pesticides and the individual pesticide residues can therefore be diluted to below the analytical limit of detection.

Published data on pesticide residues in cereals are relatively scattered. The major part of the data found and presented later, covering the period from 2000 onward, are from the United States and Europe. No data were found either from South America, Africa, or Australia. Data from Asia are from the two biggest nations, India and China, and cover therefore the majority of the population of this region.[35–38] However, the results consist only of data on DDTs and HCHs. This reflects most likely the usage pattern of these compounds, which are effective and cheap, but also that the laboratories have not, due to lack of capacity, included the newer developed pesticides in their monitoring program. From Table 11.2, it is seen that in India and China, DDTs and HCHs were frequently found in rice and wheat.

Since 1991, the U.S. Department of Agriculture (USDA) has been responsible for the pesticide residues testing program in cereals produced in the United States. The data for 2000–2003 for the five major cereal types, barley, corn, oats, rice, and wheat, are shown in Figure 11.2.[39–42]

TABLE 11.2
Pesticide Residue Results in Cereals from Asia

	ΣDDT g/kg	ΣHCH mg/kg	Other Pesticides mg/kg	Number of Samples	Year of Sampling	References
India—rice	0.023	0.066		30		[35]
India—rice	0.01 (57.7%)	0.013 (64.4%)		2000		[36]
India—wheat	0.22	2.99	7.9-ΣHeptachlor, 0.17-Aldrin	150		[37]
China—cereals	0.0045 (5.0%)	0.0011 (53.0%)		60	2002	[38]
China—cereals	0.0252	0.0053			1999	[38]
China—cereals	0.0019	0.0048			1992	[38]

Note: Numbers in brackets are percentage of samples with residues.

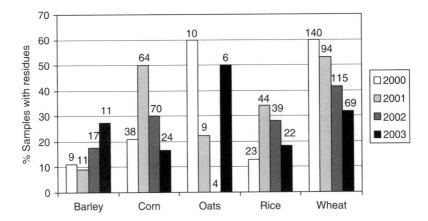

FIGURE 11.2 Frequencies of samples with residues from 2000 to 2004 in barley, corn, rice, and wheat produced in the United States. The values above the bars are the numbers of samples analyzed. (Data from US FDA, http://www.cfsan.fda.gov/~dms/pes00rep.html#table_1; http://www.cfsan.fda.gov/~dms/pes01rep.html#table_1; http://www.cfsan.fda.gov/~dms/pes02rep.html#table_1; http://www.cfsan.fda.gov/~dms/pes03rep.html#table_1).

Approximately 200 (ranging from 132 to 245) samples are analyzed each year; most of them are wheat and corn. Residues above the MRLs were found in 2.3% of the rice samples in 2001 and in 5.7% of the corn samples in 2002. Figure 11.2 shows that sample type with the highest frequency of residues was wheat and that barley had the lowest frequency of residues. For corn, rice, and wheat, a decrease has been detected from 2001 to 2003. No information is given on the specific pesticides found in the cereals as only a common list for all the commodities is given. Nevertheless, the growth regulator, chlormequat, and the herbicide, glyphosate, were not included in the list and were probably frequently used. Therefore, the number of samples with residues may have been higher than reported.

Reynolds et al.[24] investigated the levels of chlormequat in UK-produced cereals from 1997 to 2002 and found residues in 50% of 59 wheat samples at 0.05–0.7 mg/kg and in 41% of 45 barley samples at 0.06–1.1 mg/kg; none of the samples exceeded the UK MRLs for grain. A similar study from Denmark showed that chlormequat was found in wheat, rye, and oat in 71%, 60%, and 100%, respectively.[43] Glyphosate was found in more than half of the monitored cereal samples produced in Denmark from 1988 to 1999.[44]

In 2004, the number of cereal samples analyzed and compiled by the European Commission were 2719 and the percentages of samples with residues ≤MRL and exceeding the MRL (national or EC-MRL) were 29% and 1.1%, respectively. The data do not include information on the type of cereals analyzed. In 2004, the most frequently found pesticides were pirimiphos-methyl,* malathion,*

---

* Insecticide.

TABLE 11.3

Results from the EU-Coordinated Monitoring Program in Relation to Cereals

	Year of Sample Collection	Number of Samples	% Samples with Residues below MRL	% Samples above MRL	Number of Pesticides Included in the Analytical Program
Wheat	1999	1159	21	0.5	20
Wheat	2003	1021	22	0.3	41
Rice	2000	869	8.7	1.7	20
Rice	2003	635	12	1	41
Oats/rye	2004	775	19	0.6	47

Source: From http://ec.europa.eu/food/fvo/specialreports/pesticides_index_en.htm

chlorpyriphos-methyl,* chlormequat,§ deltamethrin,* chlorpyriphos,* glyphosate,†
bromides,‡ dichlorvos,* and mepiquat.§ The pesticides found were mainly insecticides and the list confirms the findings of previous years.

Apart from the national monitoring program, the Commission conducts a coordinated monitoring program, where cereals are included regularly. Wheat has been
included in 1999 and 2003, rice in 2000 and 2003, and rye/oats in 2004. The data show
that residues were found in 21%–22% of the wheat samples (Table 11.3) and the three
most frequently found pesticides were the insecticides pirimiphos-methyl, chlorpyriphos-methyl, and deltamethrin. For rice, the samples with residues were 8.7%–12%
and the three most frequent pesticides were the insecticides pirimiphos-methyl, deltamethrin, and the dithiocarbamate fungicides. The results for 2004 do not distinguish
between oats and rye, but all together residues were found in 19% of the samples.

Table 11.4 shows details from the coordinated program of residues found in
cereals from the two major cereal-producing countries in the EU, France, and
Germany.[15,16] The residues found in oat/rye and wheat were mainly organophosphorus insecticides such as malathion, pirimiphos-methyl, and chlorpyriphos-methyl.
However, the pyrethroid insecticide deltamethrin and different fungicides like imazalil, dithiocarbamates, and thiabendazol were also present. Since chlormequat and
glyphosate were not included in the monitoring program, the frequency of samples
with residues was, like in the samples from the United States, probably higher. In
Germany, exceedances of MRL were found for pirimiphos-methyl, thiabendazole,
and metalaxyl.

Storage may have only very small effect on the degradation of pesticide residues.
However, the temperature and humidity influence the degradation during storage
and malathion residues can be decreased by 30%–40% over 32 weeks at 30°.[45]

---

* Insecticide.
† Herbicide.
‡ Indicator of bromofumigants.
§ Growth regulator.

TABLE 11.4

Residues in Wheat (2003) and Oat/Rye (2004) Found in Samples from France and Germany

Pesticide	No. of Samples	With Residues ≤ MRL (%)	Residues > MRL (%)	Highest Concentration mg/kg	MRL mg/kg
Oat/rye					
France					
Chlorpyriphos-methyl	106	8		0.02	3
Malathion	106	22		2.4	8
Parathion	106	1		0.01	0.05
Pirimiphos-methyl	106	2		0.26	5
Total	106	26			
Germany					
Chlorpyriphos	180	1		0.012	0.05
Imazalil	155	1		0.01	0.02
Dithiocarbamates	103	7		0.46	1
Pirimiphos-methyl	180	18	0.6	7.9	5
Thiabendazol	132	1	0.8	0.056	0.05
Tolylfluanid	180	1		0.051	0.1
Total	180	23			
Wheat					
France					
Chlorpyriphos	131	12		0.04	0.05
Chlorpyriphos-methyl	131	18		0.15	3
Deltamethrin	131	23		0.12	1
Malathion	131	28		3.1	8
Pirimiphos-methyl	131	43		2.45	
Total	131	59			
Germany					
Benomyl	159	1		0.034	0.1
Chlorpyriphos	234	0		0.03	0.05
Dimethoate	238	0		0.007	0.3
Imazalil	230	3		0.014	0.02
Malathion	231	1		0.018	8
Dithiocarbamates	110	5		0.03	1
Metalaxyl	196	1	0.5	0.02	0.01
Parathion	198	1		0.005	0.1
Pirimiphos-methyl	235	11		0.76	5
Procymidone	195	1		0.013	0.02
Propyzamide	192	1		0.018	0.02
Total	238	16			

Sources: From EU, Monitoring of Pesticide Residues in Products of Plant Origin in the European Union, Norway, Iceland and Liechtenstein 2003, Report Summary, 2005. Available at http://www .Europa.eu.int/comm/food/plant/protection; EC, Monitoring of Pesticide Residues in Products of Plant Origin in the European Union, Norway, Iceland and Liechthenstein 2004. Report from the European Commission, SEC(2006)1416, 2006. Available at http://ec.europa.eu/food/fvo/ specialreports/pesticide_residues/report_ 2004_en.pdf

For wheat, Uygun et al.[46] have reported 50% degradation over 127 days of malathion and 30% of fenitrothion over 55 days. Residues of pesticides are greatly reduced by milling. Most residues are present in the outer part of the grain, and consequently the reduction, for example, from wheat to sifted flour can be as high as 90%, whereas the concentration in the bran increase compared with the whole grain. Further cooking reduced malathion and its degradation compounds if the grains were boiled in water.[47]

11.2.3.4   Food of Animal Origin

Pesticide residues occur in animals as a result of both previous and present uses of pesticides for agricultural purposes. The residue levels in products of animal origin are, however, generally low or nondetectable (<0.01 mg/kg). The residues ingested by, for example, livestock via the feed are metabolized by the animals and for most pesticides, in particular, in the case of the more modern pesticides the major part of the pesticides/metabolites is excreted. The highest levels of pesticide residues (including metabolites) are most often observed in organs involved in the metabolism and excretion of the pesticides, for example, liver and kidney.

In reference to products of animal origin, the focus has mainly been on persistent pesticides, authorized and used in large amounts and for a wide range of purposes from the 1950s to 1970s. The very efficient pesticides such as OCPs were also later found to be very stable in the environment, to bioaccumulate through the food chain, and to pose a risk of causing adverse human health effects.

In most parts of the world, the use of the environmentally persistent pesticides has been reduced dramatically during the last decades. The use of, for example, DDT was restricted in the United States, Canada, and most European countries in the early 1970s. In several developing countries with the need for malaria control, DDT has been used until the end of the 1990s. Other OCPs like dicofol were still in use in 2003 in, for example, China, and DDT is still authorized in different parts of the world for, for example, malaria control.[48,49]

OCPs are detectable in most matrices of animal origin especially matrices with high fat content like butter[50] (Table 11.5), cheese,[55] milk,[54,56,57] and meat.[58–60] The levels are dependent on the age of the animals at the time of slaughter[58] and the fat content of the product,[59] that is, the older the animal and the higher the fat content, the higher is in general the residue level of OCPs. In areas where the organochlorine compounds were recently or are still in use, legally or illegally, the residue levels are in some cases at or above the MRLs (Table 11.5).

Results from the Danish Monitoring program (1995–1996[43] and 1998–2003[51]) have shown that OCPs are detectable, but below the MRL, in more than half of the animal product samples analyzed ($n = 1408$). The animal products include meat, fish, butter, mixed products of butter and vegetable oils, cheese, animal fat, and eggs. In animal fat, ΣDDT was detected in the majority of the samples (about 65%) but at low levels (mean levels ≤ 15 μg/kg fat). α-HCH was detected in $<1\%$ of the samples of animal origin (excluding sea food) and at mean levels ≤ 0.5 μg/kg fat. Dieldrin was detected in $<10\%$ of the samples of animal origin at mean levels

TABLE 11.5

Residue Levels (µg/kg Fat Unless Otherwise Stated) of the Sum of DDT and Its Degradation Products DDE and DDD (ΣDDT) and Sum of HCH Isomers (ΣHCH) in Bovine Butter

Country (Year of Sampling)	ΣDDT	ΣHCH	Number of Samples/ Number of Samples with Detectable Residues	References
Denmark (1998–2003)	2	α-HCH: n.d.	DDT: 126/26 HCH: 126/0	[51]
Spain (≤2000)	p,p′-DDE: 7.3 (0.02–52.5) µg/kg wet weight	γ-HCH: 10.8 (0.0039–19.59) β-HCH: 3.2 (0.01–9.1) µg/kg wet weight	γ-HCH: 36/36 β-HCH: 36/34 p,p′-DDE: 36/35 HCB: 36/32	[52]
Turkey (~2000)	p,p′-DDT, p,p′-DDE, and p,p′-DDD all <0.001 mg/kg	γ-HCH < 0.001 mg/kg	100/0	[53]
Canada (≤2000)	5.77 (0.38–16.92)	1.21 (0.13–2.10)	6	[50]
United States (≤2000)	23.61 (0.41–141.26)	1.33 (0–2.17)	18	[50]
Australia (≤2000)	5.96 (1.44–13.78)	0.31 (0–0.86)	5	[50]
India (~2004)	120 µg/kg	0.132 µg/kg	46	[54]

Note: Mean values are presented and minimum and maximum values are presented in brackets.

≤6 µg/kg fat. None of six organophosphorus pesticides included in the analysis were detected in the 231 analyzed samples of pork and bovine meat.

The frequencies of DDT, HCH, and dieldrin found in Japanese samples during 2000–2004[61] are higher compared with the Danish data. ΣDDT was detected in 64%, 90%, and 90% of beef ($n = 25$), pork ($n = 30$), and poultry ($n = 20$) samples. ΣHCH was detected in 24%, 23%, and 20% of the Japanese samples, respectively. Dieldrin was detected in 24%, 23%, and 45% of the samples.

In several studies, butter has been analyzed as a representative of animal products with high fat contents and the levels found can be used as an indicator of the general OCP levels in animal products. The results show that the levels found in butter originating from countries like India and Mexico are higher compared with butter originating from countries such as Denmark and Germany. The results are in good agreement with the fact that the persistent OCPs were banned earlier in the latter countries than they were in the former countries.

Weiss et al.[49] have performed a worldwide survey of, among other compounds, DDT and HCB. One sample of butter was sampled from 39 European countries and from 25 non-European countries. It was found that the average level of ΣDDT in butter from all the participating countries was 10.8 μg/kg fat. The average level of HCB in butter from all the participating countries was 3.5 μg/kg fat.

11.2.3.4.1 Organochlorine pesticides in fish
The residue levels of OCPs in fish vary greatly depending on the origin. In general, higher levels are observed in seafood caught in waters close to pollution sources, for example, some coastal waters. The levels are also, in general, positively correlated with the age and the fat content of the organism.

OCPs can be found in large fractions of seafood even from waters of countries, where the compounds have been banned for several decades. Table 11.6 presents some reported levels of DDT, HCH, and dieldrin in different seafood samples caught in different parts of the world. A large study has been performed on the levels of OCPs in seafood from Taiwan, showing that OCPs were detectable in 24% of the fish samples and organophosphorus compounds in 11% of the fish samples ($n = 607$). The detection rate was lower in shellfish, that is, OCPs in 6% and organophosphorus compounds in none ($n = 62$). The mean residue level of ΣDDT in all of the sampled seafoods with detected residues was 32.5 μg/kg fresh weight.

OCPs occur in seafood samples from all over the world but the residue level of DDE, DDD, and HCHs has been reported to decrease with time. The level of, for example, DDE and DDD in cod liver from the Arctic has been reported to decrease from a level of 60 and 45 μg/kg fat in 1987/1988 to levels of 40 and 15 μg/kg fat in 1995/1998, respectively.[67]

11.2.3.5 Infant and Baby Food

Infants and children consume more foods per kilo body weight per day than adults do. Furthermore, the detoxification systems of the infants are not fully developed. These are some of the factors that make infants and young children a sensitive group of consumers. The primary food intake for infants (0–6 months of age) is accounted for by either human breast milk or formulae. As the child gets older, an increasing proportion of the daily food intake is accounted for by vegetables, fruits, and cereals and to some extent also food of animal origin, either prepared at home from raw products or as preprocessed products. Different preferences are expected in regard to which types of foods are introduced to young children and at what age, depending on the different traditions in different population groups and different countries.

Special attention has been directed toward pesticide residues in infant and weaning foods marketed as such. In 1999, Directive (99/39/EC)[68] was adopted by the European Commission to insure low residues of pesticides in these products. By this directive, the MRLs for individual pesticides in baby foods were set at 0.01 mg/kg (in many cases, corresponding to the detection level) and the use of certain pesticides for treatment of crops intended for the production of baby foods were banned. The directive only applies to infant and baby food products on the European market.

TABLE 11.6
Examples of Reported Residue Levels (μg/kg Fresh Weight) of the Sum of DDT and Its Degradation Products DDE and DDD (ΣDDT) and Sum of HCH Isomers (ΣHCH) in Fish and Seafood Samples (Unless Otherwise Stated)

Place and Year of Sampling	Matrix	ΣDDT μg/kg	ΣHCH μg/kg Fresh Weight	Dieldrin	Number of Samples/ with Detectable Residues	References
Greenland (1994–1995)	Polar cod liver	44 (16–122)	33 (13–55)		16	[62]
Greenland (1994)	Polar cod liver	(12–83)	(6–16)		77	[63]
	Cod liver	(60–98)	(7–9)		25	
Danish waters (1995–1996)	Cod liver	(12–918) μg/kg fat	(3–33) (α-HCH)	(4–51)	40	[43]
	Herring	(2–61)	(2–4) (α-HCH)	(2–9)	66	
	Mackerel	(2–5)	(2–3) (α-HCH)	(2–3)	41	
Danish waters (1998–2003)	Cod liver	260 μg/kg fat	2 (α-HCH)	18	111[a]	[51]
	Herring	8	0.4 (α-HCH)	3	219[b]	
	Mackerel	4	0.6 (α-HCH)	3	20[c]	
Adriatic Sea (1997)	Mackerel	(19–33) μg/kg fat	(0.20–0.83) (γ-HCH) μg/kg fat	(0.6–1.2)	52	[64]
China (1995/1996)	Cod	(281–399) μg/kg fat	6 (γ-HCH) μg/kg fat	(10–12)	49	[65]
	Shellfish	138 (4–479)[d]	4.6 (ND–17.2)[e]		12 sites and 50–80 specimens per site	
Taiwan (2001–2003)	Fish, shellfish, bivalve, crustacean, and cephalopod	32.5 (0.6–169.1)		0.9 (0.6–31)[f]	920[g]	[66]

Note: Mean values are presented and minimum and maximum values are presented in brackets.

[a] DDT, HCH, and dieldrin were detectable in 100%, 70%, and 95% of the cod liver samples, respectively.

[b] DDT, HCH, and dieldrin were detectable in 100%, 41%, and 94% of the herring samples, respectively.

[c] DDT, HCH, and dieldrin were detectable in 100%, 75%, and 100% of the mackerel samples, respectively.

[d] The levels were presented in microgram per kilogram dry weight (692 (22–2396)) and has therefore for comparison been divided by five, assuming a water content of 80%.

[e] The levels were presented in microgram per kilogram dry weight (23 (ND–85.9)) and has therefore for comparison been divided by five, assuming a water content of 80%.

[f] Dieldrin was detectable in 9 of 920 samples.

[g] Residues were detectable in 176 of 920 samples.

11.2.3.5.1   Residues in human breast milk

Human breast milk has a high fat content and for that reason a major concern in relation to pesticide residues in human breast milk worldwide is the environmentally stable pesticides, for example, OCPs. During breast-feeding, OCPs from the mother are excreted via the milk to the baby.

The levels of OCPs vary and depend on the age of the mother,[69,70] whether the mother has been breast-feeding before,[70] her eating habits (e.g., the amount of fatty fish),[70,71] and place of living, that is, whether there are OCPs in the local environment including the food.[71] The OCPs are of concern since they are under suspicion for having the potential to affect, for example, the birth weight of infants, the risk of cancer, and the neurodevelopment of infants.

The levels of OCPs have been shown to be higher in human breast milk from the population of Asian countries such as China, India, Cambodia, and Indonesia compared with European/North American countries such as UK, Germany, Sweden, Spain, and Canada.[69] Wong[69] have reported that the levels of DDT, DDE, and β-HCH in human breast milk are 2–15-fold higher in samples from China compared with samples from several European countries.[69] Examples of residue levels of ΣDDT and ΣHCH are shown in Table 11.7.

TABLE 11.7

Examples of Reported Residue Levels (μg/kg Fat) of the Sum of DDT and Its Degradation Products DDE and DDD (ΣDDT) and Sum of HCH Isomers (ΣHCH) in Human Breast Milk

Origin of Samples (Year of Sampling)	Level of ΣDDT μg/kg Fat Unless Other Ways Stated	Level of ΣHCH	Number of Samples Analyzed	References
Industrialized countries				
Finland and Denmark (1997–2001)	129 (31–443)	β-HCH: 13 (2.7–66) μg/kg fat	130	[72]
Germany (1995–1997)	DDT: 240 (27–1,540)	β-HCH: 40 (4–50)	246	[73]
UK (1997–1998)	DDT: 40, DDE: 430		168	[69]
Developing countries				
China (1999–2000)	DDT: 545 DDE: 2,665	β-HCH: 1,030	169	[74]
Vietnam (2000–2001)	DDT: 218 (34–6,900) DDE: 1,950 (340–16,000) ΣDDT: 2,200 (440–17,000)	β-HCH: 36 (4–160)	86	[70]
Zimbabwe (1999)	p,p'-DDE: 4,863 p,p'-DDT: 1,149 ΣDDT: 6,314	β-HCH: 216 γ-HCH: 99 ΣHCH: 383	116	[75]

Note: Mean values are presented and minimum and maximum values are presented in brackets.

The levels of OCPs in breast milk are, in general, decreasing as a result of the banning of the compounds and/or the restrictions on the uses.[71,73,74] The ΣDDT levels in breast milk from women in Taiwan sampled in 2001 have, for example, been found on average to contain 333 μg/kg milk fat (36 samples), whereas the levels of ΣDDT in breast milk sampled in Taiwan in the previous two decades on average amounted to 3595 μg/kg milk fat.[48] In breast milk from German women, the level of DDT has been found to be ~81% lower in 1995/1997 (240 μg/kg milk fat) than it was 10 years earlier.[73]

In milk from women in Indonesia (sampled 2001–2003), great differences in the levels of ΣDDT and ΣHCH have been observed. The higher levels were observed in suburban and rural areas and the lower levels in the urban areas.

Even though OCPs occur in human breast milk and therefore is consumed by infants, a literature search and Web search do not reveal any authorities or researchers that recommend avoiding breast-feeding. Thus the benefits counterbalance any possible health risk in connection with pesticide residues.

11.2.3.5.2 Residues in formulae and weaning products
Formulae and weaning foods are highly processed foods and processing most often reduces the levels of the pesticides. Especially, thermolabile pesticides are not expected to be detectable in infant formulae or weaning foods, since these products have been heat-treated during processing and for preservation. Furthermore, raw products for the production of weaning foods are washed and perhaps also peeled. The formulae available on the market are based on cow's milk or soya or a combination of the two. Weaning foods are, for example, fruit and vegetable puree, fruit juices, cereal-based meals, complete meals composed of, for example, vegetables, pasta and meat, and biscuits.

Formulae has a relatively high fat content (~25 g/100 g) and OCPs are therefore of relevance in reference to pesticide residues also in these products. Pesticides such as the organophosphorus, carbamates, and pyrethroids have not been found to accumulate in fat and milk of livestock to any significant degree and no residues of these pesticides have been detected in 1008 samples of U.S. manufactured milk-based infant formulae samples.[76]

Lackmann et al.[77] have shown that the intake of organochlorine compounds, for example, DDT and DDE is significantly higher for breast-fed infants than for bottle-fed infants in Germany. The serum concentration of DDE in breast-fed infants were about 6 times higher after 6 weeks of feeding compared with the serum concentration in bottle-fed infants. Whether this relatively large difference is maintained, also after a longer period of breast-feeding, has not yet been reported.

Higher levels of OCPs are, in general, found in formulae and weaning products produced in developing countries and lower levels in products from developed countries. In infant formulae collected from the Indian marked during 1989 residues of ΣDDT and ΣHCH were found in 94% and 70% of the samples, respectively. A total of 186 samples of 20 different brands were analyzed. The mean level of ΣDDT was found to be 300 μg/kg fat and the mean level of ΣHCH was found to be 490 μg/kg fat.

In weaning foods, it is likely to detect pesticide residues due to use of pesticides during cultivation of the raw products. In the U.S. Total Diet Studies (2000), a total of 78 items of different baby foods were analyzed. The most frequently found pesticide residues were the insecticides carbaryl (18%), endosulfan (17%), malathion (12%), and chlorpyriphos-methyl (10%) and the fungicide iprodione (12%). The highest level of 0.096 mg/kg was found for iprodione.[78]

Residues were only detected in one of a total of 181 samples of baby food collected within the Danish Food Monitoring Programme 1998–2003.[79] The sample with a residue of chlormequat (0.025 mg/kg) and mepiquat (0.019 mg/kg) was a cereal powder.

Cressey and Vannoort[80] analyzed 25 infant formulae and 30 weaning foods, commercially available in New Zealand in 1996. Soy-based formulae and weaning products were screened for about 140 pesticides and the milk-based formulae for OCPs (p,p'-DDE, p,p'-DDT, and dieldrin). p,p'-DDE was found in 7 of 20 milk-based infant formulae and residues of p,p'-DDT were found in one milk-based infant formulae. Dieldrin was detected in four of five soy-based formulae. Dithiocarbamates (LOD of 100 μg/kg) were not found in any of the soy-based formulae or any of the weaning foods. Cressey and Vannoort did not analyze for ETU, the degradation product of dithiocarbamates. Two organophosphorus pesticides, azinphos-methyl and pirimiphos-methyl, were detected in one soy-based formulae and in two out of nine cereal-based weaning foods, respectively.

In the Australian 19th total diet survey, residues were found in cereal-based infant foods but not in formulae, infant desserts, or dinners. The pesticides detected in the cereal-based products were chlorpyrifos-methyl (4 μg/kg), fenitrothion (2 μg/kg), iprodione (4 μg/kg), and piperonyl butoxide (8 μg/kg). The residue levels (nine samples) were low and the mean level ranged from 2 to 8 μg/kg. Thus, no residues of DDT, DDE, or other OCPs were detected in formulae or infant foods but some organophosphorus pesticides were found in cereal-based products.[81]

11.3 CONSUMER EXPOSURE AND RISK ASSESSMENT

11.3.1 Dietary Intake Estimation

To perform a dietary intake estimation is basically easy. The consumption is multiplied with the content:

$$\text{Intake} = \text{consumption} \times \text{content}.$$

The question is, however, what consumption and which content should be used. Should it be mean values? Should it be high values, a kind of worst-case situation? And how is the consumption respectively the content estimated? There is not one way of performing dietary intake estimations and in the literature different ways of performing the estimations have been used and in addition the data collection has been very diverse. Therefore, it is also often very difficult to compare the dietary estimations directly.

Dietary surveys can be performed in many ways. In some surveys, participants are asked to fill out a diary about what they have been eating and the amounts; in others, people are interviewed about what they have been eating, for example, yesterday. In some surveys, the food bought in the household is used for the estimation of the consumption. Here, the total amount of, for example, potatoes is divided by the number of people in the household and number of meals, where the potatoes are eaten. Both number of participants, number of days, and the details concerning the food eaten differ between dietary surveys. In many circumstances, the food as eaten are calculated back by using recipes to ingredients or RAC; for example, an apple pie is divided into flour (grain), apple, and other ingredients.

For total diet studies and duplicate diet studies, however, the content is directly determined in the food eaten. In total diet studies, a certain number of raw and prepared foods are chosen to represent the total diet of the population. The foods are then bought and prepared according to recipes and the content of the pesticides or other substances are directly determined in the foods. In the duplicate diet studies, the participants collect exactly the same amount of food as they eat and the pesticides are then determined in the collected foods.

Data concerning the content of the pesticides often comes from monitoring or surveillance. These studies differ widely in regard to which pesticides that are included, the number of pesticides included, and the number of commodities included.

In the calculation of the dietary exposure, other factors such as correction for undetectable residues or processing also influence the result. Although a pesticide is not detected in a commodity, this does not necessarily mean that it is not present; just that the level could be lower than the analytical LOD. In some calculations, the undetectable residues are set at, for example, $\frac{1}{2}$ LOD or another factor. It is known that for examples peeling or boiling can reduce the amount of pesticides, whereas drying (e.g., grapes to raisins) can increase the content of the pesticides. To perform the most reliable estimation of the dietary exposure, processing factors should be included if available.

Dietary exposure calculations can be performed with different approaches, deterministic or probabilistic, and for both chronic and acute intake. The chronic intake or the long-term intake is the possible intake over a long time, for example, a whole life and in the risk assessment this intake is compared with the ADI. The acute intake or the short-term intake, on the other hand, is the intake within 24 h or less, for example, a meal. The acute intake is compared with the ARfD in the risk assessment.

11.3.1.1 Deterministic Approach (Chronic and Acute Intake)

A calculation of the chronic intake by the deterministic approach yields only a single value for the intake and is also called a point estimate. In this approach, a single value of the consumption of a commodity is multiplied with a single value of the concentrations of residues. Often consumption and concentrations are average values, but they can also be high percentiles if a worst-case calculation is performed. If the chronic intake for a certain pesticide from all commodities is calculated, the single intakes for each possible commodity are summed.

The acute intake in the deterministic approach is always calculated for a single commodity. Depending on the commodity, different equations defined by JMPR[82] are used for calculation of the acute intake. In the two most often used equations, the so-called variability factor is included. This factor is based on the variation of the residues in a composite sample. In monitoring, usually analyzed samples are composite samples, while all the content of a pesticide found can be from just one sample. In an estimation of the acute intake, the intake from this one sample is of interest and the variability factor is an expression used to estimate the content in a single sample from the content in a composite sample.

The deterministic approach is the absolute most often used method for the calculations of pesticide intakes. The advantages of the approach are that the approach is easy and simple to perform and the results are easy to interpret. The drawback of the approach is that the exposure is expressed as single values because single values are used for both consumption and content regardless of the variability in both variables. Thus, intakes determined by the deterministic approach are generally highly overestimated.

11.3.1.2   Probabilistic Approach

Probabilistic modeling is called so because this approach yields the probability for an intake. In this approach, the whole distribution of consumption data and concentrations are used in the calculations, resulting in a new distribution for the intake; a consumption of a commodity is chosen; and a residue in this commodity is chosen; and the two values are multiplied to yield an intake. Then a new consumption value and a new residue value are chosen. This is done several times, for example 100,000 times, resulting in a distribution of intakes[83] (Figure 11.3). In this way, percentiles of the intake can be determined. The probabilistic modeling determines the acute or short-term intake, if it is the consumption for a meal or a day that is used in the calculation. Algorithms to calculate the chronic intake have become a part of some programs and the chronic intake can be compared with the ADI. The advantage of this approach is that all data of both consumptions and concentrations are used, the whole distribution of the intake is shown, and the uncertainties in the calculation can be estimated.

Probabilistic modeling is, at the moment, not widely used. In connection with authorization of pesticides, EPA in the United States use probabilistic modeling as

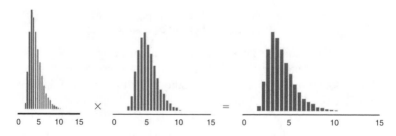

FIGURE 11.3 How a distribution of intake is performed from a distribution of consumption and of contents. (From Pieters, M.N. et al., Probabilistic Modelling of Dietary Intake of Substances, The Risk Management Question Governs the Method, RIVM, Report 3200110012005, 2005. With permission.)

part of their evaluations of the pesticides and have published guidelines for the work.[84] Both in Codex and in the EU, the use of probabilistic modeling is discussed and projects concerning the subject have been initiated.[85–87]

11.3.1.3 Cumulative Exposure

The term "cumulative exposure" can be used in different ways. Some use it as the total intake of a single pesticide from all commodities. The most often applied definition of the term, and the definition used in this book, is that the cumulative exposure is the total intake of all pesticides or a group of pesticides from all commodities. Several approaches can be used[88] but at the moment there is no common agreement on which approach to use to calculate the cumulative exposure for pesticides in our food.

Examples using two different approaches are summarized here, namely the so-called TEF approach for the Danish and Brazilian population and the margin of exposure (MOE) for the U.S. population. In both examples, the cumulative dietary exposure is calculated for choline esterase inhibiting substances (organophosphates and carbamates).

Using the TEF approach, exposures of a group of common mechanism chemicals with different potencies are normalized to yield a total equivalent exposure to one of the chemicals, the so-called index compound (IC). TEFs are obtained as the ratio of the toxic potency at the chosen toxicological end point of the IC to that of each of the other members in the group. This means that a substance with a toxic potency 10 times the IC is assigned a TEF value of 10. The exposure to each chemical is then multiplied by the appropriate TEF for example, 10 to express all exposures in terms of the IC. Summation of these values provides a total combined exposure to all chemicals in terms of the IC.

To assess the cumulative risk of the exposure in the United States, the total MOE[75] is used. MOE for a single chemical is the ratio of the effect dose level (ED) at the chosen toxicological end point to the level of dietary exposure.

$$MOE = \frac{ED}{exposure}.$$

The combined MOE is

$$Combined\ MOE = \frac{1}{1/MOE1 + 1MOE2 + 1/MOE3}, etc.$$

The greater the MOE, the lesser is the risk. In the assessments, a target value of 100 is acceptable. MOEs <100 are undesirable.

11.3.2 INTAKE CALCULATIONS OF PESTICIDE RESIDUES

In this section, intake calculations or dietary exposure from different parts of the world using different approaches are presented.

11.3.2.1   Deterministic Approach

The results from the EU-coordinated monitoring program are used to calculate both the chronic and acute intake every year.[16] For the chronic intake, the 90th percentiles of all the results was used as concentration giving a worst-case situation. Consumption figures were taken from the WHO Standard European Diet, using 60 kg as body weight. The exposure remains well below the ADI for all combinations of pesticide and commodity, ranging from 0.009% of ADI for fenhexamid in strawberry to 5.36% of the ADI for the dithiocarbamates in lettuce. For the acute intake, the data for the high or acute consumption from UK were used and the calculation was performed with the highest residue found. For eight pesticides (deltamethrin, dimethoate, lambda-cyhalothrin, dithiocarbamates, methamidophos, methidation, methomyl, and oxydemeton-methyl), this gave results exceeding the ARfD. The highest intake was 47 times the ARfD for oxydemeton-methyl in apples.

Using different figures for the residue contents, for example, MRLs or data from monitoring programs can have a great impact on the intake, which are shown by an example from a Korean intake estimate.[89] Data for consumption was from a 1998, Nation Nutrition Survey and data for residue levels were found analyzing 6164 samples representing 107 different kinds of food commodities. All the samples were collected at the same commercial market in 2001. Using MRLs as residues caused that for 16% of the analyzed pesticides the intake exceeded the ADIs (Codex values). When residue levels from the analyzed samples were used, the intake decreased dramatically to <0.15% of ADI.

For some studies, the pesticide intakes are limited to include the OCPs that bioaccumulate. From Uzbekistan, estimated monthly intake for the different isomers of HCH were 1–60 μg/month for α-HCH; 2–140 μg/month for β-HCH depending on food type.[90] The samples used for analyses were collected from three towns in February 2001. Twelve different food types were collected among them: animal products, fish, oils, vegetables, and cereals. The consumption data were collected in one city and comprised of 101 households. Information was gathered about the food consumption in the previous month.

11.3.2.2   Total Diet and Duplicate Diet Studies

In India, during 1999–2002, a kind of duplicate diet study was performed for men aged 19–24 years.[91] Every month vegetarian and nonvegetarian total diet samples comprising breakfast, lunch, and dinner were collected. Lindane was the pesticide most widely found, but the frequency decreased throughout the study from about 90% in 1999 to about 25% in 2002. The Codex ADI for lindane of 0.008 mg/kg body weight was exceeded in 1999 for the vegetarian diet and in 1999 and 2000 for the nonvegetarian diet. An explanation for the high contribution from lindane could be that about 21% of the consumption in the study came from milk and milk products, which another study showed could be highly contaminated with lindane.

In Kuwait, the dietary exposure to organophosphate pesticides was determined in the total diet and the food consumption survey was conducted as a 24 h dietary recall study.[92] The 6700 participants were interviewed about what they had consumed the previous day. All together 140 food items, divided into 11 food categories, were selected to represent the list of the Kuwaiti total diet study. The Kuwaiti diet is characterized by a high intake of cereals (grain), vegetables, and fruits and >90% of the food is imported. Intakes for 19 different age groups were calculated and all amounted to <6% of the ADIs. For males 15–19 years, diazinon contributed most to the intake. Grain and vegetables were the food groups that contributed most to the intake.

In the period 1986–1991, the US FDA has performed a Total Diet Study.[93] The consumption data were based on two nationwide surveys covering about 50,000 participants. In the studies, over 5000 types of foods were identified but to the Total Diet Study 234 foods were selected to represent all 5000 foods. The dietary intakes of pesticide residues were estimated for eight age/sex groups. In Table 11.8, the daily intakes per kilogram body weight for males 14–16 years are compared with the relevant ADI (Codex values). The six pesticides that contributed most to the daily intake and the pesticides that contribute most to the ADI are shown. The overall conclusion was that the intakes of pesticides are well below the ADI for all age groups.

11.3.2.3 Cumulative Exposure

EPA has performed cumulative risk assessment for four groups of pesticides[94] namely organophosphates, N-methylcarbamates, triazines, and chloroacetanilides. In these assessments, not only the dietary exposure is calculated but also the exposure from water and residential uses. For triazines, exposure through food was not considered as relevant and for chloroacetanilides only two pesticides were included so these assessments are not summarized.

TABLE 11.8
Six Pesticides That Contribute Most to the Intake (μg/kg bw) and ADI (%) for Males 14–16 Years in the U.S. Total Diet Study[a]

Pesticides That Contribute Most to the Intake	Intake μg/kg bw	Pesticides That Contribute Most to the ADI	% of ADI
Chlorpropham (no ADI given)	0.2899	Dieldrin	3
Thiabendazole	0.1655	Heptachlor	1
Malathion	0.0965	Omethoate	0.53
Dichloran, total	0.0505	Malathion	0.48
Permethrin, total	0.0415	Dicofol (sum)	0.40
Carbaryl	0.0306	Carbaryl	0.31

[a] Values taken from the published paper.

TABLE 11.9

**Exposure and MOE at the 99.9th Percentiles for Children
1–2 Years and Children 3–5 Years Which Have
the Lowest MOEs as Well as for Adults 20–49 Years**

	Organophosphates		Carbamates	
	Exposure (μg/kg bw)	MOE	Exposure (μg/kg bw)	MOE
Children 1–2 years	2.6	30	3.8	37
Children 3–5 years	2.3	34	3.7	42
Adults 20–49 years	1.1	75	1.3	110

For both organophosphates and the carbamates, residue data were primarily obtained from the USDA PDP program collected from 1993 to 2003 or from 1993 to 2004. The consumption data are from the USDA's Continuing Survey of Food Intakes by Individuals, 1994–1996/1998. In this survey, ca. 21,000 participants were interviewed over two discontinuous days. Processing factors are included in the estimates; undetectable residues were set to zero.

In the study, the MOE approach as well as the probabilistic approach was used. For the organophosphates, the IC was methamidophos, whereas for the carbamates oxamyl was chosen as IC (Table 11.9).

The foods that contributed most to the children's (1–2 year) exposure to carbamates were strawberry and potato, whereas the pesticides that contributed most were methomyl and aldicarb. For organophosphates, the foods that contributed most to the intake for children of 3–5 years were snap bean and the pesticides that contributed most were methamidophos and phorate.

In a study from Brazil,[95] the TEF as well as the probabilistic approach were used. In this study, methamidophos and acephate were used as index compounds.

TABLE 11.10

**Cumulative Intake (μg/kg bw/day) of Choline Esterase Inhibitors
at the 99.9% and 99.99% Percentile**

		Brazil			
		Children (0–6 Years)		Total Population	
	ARfD (μg/kg bw)	99.9% Percentile	99.99% Percentile	99.9% Percentile	99.99% Percentile
Methamidophos	10	8.02	30.7	3.36	13.5
Acephate	50	84.5	359	35.1	134

Source: From Caldas, E.D., Boon, P.E., and Tressou, J., *Toxicology*, 222, 132, 2006.

TABLE 11.11

Cumulative Intakes of Choline Esterase Inhibitors for the Adults and Children in Denmark Using Methamidophos and Chlorpyrifos as Index Compounds

	Chlorpyrifos Equivalents, TEF NL				Methamidophos Equivalents, TEF United States			
	0 LOD Adult	0 LOD Child	½ LOD Adult	½ LOD Child	0 LOD Adult	0 LOD Child	½ LOD Adult	½ LOD Child
Fruits, Vegetables, and Cereals								
Average intake µg/kg bw/day	0.0790	0.2029	0.1890	0.4870	0.0011	0.0029	0.4200	1.0800
% of ADI	0.8	2	2	5	0.03	0.07	11	27

Source: From Jensen, A.F., Petersen, A., and Granby K., *Food Addit. Contam.*, 20, 776, 2003. With permission.

The food consumption data used in this study were obtained from the Brazilian Household Budget Survey, 2002–2003. Data were collected from 45,348 households corresponding to 174,378 individuals. Each household recorded the amount of food entering in a diary over seven consecutive days and this was considered as eaten. For each individual, the week consumption was decomposed into daily consumption patterns over 7 days. Residue data were obtained from the Brazilian national program on pesticide residues. A total of 4001 samples of tomato, potato, carrot, lettuce, orange, apple, banana, papaya, and strawberry were analyzed for their contents of pesticides. Samples with nondetectable levels were assigned a value of ½ LOQ for the index compound. Processing factors were included. In Table 11.10, the intake at the 99.9% and 99.99% percentile for both index compounds are shown. Tomato was independent of IC definitely, the crop that contributed most to the total intake (>65%).

Cumulative risk assessment has been made for the Danish population for the chronic intake.[96] Residue data were from the Danish monitoring program 1996–2001, whereas the consumption data were from the Danish nationwide food consumption survey in 1995. Average values were used for residue levels. Two different IC compounds were used namely chlorpyrifos and methamidophos. The TEF values were taken, respectively, from a Dutch paper and the US EPA. Processing factors were included in the calculations. The intake was calculated both with nondetectable residues as zero and as ½ LOD. This affected not only the intake (Table 11.11) but also the pesticides or commodities that contributed most to the intake (Figure 11.4).

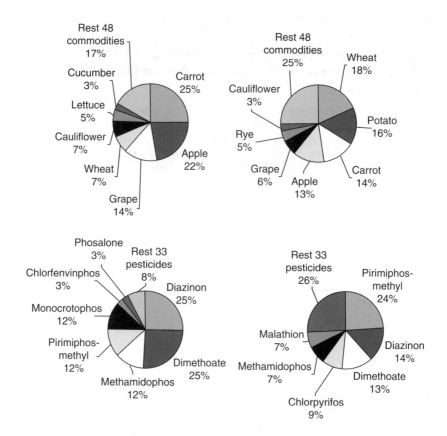

FIGURE 11.4 Commodities (two *upper* diagrams) or pesticides (two *lower* diagrams) that contributed most to the cumulative intake for the Danish population of choline esterase inhibitors in chlorpyrifos equivalents. *To the left*: nondetectable = 0; *to the right*: non-detectable = ½ LOD. (From Jensen, A.F., Petersen, A., and Granby K., *Food Addit. Contam.*, 20, 776, 2003. With permission.)

REFERENCES

1. http://www.fao.org/ag/agp/agpp/Pesticid/JMPR/PM_JMPR.htm
2. http://ec.europa.eu/food/plant/protection/pesticides/legislation_en.htm
3. http://www.epa.gov/pesticides/regulating/tolerances.htm
4. http://www.pmra-arla.gc.ca/english/legis/legis-e.html
5. http://www.daff.gov.au/agriculture-food/nrs/industry-info/mrl/australian-international
6. http://www.nzfsa.govt.nz/policy-law/legislation/food-standards/mrl-2006/nzmrlfs2006
   -consolidation.pdf
7. http://www.mohfw.nic.in/pfa%20acts%20and%20rules.pdf
8. http://www.ffcr.or.jp/zaidan/FFCRHOME.nsf/pages/eng.h-page
9. http://www.doh.gov.za/department/dir_foodcontr-f.html
10. http://www.deqp.go.th/regulation/l264.jsp?languageID = en

11. http://www.kfda.go.kr
12. New Zealand: Food Safety Authority: http://www.nzfsa.govt.nz/plant/subject/horticulture/residues/#MRLDatabase
13. USA: USDA, Foreign Agricultural Service: http://www.fas.usda.gov/htp/MRL.asp http://mrldatabase.com/query.cfm?CFID = 705&CFTOKEN = 57811986
14. FDA, Food and Drug Administration Pesticide Program Residue Monitoring 2003, 2003. Available at http://www.cfsan.fda.gov
15. EU, Monitoring of Pesticide Residues in Products of Plant Origin in the European Union, Norway, Iceland and Liechtenstein 2003, Report Summary, 2005. Available at http://www.Europa.eu.int/comm/food/plant/protection.
16. EC, Monitoring of Pesticide Residues in Products of Plant Origin in the European Union, Norway, Iceland and Liechthenstein 2004. Report from the European Commission, SEC (2006)1416, 2006. Available at http://ec.europa.eu/food/fvo/specialreports/pesticide_residues/report_2004_en.pdf
17. http://ec.europa.eu/food/fvo/specialreports/pesticides_index_en.htm
18. Dogheim, S.M. et al., Monitoring of pesticide residues in Egyptian fruits and vegetables during 1997. *Food Addit. Contam.*, 19, 1015, 2002.
19. Abbassy, M.S., Pesticide residues in selected vegetables and fruits in Alexandria City, Egypt, 1997–1998. *Bull. Environ. Contam. Toxicol.*, 67, 225, 2001.
20. Caldas, E.D. et al., Dithiocarbamates residues in Brazilian food and the potential risk for consumers. *Food Chem. Toxicol.*, 42, 1877, 2004.
21. Chang, J.M., Chen, T.H., and Fang, T.J., Pesticide residue monitoring in marketed fresh vegetables and fruits in Central Taiwan (1999–2004) and an introduction to the HACCP system. *J. Food Drug Anal.*, 13, 368, 2005.
22. Kumari, B. et al., Monitoring of seasonal vegetables for pesticide residues. *Environ. Monitor. Assess.*, 74, 263, 2002.
23. Parveen, Z., Knuhro, M.I., and Rafiq, N., Monitoring of residues in vegetables (2000–2003) in Karachi, Pakistan. *Bull. Environ. Contam. Toxicol.*, 74, 170, 2005.
24. Reynolds, S.L. et al., Occurrence and risks associated with chlormequat residues in a range of foodstuffs in the UK. *Food Addit. Contam.*, 21, 457, 2004.
25. Holland, P.T. et al., Effects of storage and processing on pesticide residues on plant products, *Pure Appl. Chem.*, 66, 335, 1994.
26. FAO/WHO, Pesticides in Food 2005, Report of the joint meeting of the FAO panel of Experts on Pesticide Residues in Food and the Environment and the WHO Core Assessment Group. FAO Plant Production and Protection Paper, p. 183. Available at http://www.who.int/ipcs/publications/jmpr/en
27. http://www.efsa.europa.eu/en/science/praper/draft_assessment_reports.html
28. Andersson, A. et al., Beräknat intag av bekämpningsmedel från vissa frukter och grönsager. Report No. 7 (Stockholm: National Food Administration), 1998.
29. Rasmussen, R.R., Poulsen, M.E., and Hansen, H.C.B., Distribution of multiple residues in apple segments after home processing. *Food Addit. Contam.*, 20, 1044, 2003.
30. Chavarri, M.J., Herrera, A., and Ariño, A., The decrease in pesticides in fruit and vegetables during commercial processing. *Int. J. Food Sci. Tech.*, 40, 205, 2005.
31. Nitz, S. et al., Fate of ethylenebis(dithiocarbamates) and their metabolites during the brew process. *J. Agric. Food Chem.*, 32, 600, 1984.
32. Ankumah, R.O. and Marshall, W.D., Persistence and fate of ethylenthiourea in tomato sauce and paste. *J. Agric. Food Chem.*, 32, 1194, 1984.
33. Cabras, P. and Conte, E., Pesticide residues in grapes and wine in Italy. *Food Addit. Contam.*, 18, 880, 2003.

34. Bar-L'Helgouach'H, C. Quality control of cereals and pulses, Arvalis, Paris. ISBN 2.86492.618.0, 2004.
35. Babu, G.S. et al., DDT and HCH residues in basmati rice (*Oryza sativa*) cultivated in Dehradun (India). *Water Air Soil Pollut.*, 144, 149, 2003.
36. Toteja, G.S., Mukherjee, A., Diwakar, S., Singh, P., and Saxena, B.N., Residues of DDT and HCH pesticides in rice samples from different geographical regions of India: a multicentre study. *Food Addit. Contam.*, 20, 933, 2003.
37. Bakore, N., John, P.J., and Bhatnagar, P., Organochlorine pesticide residues in wheat and drinking water samples from Jaipur, Rajasthan, India. *Environ. Monitor. Assess.*, 98, 381, 2004.
38. Bai, Y., Zhou, L., and Li, J., Organochlorine pesticide (HCH and DDT) residues in dietary products from shaanxi province, people's republic of china. *Bull. Environ. Contam. Toxicol.*, 76, 422, 2006.
39. http://www.cfsan.fda.gov/~dms/pes00rep.html#table_1
40. http://www.cfsan.fda.gov/~dms/pes01rep.html#table_1
41. http://www.cfsan.fda.gov/~dms/pes02rep.html#table_1
42. http://www.cfsan.fda.gov/~dms/pes03rep.html#table_1
43. Juhler, R.K. et al., Pesticide residues in selected food commodities: results from the Danish national pesticide monitoring program 1995–1996. *J. AOAC Int.*, 82, 337, 1999.
44. Granby, K. and Vahl, M., Investigation of the herbicide glyphosate and the plant growth regulators chlormequat and mepiquat in cereals produced in Denmark. *Food Addit. Contam.*, 18, 898, 2001.
45. Holland, P.T., Hamilton, D., Ohling, B., and Skidmore, M.W., Effects of storage and processing on pesticide residues in plant products. IUPAC Reports on Pesticides (31), 1994.
46. Uygun, U., Koksel, H., and Atli, A., Residue levels of malathion and its metabolites and fenitrothion in post-harvest treated wheat during storage, milling and baking. *Food Chem.*, 92, 643, 2005.
47. Lalah, J.O. and Wandiga, S.O., The effect of boiling on the removal of persistent malathion residues from stored grains. *J. Stored Prod. Res.*, 38, 1, 2002.
48. Chao, H.W. et al., Levels of OCPs in human milk from central Taiwan. *Chemosphere*, 62, 1774, 2006.
49. Weiss, J., Päpke, O., and Bergman, Å., A worldwide survey of polychlorinated dibenzo-*p*-dioxins, dibenzofurans, and related contaminants in butter. *Ambio*, 34, 589, 2005.
50. Kalantzi, O.I. et al., The global distribution of PCBs and PCPs in butter. *Environ. Sci. Technol.*, 35, 1013, 2001.
51. Fromberg, A. et al., *Chemical Contaminants—Food Monitoring, 1998–2003*, Part 1, 1st Edition, 2005. Available at http://www.dfvf.dk/Default.aspx?ID = 10875
52. Badia-Vila, M. et al., Comparison of residue levels of persistent organochlorine compounds in butter from Spain and from other European countries. *J. Environ. Sci. Health Part B*, 35, 201, 2000.
53. Yentür, G., Kalay, A., and Öktem, A.B., A survey on OCP residues in butter and cracked wheat available in Turkish markets. *Nahrung*, 45, 40, 2001.
54. Kumar, A. et al., Persistent organochlorine pesticide residues in milk and butter in Agra City, India: a case study. *Bull. Environ. Contam. Toxicol.*, 75, 175, 2005.
55. Mallatou, M. et al., Pesticide residues in milk and cheeses from Greece. *Sci. Total Environ.*, 196, 111, 1997.
56. Pandit, G.G., Persistent OCP residues in milk and dairy products in India. *Food Addit. Contam.*, 19, 153, 2002.

57. Battu, R.S., Singh, B., and Kang, B.K., Contamination of liquid milk and butter with pesticide residues in the Ludhiana district of Punjab state, India. *Ecotoxicol. Environ. Saf.*, 59, 324, 2004.

58. Glynn, A.W. et al., PCB and chlorinated pesticide concentrations in swine and bovine adipose tissue in Sweden 1991–1997: spatial and temporal trends. *Sci. Total Environ.*, 246, 195, 2000.

59. Frenich, A.G. et al., Multiresidue analysis of organochlorine and organophosphorus pesticides in muscle of chicken, pork and lamb by gas chromatography-triple quadrupole mass spectrometry. *Anal. Chim. Acta.*, 558, 42, 2006.

60. Naccari, F. et al., OCPs and PCBs in wild boars from Calabria (Italy). *Environ. Monitor. Assess.*, 96, 1991, 2004.

61. Matsumoto, H. et al., Survey of PCB and organochlorine pesticide residues in meats and processed meat products collected in Osaka, Japan. *J. Food Hyg. Soc. Jpn.*, 47, 127, 2006.

62. Cleemann, M. et al., Organochlorine in Greenland marine fish, mussels and sediments. *Sci. Total Environ.*, 245, 87, 2000.

63. Fromberg, A., Cleemann, M., and Carlsen, M., Review of persistent organic pollutants in the environment of Greenland and Faroe Islands. *Chemosphere*, 38, 3075, 1999.

64. Stefanelli, P. et al., Estimation of intake of organochlorine pesticides and chlorobiphenyls through edible fishes from the Italian Adriatic Sea during 1997. *Food Control*, 15, 27, 2004.

65. Chen, W. et al., Residue levels of HCHs, DDTs and PCBs in shellfish from coastal areas of east Xiamen Island and Minjiang Estuary, China. *Mar. Pollut. Bull.*, 45, 385, 2002.

66. Sun, F. et al., A preliminary assessment of consumer's exposure to pesticide residues in fisheries products. *Chemosphere*, 62, 674, 2006.

67. Sinkkonen, S. and Paasivirta, J., Polychlorinated organic compounds in the Arctic cod liver: trends and profiles. *Chemosphere*, 40, 619, 2000.

68. Commission Directive 1999/39/EC of May 1999. Available at http://eur-lex.europa .eu/LexUriServ/site/en/oj/1999/l_124/l_12419990518en00080010.pdf

69. Wong, M.H., A review on the usage of POP pesticides in China, with emphasis on DDT loadings in human milk. *Chemosphere*, 60, 740, 2005.

70. Minh, N.H. et al., Persistent organochlorine residues in human breast milk from Hanoi and Hochiminh City, Vietnam; contamination, accumulation kinetics and risk assessment for infants. *Environ. Pollut.*, 129, 431, 2004.

71. Sundaryanto, A. et al., Specific accumulation of organochlorines in human breast milk from Indonesia: levels, distribution, accumulation kinetics and infant health risk. *Environ. Pollut.*, 139, 107, 2006.

72. Damgaard, I.N. et al., Persistent pesticides in human breast milk and cryptorchidism. *Environ. Health Perspect.*, 114, 1133, 2006.

73. Schade, G. and Heinzow, B., Organochlorine pesticides and polychlorinated biphenyls in human milk of mothers living in northern Germany: current extent of contamination, time trend from 1986 to 1997 and factors that influence the levels of contamination. *Sci. Total Environ.*, 215, 31, 1998.

74. Wong, C.K.C. et al., Organochlorine hydrocarbons in human breast milk collected in Hong Kong and Guangshou. *Arch. Environ. Contam. Toxicol.*, 43, 364, 2002.

75. Chikuni, O. et al., Effects of DDT on paracetamol half-life in highly exposed mothers in Zimbabwe. *Toxicol. Lett.*, 134, 147, 2002.

76. Gelardi, R.C. and Mountford, M.K., Infant formulas: evidence of the absence of pesticide residues. *Reg. Toxicol. Pharmacol.*, 17, 181, 1993.

77. Lackmann, G.M., Schaller, K.H., and Angerer, J., Organochlorine compounds in breast-fed vs. bottle-fed infants: preliminary results at six weeks of age. *Sci. Total Environ.*, 329, 289, 2004.

78. U.S. Food and Drug Administration Center for Food Safety and Applied Nutrition Pesticide Program: Residue Monitoring 2000, May 2002. Food and Drug Administration Pesticide Program—Residue Monitoring 2000. Available at http://www.cfsan.fda .gov/~dms/pes00rep.html#surveys

79. Poulsen, M.E. et al., Pesticides—Food Monitoring, 1998–2003, Part 2, 1st Edition, 2005. Available at http://www.dfvf.dk/Default.aspx?ID = 9410

80. Cressey, P.J. and Vannoort, R.W., Pesticide content of infant formulae and weaning foods available in New Zealand. Food Addit. Contam., 20, 57, 2003.

81. Australia New Zealand Food Authority, The 19th Australian Total Diet Survey—A Total Diet Survey of Pesticide Residue and Contaminants. Available at http://www .foodstandards.gov.au/_srcfiles/19th%20ATDS.pdf

82. http://www.fao.org/ag/agp/agpp/pesticid/jmpr/Download/2002jmprreport2.doc

83. Pieters, M.N. et al., Probabilistic Modelling of Dietary Intake of Substances, The Risk Management Question Governs the Method, RIVM, Report 3200110012005, 2005.

84. EPA, Guidance for Submission of Probabilistic Human Health Exposure Assessments to the Office of Pesticides Program, 1998. Available at http://www.epa.gov/fedrgstr/EPA -PEST/1998/November/Day-05/o-p29665.htm

85. Development, validation and application of stochastic modelling of human exposure to food chemicals and nutrients. Food Addit. Contam., 20 (Suppl. 1), S73, 2003.

86. Boon, P.E. et al., Probabilistic intake calculations performed for the Codex Committee on pesticide residues, Report 2004.005, 2004. Available at http://www.rikilt.wur.nl/ NR/rdonlyres/BDEEDD31-F58C-47EB-A0AA-23CB9956CE18/10729/R2004005.pdf

87. Boon, P.E. et al., Estimating of The Acute Dietary Exposure to Pesticides Using the Probabilistic Approach and the Point Estimate Methodology; B1-3330SANCO2002584, 2003. Available at http://www.rikilt.wur.nl/NR/rdonlyres/BDEEDD31-F58C-47EB -A0AA-23CB9956CE18/10727/R2004008.pdf

88. Wilkinson, C.F. et al., Assessing the risks of exposures to multiple chemicals with a common mechanism of toxicity: how to cumulate? Regul. Toxicol. Pharmacol., 31, 30, 2000.

89. Chun, O.K. and Kang, H.G., Estimation of risks of pesticide exposure, by food intake, to Koreans. Food Chem. Toxicol., 41, 1063, 2003.

90. Muntean, N. et al., Assessment of dietary exposure to some persistent organic pollutants in the republic of Karakalpakstan of Uzbekistan. Environ. Health Perspect., 1, 1306, 2003.

91. Battu, R.S. et al., Risk assessment through dietary intake of total diet contaminated with pesticide residues in Punjab, India, 1999–2002. Ecotoxicol. Environ. Saf., 62, 132, 2005.

92. Sawaya, W.H. et al., Dietary intake of organophosphate pesticides in Kuwait. Food Chem., 69, 331, 2004.

93. Gunderson, E.L., FDA Total Diet Study, July 1986–April 1991, Dietary intakes of pesticides, selected elements and other chemicals. J. AOAC Int., 78, 1353, 1995.

94. EPA, Assessing Pesticide Cumulative Risk. Available at http://www.epa.gov/pesticides/ cumulative/

95. Caldas, E.D., Boon, P.E., and Tressou, J., Probabilistic assessment of the cumulative acute exposure to organophosphorus and carbamates insecticides in the Brazilian diet. Toxicology, 222, 132, 2006.

96. Jensen, A.F., Petersen, A., and Granby K., Cumulative risk assessment of the intake of organophosphorus and carbamate pesticides in the Danish diet. Food Addit. Contam., 20, 776, 2003.

12 Monitoring of Pesticides in the Environment

Ioannis Konstantinou, Dimitra Hela, Dimitra Lambropoulou, and Triantafyllos Albanis

CONTENTS

12.1 INTRODUCTION

Worldwide pesticide usage has increased dramatically during the last three decades coinciding with changes in farming practices and the increasing intensive agriculture. This widespread use of pesticides for agricultural and nonagricultural purposes has resulted in the presence of their residues in various environmental matrices. Numerous studies have highlighted the occurrence and transport of pesticides and their metabolites in rivers [1], channels [2], lakes [1,3,4], sea [5,6], air [7–10], soils [11,12], groundwater [13,14], and even drinking water [15,16], proving the high risk of these chemicals to human health and environment.

In recent years, the growing awareness of the risks related to the intensive use of pesticides has led to a more critical attitude by the society toward the use of agrochemicals. At the same time, many national environmental agencies have been involved in the development of regulations to eliminate or severely restrict the use and production of a number of pesticides (Directive 91/414/EEC) [17]. Despite these actions, pesticides continue to be present causing adverse effects on human and the environment. Monitoring of pesticides in different environmental compartments has been proved a useful tool to quantify the amount of pesticides entering the environment and to monitor ambient levels for trends and potential problems and different countries have undertaken, or currently undertaking, campaigns with various degrees of intensity and success [18]. Although numerous local and national monitoring studies have been performed around the world providing nationwide patterns on pesticide occurrence and distribution, there are still several gaps. For example, only limited retrospective monitoring data are available in all compartments and there is a lack of monitoring data for many pesticides both in space and time [5,19]. In addition, there is little consistency in the majority of these studies in terms of site selection strategy, sampling methodologies, collection time and duration, selected analytes, analytical methods, and detection limits [18,20]. Therefore, dedicated efforts are needed for comprehensive monitoring schemes not only for pesticide screening but also for the establishment of cause–effect relationships between the concentration of pesticides and the damage, and to assess the environmental risk in all compartments.

12.2 MONITORING PROGRAMS

Environmental monitoring programs are essential to develop extensive descriptions of current concentrations, spatiotemporal trends, emissions, and flows, to control the compliance with standards and quality objectives, and to provide early warning detection of pollution. Furthermore, environmental monitoring provides a viable basis for efficacious measures, strategies, and policies to deal with environmental problems at a local, regional, or global scale. Similar terms often used are "surveys" and "surveillance." A survey is a sampling program of limited duration for specific pesticides such as an intensive field study or exploratory campaign. Surveillance is a more continuous specific study with the aim of environmental quality reporting (compliance with standards and quality objectives) and/or operational activity reporting (e.g., early warning and detection of pollution) [19].

12.2.1 Purpose and Design of Pesticide Monitoring Programs

In general, pesticide monitoring is used to investigate and to gain knowledge that allows authorities tentatively to assess the quality of the environment, to recognize threats posed by these pollutants, and to assess whether earlier measures have been effective [18,21]. Whichever the objectives of a monitoring program may be, it is important that they are well defined before sampling takes place to select suitable sampling and analysis methods and to plan the project adequately. Another important characteristic of a monitoring program is that data produced are often used to implement and regulate existing directives concerning pesticides in the environment [5].

Because of the great number of parameters (pesticide physicochemical properties, climatic and environmental factors) affecting the exposure of pesticides, monitoring of a single medium will not provide sufficient information about the occurrence of pesticides in the environment. A multimedia approach that involves tracking pesticides from sources through multiple environmental media such as air, water, sediment, soil, and biota provides with data for understanding the fate and partitioning of pesticides and for the validation of environmental models [19].

A basic problem in the design of a pesticide monitoring program is that each of the earlier reasons for carrying out monitoring demands different answers to a number of questions. Thus, when a monitoring program consists of sampling, laboratory analysis, data handling, data analysis, reporting, and information exploitation, its design will necessarily have to include a wide range of scientific and management concepts, thus making a large and difficult task [21]. Therefore, cost-effective monitoring programs should be based on clear and well thought-out aims and objectives and should ensure, as far as possible, that the planned monitoring activities are practicable and that the objectives of the program will be met. There are a number of practical considerations to be dealt with when designing a monitoring program that are generic regardless of the compartment getting monitored (Figure 12.1). For pesticide monitoring programs, some general guidelines should

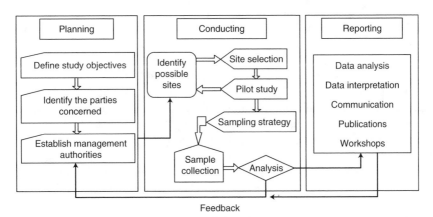

FIGURE 12.1 Phases in planning, conducting, and reporting of a monitoring program. (From Calamari, D., et al., *Evaluation of Persistence and Long Range Transport of Organic Chemicals in the Environment*, G. Klecka et al., eds., SETAC, 2000.)

be taken into consideration including the clear statement of the objectives, the complete description of the area as well as the locations and frequency of sampling, and the number of the samples. The geographical limits of the area, the present and planned water or land uses, and the present and expected pesticide pollution sources should be identified. Background information of this type is of great help in planning a representative monitoring program covering all the sources of the spatial and temporal variability of the pesticide environmental concentration. Appropriate statistical analysis can be used to determine probability distributions that may be used to select locations for further sampling programs and for risk assessment. The fieldwork associated with the collection and transportation of samples will also account for a substantial section of the plan of a monitoring program. The development of meaningful sampling protocols has to be planned carefully taking into account the actual procedures used in sample collection, handling, and transfer [22]. The design of a sampling should target the representativeness of the samples that is related to the number of samples and the selection of sampling stations intended within the objectives of the study. The sampling process of taking random grab samples and individually analyzing each sample is very common in environmental monitoring programs and is the optimal plan when a measurement is needed for every sample. However, the process of combining separate samples and analyzing this pooled sample is sometimes beneficial. Such composite sampling process is generally used under flow conditions and in situations where concentrations vary over time (surface water or air sampling), when samples taken from varying locations as well as when representativeness of samples taken from a single site need to be improved by reducing intersample variance effects. Composite sampling is also used to increase the amount of material available for analysis, as well as to reduce the cost of analysis. However, certain limitations must be taken into account and it should be used only when the researcher fully understands all aspects of the plan of choice [18,22].

Apart from sampling, the selection and the performance of the analytical method used for the determination of pesticides is a very critical subject. Earlier chapters of this book discuss the various methods that can be successfully applied to monitor pesticides in various environmental compartments. Another point that should be considered in the planning stage concerns the quality assurance/quality control (QA/QC) procedures to produce reliable and reproducible data. These quality issues relate to the technical aspects of both sampling and analysis. The quality of the data generated from any monitoring program is defined by two key factors: the integrity of the sample and the limitations of the analytical methodology. The QA/QC procedures should be designed to establish intralaboratory controls of sample collection and preparation, instrument operation, and data analysis and should be subjected to "Good Analytical Practices" (GAP). Laboratories should participate in a series of intercalibration exercises and chemical analysis cross-validations to avoid false positives [19,23].

As already mentioned, the whole planning of a monitoring program is aimed at the generation of reliable data but it is acknowledged that simply generating good data is not enough to meet monitoring objectives. The data must be proceeded and presented in a manner that aids understanding of the spatial and temporal patterns, taking into consideration the characteristics of the study areas, and that allows the human impact to be understood and the consequences of management action to be predicted. Thus, different statistical approaches are usually applied to designing, adjusting, and quantifying the informational value of monitoring data [20]. However, because data are often collected at multiple locations and time points, correlation among some, if not all, observations is inevitable, making many of the statistical methods taught to be applied. Thus, in the last decade geographic information systems (GIS) and computer graphics are used that have enhanced the ability to visualize patterns in data collected in time and space [24]. In summary, statistical methods, including chemometric methods, coupled to GIS are used in recent years to display the most significant patterns in pesticide pollution [18].

Finally, one of the major parameters of the monitoring plan should be the cost of the program. A cost estimate should be prepared for the entire program, including laboratory and field activities. The major cost elements of the monitoring program include personnel cost; laboratory analysis cost; monitoring equipment costs; miscellaneous equipment costs; data analysis and reporting costs.

As a conclusion based on the earlier arguments, monitoring activities must imply a long-term commitment and can be summarized as follows [18–20]: (1) establishment of monitoring stations for different environmental compartments to fill spatiotemporal data; (2) intensive monitoring over wider areas, and continuation of existing time trend series; (3) establishment of standardized sampling and analytical methods; (4) follow-up of improved quality assurance/quality control protocols; (5) adequate reporting of the results in the more meaningful manner; and (6) estimation of the monitoring program cost.

12.2.2 SELECTION OF PESTICIDES FOR MONITORING

The number and nature of pesticides monitored depended on the objectives of the monitoring study. Some studies concentrated on a limited number of target pesticides, whereas others performed a broad screening of different compounds. Research has usually been focused on the most commonly used pesticides either in the agricultural area around the studied sites or in the country concerned. The selection of pesticides for monitoring has also been based on pesticide properties (e.g., toxicity, persistence, and input), the cost, as well as on special directives and regulations [25].

The diversity of aims and objectives for the various monitoring programs has resulted in a variety of active ingredients and metabolites monitored in the studies performed.

For instance, until the beginning of the 1990s, halogenated, nonpolar pesticides were the focus of interest. As the environmental fate of hydrophobic pesticides became more generally understood and new, more environmental-friendly, pesticide products are introduced in the market, there has been an increase in monitoring studies that focused on currently used pesticides known to be present in the environment. Whereas environmental concentrations of halogenated, nonpolar pesticides have generally declined during the past 20 years, and whereas current concentrations in surface water are below the drinking water standards, concerns nevertheless remain, because these substances persist in the environment and accumulate in the food chain, thus continue to be in the list for investigation. Current screening strategies have also included pesticides with endocrine disruption action due to their newly discovered ecotoxicological problems on human health and environment. Among the most studied chemical classes of pesticides are the s-triazines, acetamides, substituted ureas, and phenoxy acids from the group of herbicides and organophosphorus and carbamates from the group of insecticides. Currently, modern fungicides have gain attention since their uses have been increased and new compounds have been introduced in the market.

Although that all new compounds or new uses of existing pesticides are carefully scrutinized, the list of pesticide of interest for monitoring programs is not getting shorter and there is a continuing need for development of new criteria that allow the prediction of which pesticides could be of concern for monitoring.

12.2.3 TYPES OF MONITORING

Pesticides can occur in all compartments of the environment or in other words in any or all of the solid, liquid, or gaseous phases. The environment is not a simple system and consequently pesticide monitoring should be carrying out in a specific phase (e.g., volatile pesticides in air) or may encompass two or more phases and/or media (e.g., water and sediment in the marine environment). Primary environmental matrices that are usually sampled for pesticide investigations include water, soil, sediment, biota, and air. However, each of these primary matrices includes many different kinds of samples. A brief description of each type of monitoring is given in the next paragraphs.

12.2.3.1 Air Monitoring

Historically, water contamination has garnered the lion's share of public attention regarding the ultimate fate of pesticides. In contrast, atmospheric monitoring is less expanded since the atmospheric residence time of a pesticide is very variable. However, in recent years, air quality has become a very important concern as more and more studies have shown the great impact of atmospheric pesticide pollution on environment and health. Pesticides can be potential air pollutants that can be carried by wind, and deposited through wet or dry deposition processes. They can revolatilize repeatedly and, depending on their persistence in the environment can travel tens, hundreds, or thousands of kilometers [26]. For example, currently used organochlorine pesticides (OCPs) like endosulfans and lindane have been detected in arctic samples [9,27] where, of course, they have never been used.

The design of monitoring networks for air pollution has been treated in several different ways. For example, monitoring sites may be located in areas of severest public health effects, which involves consideration of pesticide concentration, exposure time, population density, and age distribution. Alternatively, the frequency of occurrence of specific meteorological conditions and the strength of sources may be used to maximize monitor coverage of a region with limited sources.

Air concentrations of pesticides may vary over the scales of hours, days, and seasons since they respond to air mass direction and depositional events.

The sampling methods of pesticides in air may be divided into active (pump or vacuum-assisted sampling) or passive techniques (passive by diffusion gravity or other unassisted means). The sampling interval may be integrated over time or it may be continuous, sequential, or instantaneous (grab sampling). Measurements obtained from grab sampling give only an indication of what was present at the sampling site at the time of sampling. However, they can be useful for screening purposes and provide preliminary data needed for planning subsequent monitoring strategies. Probably, the collection of pesticides by using passive air samplers (PAS) is the most common sampling method for air samples. PAS continuously integrate the air burden of pesticides and give real-time or near-time assessment of the concentration of pesticide in air [8,22,28]. Most of the passive air sampling measurements have been performed using semipermeable membrane devices (SPMDs) [28], polyurethane foam (PUF) disks [29], and samplers employing XAD-resin [30].

12.2.3.1.1   *Occurrence and pesticide levels in air monitoring studies*
Numerous investigations around the world consistently find pesticides in air, wet precipitation, and even fog. Research in the 1960s to 1980s, for example, has found the infamous pesticide DDT and other OCPs in Antarctic ice, penguin tissues, and most of the whale species [31]. Monitoring programs have been established in many countries for the spatial and temporal distribution of persistent OCPs such as DDTs, HCHs, cyclodienes [19]. While many of the newer, currently used pesticides are less persistent than their predecessors, they also contaminate the air and can travel many miles from target areas. Of these, chlorothalonil, chlorpyrifos, metolachlor, terbufos, and trifluralin have been detected in Arctic environmental samples (air, fog, water, snow) by Rice and Cherniak [32] and Garbarino et al. [27] or in ecologically sensitive regions such as the Chesapeake Bay and the Sierra Nevada mountains [33]. In general, herbicides such as *s*-triazines (atrazine, simazine, terbuthylazine), acetanilides (alachlor and metolachlor), phenoxy acids (2,4-D, MCPA, dichloprop) are among the most frequently looked for and detected in air and precipitation. Regarding the modern insecticides, organophosphorus compounds (parathion, malathion, diazinon, and chlorpyrifos) have been looked for most often. The occurrence of other groups of pesticides in air and rain has been generally poorly investigated [34]. Concentrations of modern pesticides in air often range from a few picograms per cubic meter to many nanograms per cubic meter. In rain, concentrations have been measured from few nanograms per liter to several micrograms per liter. However, concentrations in precipitation depended not only on the amount of pesticides present in the atmosphere, but also on the amounts, intensity, and timing of rainfall [34]. Concentrations in fog are even higher. Deposition levels are in the order of several milligrams per hectare per year to a few grams per hectare per year [9,10].

   In general, air monitoring studies have been conducted on an ad hoc basis and are characterized by a small number of sampling sites, covering limited geographical areas and time periods. In the United States and Canada [10], however, some large, nationwide studies have been conducted. In contrast, most European (EU) monitoring studies have been focused on rain rather than in air. So far, at least over 80 pesticides have been detected in precipitation in Europe and 30 in air [35]. However, the lack of consistency in sampling and analytical methodologies holds for both United States and European studies [7].

An example of characteristic pesticide monitoring programs in air and rainwater can be mentioned, the Integrated Atmospheric Deposition Network (IADN, Canada), based on several sampling stations on the Great Lakes [36]. The Canadian Atmospheric Network for Current Used Pesticides (CANCUP, 2003) also provides new information on currently used pesticides in the Canadian atmosphere and precipitation [37]. Last example from monitoring of pesticides in rainwater is the survey established by Flemish Environmental Agency (FEA) in Flanders, Belgium [38] that monitors >100 pesticides and metabolites at eight different locations.

12.2.3.2 Water Monitoring

The principal reason for monitoring water quality has been, traditionally, the need to verify whether the observed water quality is suitable for intended uses. However, monitoring has also evolved to determine trends in the quality of the aquatic environment and how the environment is affected by the release of pesticides and/or by waste treatment operations. Currently, spot (bottle or grab) sampling, also called as active sampling, is the most commonly used method for aquatic monitoring of pesticides. With this approach, no special water sampling system is required and water samples are usually collected in precleaned amber glass containers. Although spot sampling is useful, there are drawbacks to this approach in environments where contaminant concentrations vary over time, and episodic pollution events can be missed. Moreover, it requires relatively large number of samples to be taken from any one location over the entire duration of sampling and therefore is time-consuming and can be very expensive. In order to provide a more representative picture and to overcome some of these difficulties, either automatic sequential sampling to provide composite samples over a period of time (24 h) or frequent sampling can be used. However, the former involves the use of equipment that requires a power supply, and needs to be deployed in a secure site, and the latter would be expensive because of transport and labor costs.

In the last two decades, an extensive range of alternative methods that yield information on environmental concentrations of pesticides have been developed. Of these, passive sampling methods, which involve the measurement of the concentration of an analyte as a weighted function of the time of sampling, avoid many of the problems outlined earlier, since they collect the target analyte in situ without affecting the bulk solution. Passive sampling is less sensitive to accidental extreme variations of the pesticide concentration, thus giving more adequate information for long-term monitoring of aqueous systems. Comprehensive reviews on the use of equilibrium passive sampling methods in aquatic monitoring as well as on the currently available passive sampling devices have been recently published [39–42]. Despite the well-established advantages, passive sampling has some limitations such as the effect of environmental conditions (e.g., temperature, air humidity, and air and water movement) on analyte uptake. Despite such concerns, many users find passive sampling an attractive alternative to more established sampling procedures. To gain more general appeal, however, broader regulatory acceptance would probably be required.

Other technologies available for water sampling include continuous, online monitoring systems. In such installations, water is continuously drawn from water input and automatically fed into an analytical instrument (i.e., LC-MS). These systems provide extensive, valuable information on levels of pesticides over time; however they require a secure site, are expensive to install, and have a significant maintenance cost [42].

Finally, another approach available and already in use for monitoring water quality includes sensors. A wide range of sensors for use in pesticide monitoring of water have been developed in recent years, and some are commercially available. These are based on electrochemical or electroanalytical technologies and many are available as miniaturized screen-printed electrodes [43]. They can be used as field instruments for spot measurements, or can be incorporated into online monitoring systems. However, some of these methods do not provide high sensitivity, and in some case specificity, as they can be affected by the matrix and environmental conditions, and thus it is necessary to define closely the conditions of use [44].

12.2.3.2.1 Occurrence and pesticide levels in water samples

The majority of the pesticide monitoring effort goes into monitoring surface freshwaters (including rivers, lakes, and reservoirs) and monitoring programs for pesticides in marine waters and groundwaters have received less attention. Within Europe, the contamination of freshwaters by pesticides follows comparable concentration levels and patterns as recorded in most countries. Among the most commonly encountered herbicide compounds in European freshwaters were atrazine, simazine, metolachlor, and alachlor. s-Triazine herbicides are widely applied herbicides in Europe for pre- and postemergence weed control among various crops as well as in nonagricultural purposes. In some studies, acetamide herbicides alachlor and metolachlor (which are also used to control grasses and weeds in a broad range of crops) were also detected at levels comparable with those of the triazines. Concerning insecticide concentrations in European freshwaters mainly organophosphates and organochlorine insecticides have been detected. Diazinon, parathion methyl, malathion, and carbofuran were the most frequently detected compounds [1]. OCPs continue to be present in freshwaters, but at low levels, due to their high hydrophobicity. Among them, lindane was the most frequently detected compound. Other OCPs include α-endosulfan and aldrin. Fungicides were not generally present at high concentrations in European surface waters and usually the detected levels were below detection limits. Only sporadic runoff of certain fungicides (e.g., captafol, captan, carbendazim, and folpet) was reported in estuaries of major Mediterranean rivers [45]. Finally, for the United States, the most commonly encountered compounds also include atrazine, simazine, alachlor, and metolachlor from herbicides and diazinon, malathion, and carbaryl from insecticides [46].

The water monitoring studies around the world have routinely focused on tracing parent compounds rather than their metabolites. Thus, little data are available on the occurrence of pesticide transformation products in freshwaters, including mainly transformation products of high-use herbicides, such as acetamide and triazine compounds. For example, desethylatrazine, metabolite of atrazine, has been detected in rivers of both United States [47] and Europe [48].

Agricultural uses result in distinct seasonal patterns in the occurrence of a number of compounds, particularly herbicides, in freshwaters. Regarding rivers, critical factors for the time elapse between the period of pesticide application in cultivation and their occurrence in rivers include the characteristics of the catchment (size, climatological regime, type of soil, or landscape) as well as the chemical and physical properties of the pesticides [49]. The size of the drainage basin affects the pesticide concentration profile and Larson and coworkers showed that in large rivers the integrating effects of the many tributaries result in elevated pesticide concentrations that spread out over the summer months. In rivers with relatively small drainage basins (50,000–150,000 km^2), pesticide concentrations increased abruptly and the periods of elevated concentrations were relatively short—about 1 month—as pesticides were transported in runoff from local spring rains in the relatively small area [50]. Although for the smaller drainage basins of the Mediterranean area short periods of increased pesticide concentrations would be expected, more spread out pesticide concentration profiles are observed. This is probably due to delayed leaching from soil as a result of dry weather conditions, which is reflected by the low mean annual discharges [1,51]. Generally, low concentrations were observed during the winter months because of dilution effects due to high-rainfall events and the increased degradation of pesticides after their application. Thus, pesticides were flushed to the surface water systems as pulses in response to late spring and early summer rainfall as reported elsewhere [52].

The character of the landscape in combination with the type of cultivation in the catchment area may as well affect the temporal variations in riverine concentrations of pesticides. For example, for the relatively large basin of the river Rhone, the concentration of triazines displays a short peak from late April to late June with relatively constant concentrations during the rest of the year [53], due to the fact that herbicides are used in vineyards situated on mountain slopes which promotes rapid runoff. Finally, similar trends and temporal variations were observed also in lakes. The only difference is that residues were detected during a longer period as a result of the lower water flushing and renewal time compared with rivers.

Several pesticides and their metabolites have also been identified in groundwater [54]. However, fewer pesticide measurements are available around the world located mainly in the area of United States and Europe. In previous published studies that summarized the groundwater monitoring data for pesticides in the United States [55], researchers reported that at least 17 pesticides have been detected in groundwater samples collected from a total of 23 States. About half of these chemicals were herbicides such as alachlor, atrazine, bromacil, cyanazine, dinoseb, metolachlor, metribuzin, and simazine. The reported concentrations of these herbicides ranged from 0.1 to 700 µg/L. Cohen et al. [55] have compiled the chemodynamic properties of the detected pesticides in groundwater and concluded that most of these chemicals had aqueous solubility in excess of 30 mg/L and degradation half-lives longer than 30 days.

In EU countries, as in the case of the United States, commonly used pesticides such as triazines (atrazine and simazine) and the ureas (diuron and chlortoluron), which are used in relatively large quantities, are often detected in raw water sources. Because atrazine and simazine frequently appear in groundwater, several European countries have banned or restricted the use of products containing these active ingredients and a recent assessment revealed a statistically significant downward trend in the contamination of groundwater with atrazine and its metabolites in a number of European countries [15]. However, in Baden–Wurttemberg, Germany, where atrazine concentrations in groundwater appear to be decreasing, concentrations of another triazine herbicide, hexazinon, show an upward trend [15]. As an example of groundwater monitoring program, the Pesticides in European Groundwaters (PEGASE) is a detailed study of representative aquifers. Furthermore, the Pesticide National Synthesis Project which is a part of the U.S. Geological Survey's National Water Quality Assessment Program (NAWQA) with the aim of long-term assessment of the status and trends of water resources including pesticides as one of the highest priority issues is also a nice example for water monitoring programs (http://ca.water.usgs.gov/pnsp/).

As mentioned previously, limited monitoring data are available for the occurrence of pesticides in marine waters. Mainly estuarine environments, ports, and marinas have been monitored for pesticide loadings. Nice example of such monitoring program is the Fluxes of Agrochemicals into the Marine Environment (FAME) project, supported by the European Union, that provide information for Rhone (France), Ebro (Spain), Louros (Greece), and Western Scheldt (The Netherlands) river/estuary systems [56] and MEDPOL program for monitoring priority fungicides in estuarine areas of the Mediterranean region [44,57]. In addition, the Assessment of Antifouling Agents in Coastal Environments (ACE) project of the European Commission (1999–2002) provides data concerning contamination and effects/risks of the most popular biocides currently used in antifouling paints to prevent fouling of submerged surfaces in the sea as alternatives to tributyltin compounds. A number of booster biocides have been detected in many European countries including Irgarol 1051, diuron, sea nine, and chlorothalonil. The occurrence, fate, and toxic effects of antifouling biocides have been reviewed recently [58,59].

12.2.3.3 Soil and Sediment Monitoring

Soil and sediment compartments might also be regarded as reservoirs for many types of pesticides. Although high amounts of pesticide as well as a complex pattern of their metabolites are usually present in soils, this matrix is not generally monitored on a regular basis and there is a gap in knowledge on national and global level regarding the pesticide residue levels. The majority of the investigation studies were carried out by researchers' initiative or licensing of new substances or under the frame of founded projects. Regarding Europe, recent discussions have taken place to consider regulation of persistence of soil residues beyond the guidelines given in the Directive 91/414/EEC [17]. In this regard, stronger emphasis should be given to soil monitoring programs such as Monitoring the State of European Soils (MOSES; http://projects-2004.jrc.cec.eu.int/) and Environmental Indicators for Sustainable Agriculture (ELISA; http://www.ecnc.nl/CompletedProjects/Elisa_119.html).

In contrast to soils, sediments are usually monitored for pesticide contamination. Sediments from river, lake, and seawaters provide habitat for many benthic and epibenthic organisms and are a significant element of aquatic ecosystems. Many pesticide compounds, because of their hydrophobic nature, such as OCPs, are known to associate strongly with natural sediments and dissolved organic matter and high concentrations of pesticides are frequently found in bed sediments, both freshwater and coastal [60]. Monitoring studies using sediment core stratification also have the advantage of providing information on the chronologies of accumulation rates of persistent pesticides. This information is important to evaluate the rate of emission from probable sources, and to relate specific rates of pesticide accumulation and rates of ecosystem response. Sediment monitoring is also a task for the correct implementation of the Water Framework Directive (WFD) to assess any changes in the status of water bodies.

Soils and sediments are typically very inhomogeneous media, thus a large number of samples may be required to characterize a relatively small area. Sampling sites could be distributed spatially at points of impact, reference sites, areas of future expected changes, or other areas of particular interest. Selection of specific locations is a subject of accessibility, hydraulic conditions, or other criteria. The devices used for soil and sediment sampling are usually grab samplers or corers. Grab samplers are available for operation at surfacial depths. Box corers or multicorers can be employed if more data on the chronologies of accumulation rates of the analytes are needed.

12.2.3.3.1 Occurrence and pesticide levels in soils and sediments

In view of the current concern about the assessment of soil quality, some recent pesticide monitoring studies have been conducted within Europe [11,12,61,62]. According to the results a variety of pesticides, mainly herbicides and insecticides appeared consistently as contaminants of the tested soil samples. Concerning pesticide contamination of soils in United States pesticides such as atrazine, chlorpyrifos, and others have been detected [63].

The monitoring studies performed on sediments show a large number of detected pesticides over the last 40 years. Most of the target analytes detected were OCPs and their transformation products despite the fact that most of them were banned or severely restricted by the mid-1970s in the United States and EU. This reflects both the environmental persistence of these compounds and limited target analytes list. DDT and metabolites, chlordane compounds, α-, β-, γ-HCH, and dieldrin were the most detected pesticides in bed sediments. Other OCPs that sometimes were detected included endosulfan compounds, endrin and its metabolites, heptachlor and heptachlor epoxide, methoxychlor, and toxaphene [64].

Recent studies in sediment cores have shown that concentration levels of OCPs have a relative steady state for DDTs, with a slight decrease in the top layers, suggesting a slight decline in their concentrations due to restrictions in their usage [65]. Besides the OCPs, a few compounds in other pesticide classes were detected in some studies. Most of these pesticides contained chlorine or fluorine substituents and have medium hydrophobicity. Currently used pesticides detected in sediments included the herbicides atrazine, ametryne, prometryne, trifluralin, dicamba, alachlor, metholachlor, and diuron; the organophosphorus insecticides diazinon, chlorpyrifos, ethion, and pyrethrines such as cypermethrin, fenverate, and deltamethrin [2,3]. Of pesticides from other chemical classes, most were targeted at relatively few sites. Examples in this case include the booster biocides such as irgarol, diuron, and chlorothalonil, which were detected in coastal marine sediments [58,59].

12.2.3.4 Biological Monitoring

A lot of biological organisms, from flora and fauna to human beings, are monitored to determine amounts of these pesticides that are present in the environment and evaluate the associated hazard and risk. This type of monitoring is an essential part of pesticide pollution studies that is known as biological monitoring or biomonitoring. Another important facet of environmental biomonitoring is the emerging field of environmental specimen banking. A specimen bank acts as a bridge connecting real-time monitoring with future trends monitoring activities.

In general, biomonitoring overcomes the problem of achieving a snapshot of the quality of the environment, and can provide a more representative picture of average conditions over a period of weeks to months. However, the use of biomonitors has limitations since some compounds are metabolized or eliminated at a rate close to the rate of uptake, and thus are not accumulated. Moreover, because of cost, the monitoring may be carried out only on a limited number of species and there is no guarantee that important species will be selected. Not all pesticides are amenable to biological monitoring. Pesticides that are rapidly absorbed and are neither sequestered nor metabolized to a significant extent are usually good candidates. Pesticides that have a high tendency to bioaccumulate, such as OCPs, are the most commonly detected pesticides in biota samples.

Sample collection methods must be selected considering both the organisms to be collected and the conditions that will be encountered. Organisms that can be deployed for extended periods of time, during which they passively bioaccumulate pesticides in the surrounding environment are usually selected. Plankton, bacteria, periphyton, benthos, fish, and fish-eating birds are the most common specimens for monitoring aquatic compartment. Analysis of the tissues or lipids of the test organism(s) can give an indication of the equilibrium level of waterborne pesticide contamination. Adipose tissues, eggs, and liver have been recognized as accumulators of lipophilic pesticides and they are usually monitored to quantify the threat of pesticide contamination in species of wildlife. Apart from aquatic organisms and wildlife species, increasing attention is focused on the monitoring and assessment of human exposure to pesticides throughout the world. Urine, blood, and exhaled air are the mostly used specimens for routine biological monitoring to human beings. Other biological media include adipose tissue, liver, saliva, hair, placenta, and body involuntary emissions such as nasal accretions, breast milk, and semen. However, many of these media have some serious problems (e.g., matrix effects, insufficient dose–effect relationships) and they do not necessarily provide consistent results to that from blood, urine, or breathe [66].

12.2.3.4.1 Occurrence and pesticide levels in biota

Several studies have been conducted around the world on the general topic of biological monitoring of pesticides. As in the case of sediments, most of the studies reveal the presence of OCPs and their transformation products. These compounds have been detected in different human specimens such as human milk, saliva, urine, adipose tissues, and liver [66–69]. DDT and its metabolites are still the most frequently determined compounds, especially in samples from developing countries. Other OCPs determined were cyclodienes such as dieldrin, aldrin, endrin, heptachlor and its epoxide, chlordane as well as isomers of hexachlorocyclohexane [67]. Moreover, endosulfan I and II and the sulfate metabolite have been detected in fatty and nonfatty tissues and fluids from women of reproductive age and children in Southern Spain [69]. Apart from OCPs, currently used pesticides have also been detected in different human biological samples. Examples include bromophos in blood; fenvalerate, malathion, terbufos, and chlorpyrifos methyl in urine; paraquat, 2,4-D, and pentachlorophenol in urine and blood; carbaryl, atrazine, and ethion in saliva; and DDT in blood and adipose tissue, and so on [68]. From the currently used pesticides, organophosphorus pesticides (OPPs) are the most frequently detected in different human biological fluids. Apart from the parent compounds, the measurement of dialkyl phosphate metabolites has been frequently used to study exposure to a wide range of OPPs. These metabolites have been detected in urine samples from exposed workers as well as from people who had no occupational exposure to OPPs. In addition, metabolites of carbamates (carbaryl, carbofuran) and pyrethrines (cypermethrin, deltamethrin, permethrin) have been also detected in urine samples [66–68].

Except of human biological samples, the accumulation pattern of OCPs in aquatic organisms as well as terrestrial wildlife has been reported. For example, concentration levels of DDT and its metabolites have been detected in different species of arctic wildlife such as terrestrial animals, fish, seabirds, and marine mammals [70]. Extensive results have also reported for various bird species [4,71,72], fish, and amphibian [73,74] as well as mammals [75,76], when adipose tissues, liver, or eggs of these organisms have been analyzed. p,p'-DDE, a major metabolite of DDT, continued to be the dominating OCP burden in almost all the tested species, whereas cyclodienes and HCHs occurred at lower concentrations. Apart from OCPs, several currently used pesticides (despite their lower bioaccumulation) such as trifluralin, chlorothalonil, parathion methyl, phosalone, disulfoton, diazinon, dimethoate, and chlorpyrifos have also been detected in biota samples [6,77]. It is notable that a high variability in the concentrations of pesticides within the same species was observed and this was related to sampling location, age, and sex and with condition and stage of the life cycle (starvation/feeding, lactation, illness/disease) of the analyzed organisms.

A comparison of studies regarding the aquatic monitoring in sediments and biota suggests that pesticides were detected more often in aquatic biota than in bed sediment. In addition, the transformation products were also found at higher levels in biota samples than in associated sediment [4].

An example of monitoring program that report a range of diverse invertebrate, vertebrate, and human relevant tests is the Comparative Research on Endocrine Disrupters—Phylogenetic Approach and Common Principles focusing on Androgenic/Antiandrogenic Compounds (COMPRENDO) project [78].

12.2.4 WATER FRAMEWORK DIRECTIVE AND MONITORING STRATEGIES

The potential adverse consequences that are derived from the use of pesticides have led to the development of special regulations. For instance, in the European

Community, several directives and regulations have been issued with the aim of safeguarding human health and the environment from the undesirable effects of these chemicals (i.e., Dangerous Substances (76/464/EC) [79], Groundwater (80/68/EEC) [80], and Pesticide (91/414/EEC) [17] Directives). The newly introduced WFD (2000/60/EC) [81] is widely recognized as one of the most ambitious and comprehensive pieces of European environmental legislation. Its aim is to improve, protect, and prevent further deterioration of water quality at the river-basin level across Europe. The term "water" within the WFD encompasses most types of water bodies. Furthermore, to monitor the progressive reduction in contaminants, trend studies, whether spatial or geographical, should be envisaged through the measurement of contaminants in sediment and biota. The Directive aims to achieve and ensure "good quality" status of all water bodies throughout Europe by 2015, and this is to be achieved by implementing management plans at the river-basin level. The WFD foresees that water quality should be monitored on a systematic and comparable basis. Thus, technical specifications should follow a common approach (e.g., the standardization of monitoring, sampling, and methods of analysis). Chemical monitoring is expected to be intensified and will follow a list of 33 priority chemicals (inorganic and organic pollutants including pesticides) that will be reviewed every 4 years. The concentrations of the priority substances in water, sediment, or biota must be below the Environmental Quality Standards (EQSs) and this is expressed as "compliance checking." However, EQSs for these substances including pesticides have yet to be stated [25,82]. The derivation of EQSs through a risk assessment procedure is presented later in this chapter.

The implementation of the WFD is based on a three-level monitoring system, which will form part of the management plans and was to be implemented from December 2006 [81,83]: This include (1) surveillance monitoring aimed at assessing long-term changes in natural conditions; (2) operational monitoring aimed at providing data on water bodies at risk or failing environmental objectives of the WFD; and (3) investigative monitoring aimed at assessing the causes of such failure and the effects.

Comprehensive reviews focused on principal monitoring requirements of the WFD as well as on emerging techniques and methods for water quality monitoring have been published recently to identify and outline the tools or techniques that may be considered for water quality monitoring programs necessary for the implementation of WFD [24,83].

12.3 ENVIRONMENTAL EXPOSURE AND RISK ASSESSMENT

12.3.1 ENVIRONMENTAL EXPOSURE

12.3.1.1 Point and Nonpoint Source Pesticide Pollution

Environmental exposure of pesticides can be occurred by point and nonpoint sources. A point source can be any single identifiable source of pollution from which pesticides are discharged such as the effluent pipes, careless storage, and disposal of pesticide containers, accidental spills, and overspray. Pesticide movement away from the targeted application site is defined as nonpoint source pollution and can occur through runoff, leaching, and drift. Nonpoint source pollution occurs over broad geographical scales and because of its diffuse nature it typically yields relatively uniform environmental concentrations of pesticides in surface waters, sediments, and groundwater. Runoff is the surface movement of pesticide in water or bound to soil particles, while leaching is the downward movement of a pesticide through the soil by water percolation. Drift is the off-target movement by wind or air currents and can be in the form of spray droplet drift, vapor drift, or particle (dust) drift.

12.3.1.2   Environmental Parameters Affecting Exposure

The environmental parameters that affect pesticide exposure could be classified as follows:

1. *Soil characteristics and field topography*: Texture composition and pH are the main soil properties that affect pesticide fate and transport, whereas topographic characteristics of the fields like watershed size, slope, drainage pattern, permeability of soil layers affect greatly the potential to generate runoff water or leachates.
2. *Weather and climate*: Climatic factors such as the amount and timing of rainfall, duration, and intensity, as well as temperature and air movement influence the degree to which pesticides are mobilized by runoff, leaching, and drift. In addition, temperature and sunlight affect all abiotic and biotic transformation reactions of pesticides [84,85].

12.3.1.3   Pesticide Parameters Affecting Exposure

The pesticide factors affecting exposure could be organized on three main sets:

1. *Application factors*: These include the application site (crop or soil surface) and method, the type of use (agricultural, nonagricultural applications, indoor pest management, etc.), the formulation (e.g., granules or suspended powder or liquid) and the application amount, and frequency. In addition, the application time does affect its possible routes of transport in the environment.
2. *Partitioning and mobility of pesticides in the environment*: The main physicochemical properties of pesticides that affect their mobility are the water solubility, vapor pressure, and soil–water partition coefficient (K_{oc}). K_{oc} defines the potential for the pesticide to bind to soil particles. Off-target movement by drift also depends on the spray droplet size and the viscosity of the liquid pesticide while plant uptake from the soil is another important pathway in determining the ultimate fate of pesticide residues in the soil [84,85].
3. *Persistence in the environmental compartments*: Persistence is usually expressed in terms of half-life that is the time required for one-half of the pesticide to decompose to products other than the parent compound. The longer a pesticide persists within the environment, the greater the risk it

poses to it. Hydrolysis, direct and indirect photolysis, and biodegradation are the principal pesticide degradation processes and their rates depend on pesticide chemistry, as well as on environmental conditions [84].

12.3.1.4 Modeling of Environmental Exposure

Monitoring data and environmental modeling are interconnected to each other. Monitoring could provide the correct input data to models for calibration and validation or could be devoted to collect data on the timing and magnitude of loadings. Mathematical models that simulate the fate of pesticides in the environment are used for developing Environmental Estimated Concentrations (EECs) or Predicted Environmental Concentrations (PECs). This means "predicting exposure" in space and time, drawing on available environmental fate data, physicochemical data, and the proposed agricultural practices and usage pattern associated with the pesticide [86]. A complete presentation of environmental models describing the exposure of pesticide in the environment is outside the scope of the present chapter. Thus, only common environmental models that are used to estimate environmental exposure concentrations for aquatic systems in the context of current risk assessing techniques will be presented.

The Generic Estimated Environmental Concentration (GENEEC) model, developed by the EPA, determines generic EEC for aquatic environments under worst-case conditions (i.e., application on a highly erosive slope with heavy rainfall occurred just after the pesticide application, the treatment of the entire area—essentially 10 acres of surface area with uniform slope—with the pesticide, and the assumption that all runoff drains directly into a single pond). The model uses environmental fate parameters derived from laboratory studies under standard procedures as well as soil and weather parameters. The outputs of the model are the pesticide runoff and environmental concentration estimates [87]. This model can be used as first tier approach since it is based on a single event and a high-exposure scenario. On a higher tier approach (second and third), models that can account for multiple weather conditions and/or multiple sites are used. Such models are the Pesticide Root Zone Model (PRZM), edge of field runoff/leaching the Exposure Analysis Modeling System (EXAMS), fate in surface water, and AgDrift (spray drift) [87] that used additional parameters, more descriptive of the site studied. PRZM simulates the leaching, runoff, and erosion from an agricultural field and EXAMS simulates the fate in a receiving water body. The water body simulated is a static pond, adjacent to the crop of interest. Typical conditions of the site including the soil characteristics, hydrology, crop management practices, and weather information are used. The output of this higher tier analysis is to define the EEC that can be reasonably expected under variable site and weather conditions. The model yields an output of annual maxima distributions of peak, 96 h, 21 days, 60 days, 90 days, and yearly intervals. AgDrift includes generic data for screening level assessments including pesticide formulation, drop height, droplet size, nozzle type, and wind speed. The earlier approaches are used by pesticide registrants to address environmental exposure concerns and are frequently combined with geographical information systems (GIS) to produce regional maps.

The fugacity approach has also proven particularly suited for describing the behavior of pesticides in the environment. A tiered system of fugacity models has been introduced which distinguishes four levels of complexity, depending on whether the system is closed or in exchange with the surrounding environment. The four levels are Level I, close system equilibrium; Level II, equilibrium steady state; Level III, Nonequilibrium steady state; and Level IV, Nonequilibrium non-steady state. Levels I and II are used in lower tier approaches, whereas Level III is widely used in higher tiers to obtain exposure concentrations due to emission flux into a predefined standard environment. A detailed introduction into fugacity-based modeling can be found in Ref. [88].

For evaluating the impact of management practices on potential pesticide leaching, the Groundwater Loading Effects of Agricultural Management Systems (GLEAMS) is a widely used, field-scale model. GLEAMS assumes that a field has homogeneous land use, soils, and precipitation. It consists of four major components: hydrology, erosion, pesticide transport, and nutrients. GLEAMS estimates leaching, surface runoff, and sediment losses from the field and can be used as a tool for comparative analysis of complex pesticide chemistry, soil properties, and climate. The model output data are daily, monthly, annual pesticide mass and concentrations in runoff and sediment.

Finally, a fourth tier approach can be used based on watershed site assessments. These assessments are very complex since the landscape studied has a very high surface area, high diversity of soils and weather conditions, varied proximities of agricultural lands to receiving waters and various water bodies. Thus, GIS are commonly used to distinguish high-risk versus low-risk areas on a watershed basis. Finally, modeling and monitoring are often combined within tier 4 to provide more accurate distribution of pesticide exposure.

12.3.2 RISK ASSESSMENT

In order to evaluate the negative impact of pesticides on ecosystems, the environmental risk assessment is necessary. It is known that the environmental impact of a pesticide depends on the degree of exposure and its toxicological properties [89]. The risk assessment procedure involves three main steps: a formulation of the problem to be addressed followed by an appraisal of toxicity and exposure and concluding with the characterization of risk. A typical framework for ecological risk assessment is shown in Figure 12.2 [90]. The objective of the exposure assessment is to describe exposure in terms of source, intensity, spatial and temporal distribution, evaluating secondary stressors (metabolites) to derive exposure profiles. Usually exposure assessment involves the measured environmental concentrations (MECs) derived from monitoring studies or the developing and application of models as discussed previously.

The toxicity assessment identifies concentrations that when administered to surrogate organisms result in a measurable adverse biological response. Toxicological assessment is commonly based on laboratory studies with the aim of determination of the relationship between magnitude of exposure and extent of observed effects commonly referred as dose–response relationship. Toxicity impacts were

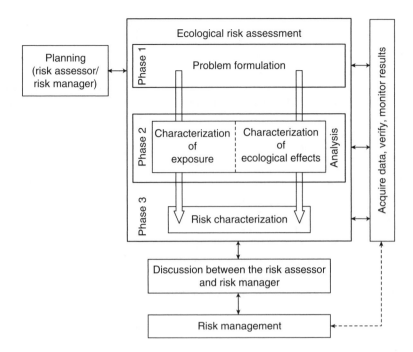

FIGURE 12.2 EPA framework for ecological risk assessment. (From USEPA, U.S. Environmental Protection Agency, *Framework for Ecological Risk Assessment*, Risk Assessment Forum, Washington, D.C., 1992.)

usually studied by indicator species selected to represent various trophic levels within an ecosystem. Representative groups of organisms are assessed for risk to pesticides, including fish, aquatic invertebrates, algae, and plants from the aquatic environment and birds, mammals, bees and beneficial arthropods, earthworms, soil microorganisms, and nontarget plants from the terrestrial environment. All these organisms are assessed in Europe under 91/414/EEC [17], whereas the USEPA concentrates on birds and mammals, bees, nontarget plants, and aquatic organisms. It is impossible and inadvisable to test every species (abundant, threatened, endangered) with every pesticide but the need for more toxicological data is acknowledged. Chosen organisms like *Daphnia* sp. for freshwater zooplankton or rainbow trout for freshwater fish categories should typically satisfy some basic criteria like the ecological significance, the abundance and the wide distribution, the susceptibility to pesticide exposure, and the availability for laboratory testing.

Stressor–response analysis can be derived from point estimates of an effect (i.e., lethal concentration or effect concentration for 50% of the organism population, LC_{50} or EC_{50}) or from multiple-point estimates (hazardous concentration for 5% of the species, HC_5) that can be displayed as cumulative distribution functions (species sensitivity distributions, SSDs). In addition, the establishment of cause–effect relationships from observational evidences or experimental data could be performed.

In a third phase, the risk characterization takes place defining the relationship between exposure and toxicity. Two different approaches are usually applied for this purpose. The first is a deterministic approach that is based on simple exposure and toxicity ratios and the second is a probabilistic approach in which the risk is expressed as the degree of overlap between the exposure and effects. Apart from these methods, numerous Pesticide Risk Indicators (PRIs) based on classification systems have been developed for fast preliminary assessments and comparative purposes. All methods will be analyzed in detail later.

The last step in the assessment of risk is the weight-of-evidence analysis. Strengths, limitations, and uncertainties as well as magnitude, frequency, and spatial and temporal patterns of previously identified adverse effects and exposure concentrations are discussed in the weight-of-evidence analysis.

The assessment of the pesticides risk usually follows a tiered approach adopt. Tiers are normally designed such that the lower tiers are more conservative, whereas the higher tiers are more realistic with assumptions more closely approaching reality. Tier 1 is essentially a screen, thereby to identify low-risk uses, or those groups of organisms at low risk [91–94]. Higher tier approaches aim to the refinement of risk, that is, a procedure (method, investigation, evaluation) performed to characterize in more depth the pesticide risks arising from the preliminary (tier 1) risk assessment. The risk refinement is triggered to increase more realistic and/or comprehensive sets of data, assumption and models, and/or mitigation options. Thus, if the assessment fails to "pass" tier 1, then a more detailed risk assessment is required.

12.3.2.1   Preliminary Risk Assessment–Pesticide Risk Indicators–Classification Systems

A preliminary estimation of the environmental impact of pesticides use could be performed through the development and use of PRIs, which are indices that combine the hazard and exposure characteristics for one or several environmental compartments that are assessed separately. PRIs make use of the physicochemical and biological properties of pesticides and have been used over the years by a large number of organizations for the purposes of selecting pesticide compounds for further regulatory actions.

Firstly, the development of a PRI is generally based on the concept of risk ratios, that is, the division of exposure concentration by effect concentration. Several approaches are based on this standard framework for risk assessment (analyzed in the following section) such as the Evaluation System for Pesticides (ESPE) [95], the Ecological Relative Risk (EcoRR) [96], the Environmental Yardstick [97], and SYNOPS [98]. Although the risk ratio approach is favored by many researchers, different methodologies have also been used such as the scoring and ranking of pesticides in terms of their environmental hazard. In general, the proposed systems are also based on factors describing the physicochemical and ecotoxicological properties of pesticides. Such indices are developed by assigning scores to the previously mentioned properties. The scores are then aggregated using different algorithms or weights of evidence finally to obtain a numerical or descriptive index useful for comparative assessment of the environmental impact of pesticide applications [99]. There are several screening tools in use that were developed for priority setting in risk assessment, which involves ordering chemicals by scoring and ranking them individually or placing them in group based on degree of concern (e.g., high, medium, low). Examples of such approaches are the Scoring and Ranking Assessment Model (SCRAM), [100], the Environmental Impact Quotient (EIQ) [101], and the Pesticide Environmental Risk Indicator (PERI) [102]. Such approaches can be useful for several management purposes such as the selection of pesticides with less environmental impact and the setting of priority list for planning environmental monitoring or further experimental research [103]. These methods are simple and fast for ecological screening assessments but are highly arbitrary [104]. In addition, some other systems use the risk ratio methodologies combined with rating and scoring approaches in aggregated indices. The short-term or long-term pesticide risk indexes for the surface water system (PRISW-1, PRISW-2) [104] belong to this category. Finally, van der Werf and Zimmer [105], in 1998, have developed an expert system using fuzzy logic (I-pest) to assess the environmental impact of a single pesticide application to rank various alternatives.

Recently, an attempt to evaluate and compare the various methodologies has been made in Europe by the Concerted Action on Pesticide Environmental Risk Indicators (CAPER) project [102]. According to the project conclusions, PRIs differed considerably with regard to several aspects such as purposes, methodologies, compartments, and effects to take into account. However, the earlier aspects barely influenced the rankings of the pesticides. Further details on all the previous and other approaches and systems are well described and compared in recently published articles and reports [102,103,105,106]. In conclusion, the present indicators leave room for users and scientists to select the most appropriate indicator, according to the considered environmental effects and the environmental specific conditions at national or regional level. However, a harmonized scientific framework is highly recommended.

12.3.2.2 Risk Quotient–Toxicity Exposure Ratio Method (Deterministic-Tier 1)

At present, the usual approaches to decide the acceptability of environmental risks are generally based on the concept of risk ratios expressed as the toxicity–exposure ratio (TER) adopted by the EU (Equation 12.1) [17] or the risk quotient (RQ) adopted by USEPA (Equation 12.2) [107]. This methodology usually involves comparing an estimate of toxicity, derived from a standard laboratory test with a worst-case estimate of exposure, EEC, or PEC from model applications or peak measured concentrations, for the US and EU, respectively.

$$\text{TER} = \frac{\text{toxicity}}{\text{exposure}}, \tag{12.1}$$

$$\text{RQ} = \frac{\text{exposure}}{\text{toxicity}}. \tag{12.2}$$

Since the term risk implies an element of likelihood which is usually reported as probabilities, it is more correct that the risk quotient should be better expressed as hazard quotient (HQ). However, both terms are used in several studies with the same meaning. Examples of toxicity measurements used in the calculation of RQs are LC_{50} (fish and amphibians, birds); LD_{50} (birds and mammals); EC_{50} (aquatic plants and invertebrates); EC_{25} (terrestrial plants); EC_{05} or nonobserved effect concentration (NOEC) (endangered plants).

According to Directive 414/91/EEC [17], one standard procedure for the risk assessment in aquatic systems is the determination of RQ method for three taxonomic groups (i.e., algae, zooplankton, fish) at two effect levels (i.e., acute level, using LC_{50} or EC_{50} values and chronic level, using NOEC or predicted noneffect concentration [PNEC] values).

For assessing the risk in sediments, if results from whole-sediment tests with benthic organisms are available, the $PNEC_{sed}$ has to be derived from these tests. In the case that not enough reliable ecotoxicological data for sediment-dwelling organisms are known, the equilibrium partitioning method can be used [108] to derive $PNEC_{sed}$ according to the following equation:

$$PNEC_{sed} = \frac{PNEC_{wat} \times K_{susp-water}}{RHO_{susp}} \times 1000, \qquad (12.3)$$

where

$PNEC_{wat}$ is the PNEC calculated for the water compartment

$K_{susp-water}$ is the sediment/water partition coefficient

RHO_{susp} is the bulk density of the sediment

The same methodology can be applied for deriving PNEC values for soil using the corresponding K_{psoil} (soil/water) partition coefficient.

For terrestrial systems, the estimate of the distribution of exposure is separated into the chemical/physical and biological components. The first component of dose estimate is the environmental and chemical variables that influence the distribution of residue levels. The major variables that influence the biological component are species-dependent including (1) food, water, and soil ingestion rates; (2) dermal and inhalation rates; (3) dietary diversity; (4) habitat requirements and spatial movement; and (5) direct ingestion rates. These variables are combined into Equation 12.4 to estimate the distribution of total dose:

$$Dose_{total} = Dose_{oral} + Dose_{dermal} + Dose_{inhal}. \qquad (12.4)$$

The oral dose can be further analyzed as follows:

$$Dose_{oral} = Dose_{food} + Dose_{water} + Dose_{soil} + Dose_{preening} + Dose_{granular}. \quad (12.5)$$

For each of these sources of oral exposure, the equations which can be used to estimate the dose are reported elsewhere [109]. Frequently for birds and mammals, it is assumed that exposure is through eating treated food items and residue concentrations (w/w) in milligram per kilogram are compared with dietary LC_{50}, NOEC.

12.3.2.2.1 The use of assessment factors for the characterization of uncertainty

For many substances, the available toxicity data that can be used to predict ecosystem effects are very limited, and thus, empirically derived assessment factors must be used depending on the confidence with which a PNEC can be derived from the existing data. The proposed assessment factors according to EC guidelines [108] are presented in Table 12.1 for water and sediment.

If the database on SSDs from long-term tests for different taxonomic groups is sufficient, statistical extrapolation methods may be used to derive a PNEC. In such methods, the long-term toxicity data are log-transformed and fitted according to the distribution function and a prescribed percentile of that distribution is used as criterion. Kooijman [110] and Van Straalen and Denneman [111] assume a log-logistic function, Wagner and Lokke [112] a log-normal function, and Newton et al. [113] a Gompertz distribution. Newman et al. [113] proposed to bootstrap the data as a nonparametric alternative whereas Van der Hoeven [114] proposed a nonparametric method to estimate HC_5 without any assumption about the distribution and without bootstrapping. Aldenberg and Jaworska [115] refined the way to estimate the uncertainty of the 95th percentile by introducing confidence levels. The 95% confidence level provides more strict values while 50% of confidence level is usually applied. According to the earlier discussions, a PNEC value can be calculated as

TABLE 12.1
Assessment Factors to Derive a PNEC$_{aquatic}$

Available Data	Assessment Factor
At least one short-term L(E)C$_{50}$ from each of three trophic levels of the base set (fish, *Daphnia*, and algae)	1000[a]
One long-term NOEC (either fish or *Daphnia*)	100[b]
Two long-term NOECs from species representing two trophic levels (fish, *Daphnia*, and/or algae)	50[b]
Long-term NOECs from at least three species (fish, *Daphnia*, and algae) representing three trophic levels	10[b]
Species sensitivity distribution (SSD) method	5–1
Field data or model ecosystems	Reviewed on case by case basis

Source: From European Commission, Technical Guidance Document on Risk Assessment in Support of Council Directive 93/67/EEC for New Notified Substances and Commission Regulation 1488/94 on Risk Assessment for Existing Substances and Directive 98/8/EC of the European Parliament and the Council Concerning the Placing of Biocidal Products of the Market, EU, JRC, Brussels, Belgium, 2002.

[a] A factor of 100 could be used for pesticides subjected to intermittent release.

[b] The same assessment factors are used for derivation of PNEC in sediments using appropriate species.

$$PNEC = \frac{5\%SSD(50\%c.i.)}{AF}. \qquad (12.6)$$

AF is an appropriate assessment factor between 5 and 1 (as proposed in Table 12.1), reflecting the further uncertainties identified. Confidence can be associated with a PNEC derived by statistical extrapolation if the database contains at least 10 NOECs (preferably >15) for different species covering at least 8 taxonomic groups [108].

Uncertainty arises from an incomplete knowledge of the system that is assessed and it is associated with the following aspects: measurement errors (accuracy), inherent variability, model error both conceptual and mathematical, assumption errors, and lack of data. As already mentioned, the characterization of risk at a first level of assessment is typically highly conservative, both from exposure and effects characterization perspective and thus it is characterized by high uncertainty. This means that even values of RQ that are below 1 are quite likely to be capable of causing an effect. Usually, a safety factor is applied to risk quotients for covering uncertainty. The factor can vary between 1 and 100, depending on the organisms that is assessed and whether the toxicity end point is acute, based on short-term effects ($LD/LC/EC_{50}$) or chronic, based on NOEC [56,94].

Therefore, as a final step in risk characterization procedure, the results of the RQ are compared with acceptable levels designed by particular jurisdiction [116]. These regulatory triggers used to categorize the potential risk are defined as levels of concern (LOC). An example of LOCs of RQ values that can be used for terrestrial and aquatic risk assessments is shown in Table 12.2. In the EU, TERs for terrestrial acute effects must be ≥ 10 and for aquatic short-term effects ≥ 100.

TABLE 12.2

EPA Established Risk Quotients and Levels of Concern for Different Environmental Applications

End Point and Scenario	Risk Quotient	Nonendangered	Endangered
Mammalian acute (granular)	$EEC/LD50/FT^2$	0.5	0.1
Mammalian acute (spray)	EEC/LC_{50}	0.5	0.1
Mammalian chronic (spray)	$EEC/NOEC$	1.0	1.0
Avian acute (granular)	$EEC/LD50/FT^2$	0.5	0.1
Avian dietary (spray)	EEC/LC_{50}	0.5	0.1
Avian chronic (spray)	$EEC/NOEC$	1.0	1.0
Aquatic acute	EEC/LC_{50}	0.5	0.05
	EEC/EC_{50}		
Aquatic chronic	$EEC/NOEC$	1.0	1.0
Terrestrial plants	EEC/EC_{25}	1.0	1.0
Aquatic plants	EEC/EC_{50}	1.0	1.0

Source: From Whitford, F. in *The Complete Book of Pesticide Management. Science, Regulation, Stewardship and Communication*, John Wiley & Sons, New York, USA, 2002.

Chronic and subchronic TERs ≥ 5 and 10 for terrestrial and aquatic species, respectively, are acceptable.

Descriptive uncertainty analysis is usually performed in the lower tiered risk assessments while sensitivity analysis and more complex model (i.e., Monte Carlo) simulation are usually completed in higher tier assessments. Monte Carlo simulations can be performed by using risk quotient approach by using randomly selected toxicity values from the generated SSDs and dividing these by the environmental concentrations randomly selected from their specified distributions to produce RQ or TER values. Such an approach when repeated thousands of times builds up a distribution of RQ or TER values and provides information on the risk assessment uncertainty, as more environmentally realistic assumptions are introduced [117].

In conclusion, if consideration of the "worst-case" scenario results in TERs or RQs that are acceptable when compared with LOC, then no further risk assessment is needed. If the tier 1 assessment does not pass the risk criteria, then the assessment needs to be refined and iterated back to the initial exposure and toxicity characterization but using a higher tier procedure.

12.3.2.2.2 Risk Refining and hazard of pesticide mixtures

Risk refinement must be a tiered process that more realistic and/or comprehensive sets of data, assumptions, and models are used to reexamine the potential risk. There is a tendency to jump straight from tier 1 to chemical monitoring in the environment and generate "real-world" data. However, this approach has its limitations since it provides only a snapshot in time and rarely gives sufficient information about concentrations over time, which is often necessary to determine exposure. For tier 1, the USEPA uses the GENEEC exposure model; and for tier 2, the PRZM/EXAMS modeling systems which is specific to a particular crop and region [94]. Currently used models that are used in risk assessment approaches were presented in Section 12.3.1.4. Refinement of toxic effects is usually obtained through the application of probabilistic approaches presented later in this chapter.

Until now, the relative risk of single pesticide compounds has been discussed. However, as already reported in the first few sections of this chapter, multiresidues of pesticides are usually detected in different environmental compartments. For the estimation of pesticide mixture effects, the quotient addition method is generally applied. The quotient addition approach assumes that toxicities are additive or approximately additive and that there are no synergistic, antagonistic, or other interactions. The additive response of a mixture of pesticides with the same toxicological mode of action can be assessed, according to the so-called Loewe additivity model [118] as described in Equation 12.7. The sum of the toxic quotients of all compounds detected gives an estimate of the total toxicity of the sample with respect to the compounds determined.

$$\mathrm{TU}_{\mathrm{mix}} = \sum_{i=1}^{n} \mathrm{TU}_i, \qquad (12.7)$$

where $\mathrm{TU}_i = C_i/\mathrm{EC}_i$ are the toxic units of individual pesticides calculated as TERs or RQs.

This assumption may be most applicable when the modes of action of chemicals in a mixture are similar (as for carbamates and phosphate esters), but there is evidence that even with chemicals having dissimilar modes of action, additive or near-additive interactions are common [92,119]. This approach provides an estimate of the contribution of the compound of interest to the total toxicity of the water sample analyzed to a certain taxonomic group.

12.3.2.2.3 Limitations of the method

The risk quotient is a useful tool because it provides the risk managers a screening method to facilitate the rapid identification of pesticides that are not likely to pose an ecological risk. However, the risk quotient cannot address issues related to magnitude, probability, and species diversity. A common error in the interpretation of RQs is the assumption that the RQ itself is proportional to the risk. Since the concept of risk incorporates an element of probability, the RQ is biased because it assumes that the conditions exist on every occasion and in every location, and that there is a 100% probability of cooccurrence of the stressor and the most sensitive organism.

Thus, major limitations of the quotient method for ecological risk assessment are that it fails to consider variability of exposures among individuals in a population, ranges of sensitivity among species, and the ecological function of species assuming that is a keystone organism in the environment.

12.3.3 PROBABILISTIC RISK ASSESSMENT (TIER 2)

The use of probabilistic approaches allows the quantification of likelihood of effects which by definition is risk. In probabilistic approaches, the risk is expressed as the degree of overlap between the exposure and the effects that is acceptable for a certain level of protection that would be attained [120]. PRA approaches use SSD combined with distributions of exposure concentrations to better describe the likelihood of exceedances of effect thresholds and thus the risk of adverse effects. The frequency of occurrence of levels of exposure (return frequencies) could be classified as follows: typical case (50th percentile), reasonable worst case (90th percentile), and extreme worst case (99th percentile). From the resulted SSDs, exposure levels that would protect 90%, 95%, or indeed any percentage of the species can be determined. Of course, there are a number of concerns such as what level, if any, of species affected might be acceptable; which species might be affected; how they might be affected; and are they economically, ecologically, or otherwise important.

Hart [121] in his summary of an EU-funded workshop on pesticide PRA identified several strengths and weaknesses of PRA within the context of EC Directive 91/414/EEC [17]. Strengths of PRA include (1) the ability to quantify the type, magnitude, and frequency of toxic effects and communicate more "meaningful" outputs to decision-makers and the public; (2) the ability to quantify variability, uncertainty, and model sensitivity; (3) the better use of available information by taking into account all available toxicity data to quantify variation between species and not just the more sensitive or representative organism for the ecosystem only; and (4) finally, probabilistic methods are also more prone to be coupled with new approaches such as GIS and population modeling. Potential weaknesses include the greater complexity that could lead to misleading results, the requirement of more toxicity data and thus the increased animal testing, the lack of available expertise and guidance, and the lack of established criteria for decision-makers [121].

PRAs could be applied for all organisms as well as for human health and has been recommended for regulatory assessment of pesticides [91]. The general concepts have been reviewed and discussed [120–123]. The different PRA methods are developed similarly, but they may be used for different purposes. Some uses include the setting of environmental quality objectives and criteria, whereas others are used for assessing risks of known exposure. As a first example of a PRA method of risk assessment, the inverse method of Van Straalen and Denneman [111] is presented. The method is based on the assumption that the frequency distribution of effect end points for different species is log-logistic [119]. The parameters describing the distribution could be estimated for the mean and the standard deviation of the ln-transformed data set of a number of toxicity end points of a given pesticide reported in the literature. From this distribution, a concentration is calculated that is hazardous for 5% of the species in an ecosystem (HC$_5$, Equation 12.8), which is an acceptable level for protecting aquatic ecosystems [56,124].

$$HC_5 = \exp(x_m - k_L s_m),$$
(12.8)

where

m = the number of the test species
x_m = the mean of the ln-transformed toxicity end points (LC$_{50}$ or EC$_{50}$ or NOEC)
s_m = the standard deviation of the ln-transformed effect levels
k_L = the extrapolation constant as reported in Ref. [125]

The 95% confidence level provides a strict or safe HC$_5$, whereas the 50% confidence level provides the most probable or mean. In most studies, the 50% confidence level was used.

The hazard or ecological risk is estimated by defining it as the probability, Φ, that a random species will be affected by the measured field concentrations (C).

$$\Phi = \left[1 + \exp\left\{ \frac{x_m - \ln C}{k_L / \ln(95/5)s_m} \right\} \right]^{-1}.$$
(12.9)

This method has been followed by several researchers in pesticide risk assessment in aquatic systems [3,56]. For estimating the combined risk ($\Sigma\Phi$) from pesticide mixtures, the equation for the addition of probabilities can be used as follows:

$$\Phi[A_1 + A_2 + \cdots + A_n]$$

$$= \sum_{i=1}^{n} \Phi[Ai] - \sum_{i_1 < i_2} \Phi[Ai_1 Ai_2] + \cdots$$

$$+ (-1)^{r+1} \sum_{i_1 < i_2 < \ldots < i_r} \Phi[Ai_1 Ai_2 \cdots Ai_r] + \cdots + (-1)^{(n+1)} \Phi[A_1 A_2 \cdots A_n].$$
(12.10)

The summation $\sum_{i_1 < i_2 < \cdots < i_r} \Phi[Ai_1 Ai_2 \cdots Ai_r]$ is taken over all of $\binom{n}{r}$ possible subsets of the ecological risk r of the compounds $\{1, 2, \ldots, n\}$. The equation does not account for synergistic or antagonistic interactions.

A second generic method of PRA is presented later. The method has been used by a number of authors [120,126–128] and is currently implemented by the USEPA [109,129]. Toxicity data for all species are combined to produce a distribution curve of effects concentration where appropriate data for all species fitted to log-normal distributions, while other models or bootstrapping models can be also used. The exposure data (measured values from monitoring programs or estimated by modeling) are plotted on the same axes as the effects data. The extent of overlapping between the two curves indicates the probability of exceeding an exposure concentration associated with a particular probability of effects of the studied pesticide. For plotting cumulative percentage (or cumulative probability) of the total distribution, both the acute and chronic toxicity data and the environmental pesticide concentrations are separately sorted into ascending order and ranked. These data are then converted to a cumulative percentage of the total distribution using the following equation [126]:

$$\text{Cumulative percent} = [\text{rank}/(n + 1)] \times 100, \qquad (12.11)$$

where n is the total number of environmental concentration or toxicity data used to perform the quantitative assessment. These percentiles were plotted against the log-transformed concentration, and a linear regression was performed to characterize each distribution (Figure 12.3A and B). Alternatively, straight-line transformations

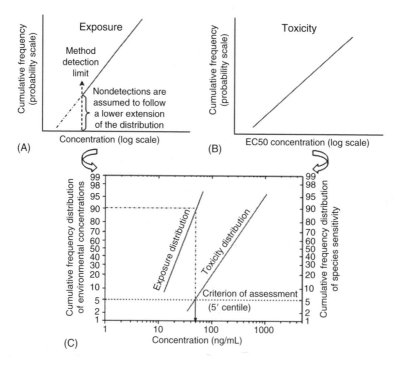

FIGURE 12.3 Graphical representation of combination of (A) exposure and (B) toxicity data expressed as linearized probability distributions (C) for the probabilistic risk estimation.

of probability functions are obtained by probit transformation according to the equation:

$$\int (x,\mu,\sigma) = \frac{1}{\sqrt{2\pi\sigma}} e^{-(x-\mu)/2\sigma}, \tag{12.12}$$

where μ is the distribution mean and σ is the distribution standard deviation [120].

Approaches for handling data below the detection limits include the assigning of values as zero or one-half the detection limit. Alternatively, nondetected concentrations are assumed to be distributed along a lower extension of the distribution (Figure 12.3A). The use of distribution curves for exposure and toxicity data allows the application of a joint probability method (Figure 12.3C) to perform the environmental risk assessment. In this way, any level of effect is associated with an exposure concentration and inversely for any concentration level a probability of exceedance of this level can be determined [120]. In the example provided (Figure 12.3C), the concentration at which 5% of species toxicity values will be exceeded is ~50 µg/L. Approximately 90% of all water concentrations would be expected to be ≤50 µg/L or in other words this concentration would be exceeded 10% of the times. The final step in the probabilistic approach is to generate a joint probability plot of the exceedance of data (exceedance profile). This can be performed by solving the functions describing the probability of exceeding both an exposure and an effect concentration with appropriate fitted regression models or from Monte Carlo modeled data [91]. The graphical representation of a joint probability curve (JPC) which describes the probability of exceeding the concentration associated with a particular degree of effect is shown in Figure 12.4. In such type of representation, the closer the JPC to the axes, the lesser the probability of adverse effect (Figure 12.5) [91].

There is a debate over which value from a range of species sensitivities is most appropriate to protect the various environmental compartments. The 5th percentile value is a generically applicable level of species protection used by USEPA [129,130], European [115,131], and Australian [132] quality criteria. While the 5th percentile is therefore the accepted norm, previous studies with pyrethroids and atrazine have proposed that the 10th percentile effect concentration is adequate [124].

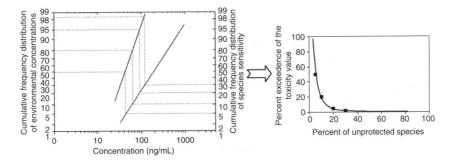

FIGURE 12.4 Graphical representation of the derivation of a joint probability curve (exceedance profile from exposure and toxicity probability functions). (Modified from Solomon, K., Giesy, J., and Jones, P., *Crop Prot.*, 19, 649, 2000.)

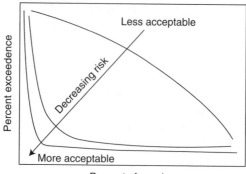

Percent of species

FIGURE 12.5 Illustration of the use of the joint probability curve in decision making. (From ECOFRAM, Ecological Committee on FIFRA Risk Assessment Methods, Aquatic and Terrestrial Final Draft Reports, USEPA, 1999, www.epa.gov/oppefed1/ecorisk/index.htm)

This level of species protection is not universally accepted, especially if the unprotected 10% are keystone species and have commercial or recreational significance. However, protection of 90% of the species in 90% of the time (10th percentile) has been recommended by the SETAC [133].

For PRA, however, a harmonization of the methods used through calibration and validation should be established for their appropriate use in environmental risk assessment. In addition, methods for dealing with spatial and temporal variation and regional scenarios are recommended to be developed and validated with the help of GIS approaches.

As a generic conclusion, the methods that can be followed for pesticide risk assessment in the environmental compartments could include comparisons between point estimates and/or distributions of exposure and toxicity data, depending on the data available and the questions that are addressed in the assessment (Table 12.3). An example of tiered approach for risk characterization is showed in Table 12.4. Once risk has been characterized, it is necessary to follow basic guidelines for risk management and communication strategies [134].

12.4 ENVIRONMENTAL QUALITY STANDARD REQUIREMENTS AND SYSTEM RECOVERY THROUGH PROBABILISTIC APPROACHES

EQSs are concentration limits that should represent the theoretical "no effect" figure, and must be elaborated in European Union for a list of priority substances including pesticides to implement the WFD [135]. Often, published papers on pesticide monitoring state that determined residue levels have exceeded or not "safe" levels or that acute/chronic effect may be expected with reference to "maximum permissible concentrations" (MPC) or "EQSs."

Since EQSs cannot be determined experimentally, some of the current procedures are based on the use of application factors to experimental NOECs [136]. However, SSDs have also been proposed to derive EQSs [137]. The extrapolation of using such

TABLE 12.3
Methods for the Risk Assessment of Pesticides in the Environment Depending on the Data Availability and the Questions that Are Addressed in the Assessment

Method	Exposure	Effects	Output
Point estimate quotients	Point estimation	Point estimation	A ratio of exposure–toxicity
Distribution–point estimation comparison	Distribution	Point estimation	Probability of exposure exceeding the effect levels
Exposure and effect distribution comparison	Cumulative frequency distribution	Distribution	Probability of certain effects occurring when a fixed exposure level is exceeded
Distribution-based quotients	Distribution (Monte Carlo simulation)	Distribution	Probability distribution of quotients (probability that exposure exceeds toxicity)
Integrated exposure and effects distribution	Distribution (Monte Carlo simulation)	Distribution (Monte Carlo simulation)	Probability and magnitude of effect occurring

TABLE 12.4
Tiered Risk Assessment Scheme

Tier Level	Exposure Assessment	Risk Assessment
Tier 1 (Deterministic)	Screening level EEC based on a high-exposure scenario	Is the EEC $<$ point estimates of toxicity for the most sensitive species (L(E)C$_{50}$ or NOEC)
	If yes $\rightarrow$ no further assessment necessary; if no $\rightarrow$ tier 2 or mitigate	
Tier 2 (Probabilistic)	Reasonable high-exposure EECs based on improved model simulations	Is the upper 10th percentile of the distribution of EEC $<$ the lower 10th percentile of the distribution of toxicity estimates (L(E)C$_{50}$ or NOEC)
	If yes $\rightarrow$ no further assessment necessary; if no $\rightarrow$ tier 3 or mitigate	
Tier 3	More specific scenarios for defining geographical and climate-driven EECs	As for tier 2
	If yes $\rightarrow$ no further assessment necessary; if no $\rightarrow$ tier 4 or mitigate	
Tier 4	Site-specific EECs (pulsed exposures) or landscape modeling confirmed by environmental monitoring	As for tier 2 or use more realistic toxicity tests
	If yes $\rightarrow$ no further assessment necessary; if no $\rightarrow$ mitigate	

Source: Modified from SETAC, Pesticide Risk and Mitigation, Final Report of the Aquatic Risk Assessment and Mitigation Dialog Group, SETAC Foundation for Environmental Education, Pensacola, FL, 220 pp., 1994.

distributions for setting EQSs originated from past studies [110,130]. PNEC values were calculated to serve as EQS by using the HC_5 values (hazardous concentration for 5% of species in the ecosystem under investigation) with a default safety factor of 5 as it is suggested in the EC Technical Guidance Document [108].

In addition, the need to develop quality objectives not only for single substances but also for mixtures of pesticides seems evident. For that purpose, the conceptual basis could be the use of the two existing biometric models: concentration addition (CA) and independent action (IA) or response addition. They may allow calculation of the toxicity of mixtures of pesticides with similar modes of action (CA) and dissimilar modes of action (IA), respectively. The research project of Prediction and Assessment of the Aquatic Toxicity of Mixtures of Chemicals (PREDICT) [138] provided results for several multiple mixtures, composed of similarly as well as dissimilarly acting chemicals and revealed that significant mixture toxicity occurs even at a mixture concentration consisting of the sum of the EC_{01} concentrations of the mixture components. Thus, NOECs should not be generally considered as concentrations of no environmental concern with respect to multiple pesticide mixtures.

The rate of recovery of a system could also be estimated based on SSDs. The approach is based on the dissipation half-life of the pesticide ($T_{1/2}$), the initial concentration (C_0), and the hazardous concentration for 5% of species (HC_5) [139]. As an example, for the recovery of the system within a year after application, the pesticide's half-life should meet the condition:

$$T_{1/2} < \frac{\ln 2}{\ln\left\{\frac{C_0}{HC_5}\right\}}. \tag{12.13}$$

Although complete dissipation will be necessary for complete recovery, under some conditions it may not be sufficient and ecological recovery may lag behind the disappearance of the pesticide [124]. Of course, the factors that influence the recovery of biota population after a significant perturbation are complex and the earlier approach could be used only for preliminary assessment.

12.5 LIMITATIONS AND FUTURE TRENDS OF MONITORING AND ECOLOGICAL RISK ASSESSMENT FOR PESTICIDES

Although many monitoring studies have been conducted in the past several gaps need to be completed. Monitoring has been preoccupied with measuring environmental levels rather than describing exposure and fate, determining the possible adverse effects, and/or evaluating the efficiency of mitigation methods. Metabolites have not been included in many monitoring programs and also novel pesticides should be studied since the patterns of pesticide use are constantly changing as the popularity of existing compounds rises and falls as new compounds are introduced into farming. In addition, available monitoring data are rarely comparable due to variability of analytical methods followed and the objectives targeted. Well-structured monitoring programs based on multimedia monitoring approaches and conducted according to QA/QC in sampling and analysis should be developed covering regional, national, and/or global patterns.

Concerning the current approaches of ecological risk assessment, one criticism is that some groups of organisms are not represented. The most commonly mentioned organisms are wild mammals, reptiles, and amphibians. Ecological risk assessment for pesticides concentrates on direct impacts on exposed species. If direct effects are anticipated, the potential for indirect effects does need to be considered and should be addressed by community level studies. In addition, research is needed to characterize the response of organisms to pulsed exposures and site-specific conditions. Furthermore, ecological risk assessment for pesticides often concentrates on single compounds, whereas the environment organisms might be exposed to more than one pesticide or other chemical mixtures including natural toxins and a variety of other stressors. Additive effects are usually considered for chemicals having the same mode of action. Greatest concern should be expressed for synergistic action, that is, the effect of the chemicals together is greater than that predicted from the parts, while the converse of this (antagonistic effect) is not likely to be of priority concern in a risk assessment. For a standard risk assessment procedure applicable for screening and comparison purposes, a database of exposure and toxicity for major pesticides and metabolites using a standard methodology is required. In many cases, risk assessments are hampered by the lack of data, especially toxic effects, thus more ecotoxicological studies which will result in less conservative hazard limits with less uncertainty are needed.

In conclusion, monitoring of the environment for pesticide residues and ecological risk assessment must continue based on harmonized methodologies and systematic studies. Monitoring data are critical elements in quantitative evaluations of environmental and human hazards and risk. We must be vigilant for early warning signs of damage of ecological systems. The ecological risk assessment approach could thus contribute to debate and give invaluable help in defining environmental guidelines for pesticides. To achieve the goal of environmental sustainability, the continuous and deeper scientific knowledge obtained from the monitoring and risk assessment procedures constitutes a powerful tool.

REFERENCES

1. Konstantinou I.K., D.G. Hela, T.A. Albanis, The status of pesticide pollution in surface waters (rivers and lakes) of Greece. Part I. Review on occurrence and levels, *Environ. Pollut.*, 141, 555, 2006.
2. Müller J.F., et al., Pesticides in sediments from Queensland irrigation channels and drains, *Mar. Pollut. Bull.*, 41, 294, 2000.
3. Hela D.G., et al., Determination of pesticide residues and ecological risk assessment in lake Pamvotis, Greece, *Environ. Toxicol. Chem.*, 24, 1548, 2005.
4. Kannan K., J. Ridal, J. Struger, Pesticides in the Great Lakes, *Hdb. Env. Chem.*, Vol. 5, DOI 10.1007/698_5_041, Springer-Verlag, Berlin, 2005.
5. Albaigés J., Persistent organic pollutants in the Mediterranean Sea, *Hdb. Env. Chem.*, Vol. 5, Part K: DOI 10.1007/b107145, Springer-Verlag, Berlin, 2005.

6. Sapozhnikova Y., O. Bawardi, D. Schlenk, Pesticides and PCBs in sediments and fish from the Salton Sea, California, USA, *Chemosphere*, 55, 797, 2004.

7. Harrie F.G., et al., *Fate of Pesticides in the Atmosphere: Implications for Environmental Risk Assessment*, Kluwer Academic Pub., The Netherlands, 1999.

8. Harner T., et al., Global pilot study for persistent organic pollutants (POPs) using PUF disk passive air samplers, *Environ. Pollut.*, 144, 445, 2006.

9. Hung H., et al., Temporal and spatial variabilities of atmospheric polychlorinated biphenyls (PCBs), organochlorine (OC) pesticides and polycyclic aromatic hydrocarbons (PAHs) in the Canadian Arctic: results from a decade of monitoring, *Sci. Total Environ.*, 342, 119, 2005.

10. Tuduri L., et al., A review of currently used pesticides in Canadian air and precipitation: Part 1: Lindane and endosulfans, *Atmos. Environ.*, 40, 1563, 2006.

11. Sánchez-Brunete C., B. Albero, J.L. Tadeo, Multiresidue determination of pesticides in soil by gas chromatography–mass spectrometry detection, *J. Agric. Food Chem.*, 52, 1445, 2004.

12. Goncalves C., et al., Chemometric interpretation of pesticide occurrence in soil samples from an intensive horticulture area in north Portugal, *Anal. Chim. Acta*, 560, 164, 2006.

13. Guzzella L., F. Pozzoni, G. Giuliano, Herbicide contamination of surficial groundwater in Northern Italy, *Environ. Pollut.*, 142, 344, 2006.

14. Sahoo G.B., et al., Application of artificial neural networks to assess pesticide contamination in shallow groundwater, *Sci. Total Environ.*, 367, 234, 2006.

15. Bartram J., Water and Health in Europe: A Joint Report from the European Environment Agency and the WHO Regional Office for Europe, World Health Organization (WHO) Regional publications European series, No. 93, 2002.

16. Maloschik E., et al., Monitoring water-polluting pesticides in Hungary, *Microchem. J.*, 85, 88, 2007.

17. European Economic Community, 1991. Council Directive concerning the placing of plant protection products on the market, 91/414/EEC, OJ L 230, Brussels, Belgium.

18. Wiersma B.G., *Environmental Monitoring*, CRC Press, Boca Raton, NY, 2004.

19. Calamari D., et al., Monitoring as an indicator of persistence and long-range transport. In Klecka G., et al., Eds., *Evaluation of Persistence and Long Range Transport of Organic Chemicals in the Environment*, SETAC, 2000.

20. Gilbert R.O., *Statistical Methods for Environmental Pollution Monitoring*, John Wiley & Sons, New York, USA, 1987.

21. Minsker B., *Long-Term Groundwater Monitoring Design: The State of the Art*, American Society of Civil Engineers, Reston, VA, USA, 2003.

22. Keith L.H., *Principles of Environmental Sampling*, The American Chemical Society, Washington, USA, 1996.

23. Quevauviller Ph., *Quality Assurance for Water Analysis*, John Wiley & Sons, Chichester, UK, 2002.

24. Usländer T., Trends of environmental information systems in the context of the European Water Framework Directive, *Environ. Modell. Softw.*, 20, 1532, 2005.

25. Coquery M., et al., Priority substances of the European Water Framework Directive: analytical challenges in monitoring water quality, *Trends Anal. Chem.*, 24, 117, 2005.

26. Shen L., et al., Atmospheric distribution and long range transport behavior of organochlorine pesticides in North America, *Environ. Sci. Tech.*, 39, 409, 2005.

27. Garbarino J.R., et al., Contaminants in Arctic snow collected over northwest Alaskan sea ice, *Water Air Soil Poll.*, 139, 183, 2002.

28. Klánova J., et al., Passive air sampler as a tool for long-term air pollution monitoring: Part 1. Performance assessment for seasonal and spatial variations, *Environ. Pollut.*, 144, 393, 2006.

29. Pozo K., et al., Passive-sampler derived air concentrations of persistent pollutants on a northsouth transect in Chile, *Environ. Sci. Technol.*, 38, 6529, 2004.

30. Wania F., et al., Development and calibration of a resin-based passive sampling system for monitoring persistent organic pollutants in the atmosphere, *Environ. Sci. Technol.*, 37, 1352, 2003.

31. Wargo J., *Our Children's Toxic Legacy*, Yale University Press, p. 254, 1996.

32. Rice C.P., S.M. Cherniak, Marine arctic fog: an accumulator of currently used pesticide, *Chemosphere*, 35, 867, 1997.

33. Thurman E.M., A.E. Cromwell, Atmospheric transport, deposition and fate of triazine herbicides and their metabolites in pristine areas at Isle Royale National Park, *Environ. Sci. Technol.*, 34, 3079, 2000.

34. Dubus I.G., J.M. Hollis, C.D. Brown, Pesticides in rainfall in Europe, *Environ. Pollut.*, 110, 331, 2000.

35. Van Dijk H.F.G., R. Guichert, Atmospheric dispersion of current-used pesticides: a review of the evidence from monitoring studies, *Water Air Soil Poll.*, 115, 21, 1999.

36. Buehler, S.S., I. Basu, R.A. Hites, Causes of variability in pesticide and PCB concentrations in air near the Great Lakes, *Environ. Sci. Technol.*, 38, 414, 2004.

37. Tuduri L., et al., *CANCUP: Canadian Atmospheric Network for Currently Used Pesticides. Managing Our Waters*, Cornwall, Canada, 2004.

38. Quaghebeur D., et al., Pesticides in rainwater in Flanders, Belgium: results from the monitoring program 1997–2001, *J. Environ. Monit.*, 6, 182, 2004.

39. Mayer P., et al., Equilibrium sampling devices, *Environ. Sci. Technol.*, 37, 185, 2003.

40. Namiesnik J., et al., Passive sampling and/or extraction techniques in environmental analysis: a review, *Anal. Bioanal. Chem.*, 381, 279, 2005.

41. Stuer-Lauridsen F., Review of passive accumulation devices for monitoring organic pollutants in the aquatic environment, *Environ. Pollut.*, 136, 503, 2005.

42. Vrana B., et al., Passive sampling techniques for monitoring pollutants in water, *Trends Anal. Chem.*, 24, 845, 2005.

43. Hanrahan G., D. Patil, J. Wang, Electrochemical sensors for environmental monitoring: design, development and applications, *J. Environ. Monit.*, 6, 657, 2004.

44. Tschmelak J., G. Proll, G. Gauglitz, Optical biosensor for pharmaceuticals, antibiotics, hormones, endocrine disrupting chemicals and pesticides in water: assay optimization process for estrone as example, *Talanta*, 65, 313, 2005.

45. Readman J.W., et al., Fungicide contamination of Mediterranean Estuarine Waters: results from a MED POL pilot survey, *Mar. Poll. Bull.*, 34, 259, 1997.

46. U.S. Geological Survey (USGS), The Quality of our Nation's Waters—Nutrients and Pesticides: U.S. Geological Survey Circular 1225, p. 82, 1999.

47. Coupe R.H., Occurrence of Pesticides in Five Rivers of the Mississippi Embayment Study Unit, 1996–1998, USGS Report 99-4159.

48. Claver A., et al., Study of the presence of pesticides in surface waters in the Ebro river basin (Spain), *Chemosphere*, 64, 1437, 2006.

49. Capel P.D., S.J. Larson, T.A. Winterstein, The behaviour of 39 pesticides in surface waters as a function of scale, *Hydrol. Process.*, 15, 1251, 2001.

50. Larson S.J., et al., Relations between pesticide use and riverine flux in the Mississippi river basin, *Chemosphere*, 31, 3305, 1995.

51. Albanis T.A., et al., Monitoring of pesticide residues and their metabolites in surface and underground waters of Imathia (N. Greece) by means of solid-phase extraction disks and gas chromatography, *J. Chromatogr. A*, 823, 59, 1998.

52. Thurman E.M., et al., Herbicides in surface waters of the midwestern United States: the effect of spring flush, *Environ. Sci. Technol.*, 25, 1794, 1991.

53. Steen R.J.C.A., Fluxes of Pesticides into the Marine Environment: Analysis, Fate and Effects, PhD thesis, Vrije Universiteit, Amsterdam, The Netherlands, 2002.

54. Fielding, M., et al., Pesticides in Ground and Drinking Water, EU, Water Pollution Research Report 27, 1991.

55. Cohen S.Z., et al., A ground water monitoring study for pesticides and nitrates associated with golf courses on Cape Cod, *Ground Water Monit. Rev.*, 10, 160, 1990.

56. Steen R.J.C.A., et al., Ecological risk assessment of agrochemicals in European estuaries, *Environ. Toxicol. Chem.*, 18, 1574, 1999.

57. FAO/IAEA/UNEP, Report of the FAO/IAEA/UNEP Consultation Meeting on the Fungicides Pilot Survey (Ioannina, Greece, 1993), UNEP, Athens, 9 pp. (mimeo), 1993.

58. Konstantinou I.K., T.A. Albanis, Worldwide occurrence and effects of antifouling paint booster biocides in the aquatic environment: a review, *Environ. Int.*, 30, 235, 2004.

59. Konstantinou I.K., Ed., Antifouling paint biocides, *Hdb. Env. Chem.*, Vol. 5, Part O, 10.1007/11555148, Springer-Verlag, Berlin, Heidelberg, 2006.

60. Long J.L.A., et al., Micro-organic compounds associated with sediments in the Humber rivers, *Sci. Total Environ.*, 210–211, 229, 1998.

61. Vega A.B., et al., Monitoring of pesticides in agricultural water and soil samples from Andalusia by liquid chromatography coupled to mass spectrometry, *Anal. Chim. Acta*, 538, 117, 2005.

62. Oldal B., et al., Pesticide residues in Hungarian soils, *Geoderma*, 135, 163, 2006.

63. Harner T., et al., Residues of organochlorine pesticides in Alabama soils, *Environ. Pollut.*, 106, 323, 1999.

64. Wurl O., J.P. Obbard, Organochlorine pesticides, polychlorinated biphenyls and polybrominated diphenyl ethers in Singapore's coastal marine sediments, *Chemosphere*, 58, 925, 2005.

65. Covaci A., et al., Polybrominated diphenyl ethers, polychlorinated biphenyls and organochlorine pesticides in sediment cores from the Western Scheldt river (Belgium): analytical aspects and depth profiles, *Environ. Int.*, 31, 367, 2005.

66. Que Hee S., *Biological Monitoring: An Introduction*, Wiley & Sons, New York, 1993.

67. Subramanian K.S., G.V. Iyengar, *Environmental Biomonitoring: Exposure Assessment and Specimen Banking*, ACS, USA, 1997.

68. Franklin C.A., J.P. Worgan, *Occupational and Residential Exposure Assessment for Pesticides*, Wiley & Sons, New York, USA, pp. 28–45, 2005.

69. Cerrillo I., et al., Endosulfan and its metabolites in fertile women, placenta, cord blood, and human milk, *Environ. Res.*, 98, 233, 2005.

70. Fisk A.T., et al., An assessment of the toxicological significance of anthropogenic contaminants in Canadian arctic wildlife, *Sci. Total Environ.*, 351–352, 57, 2005.

71. Albanis T.A., et al., Organochlorine contaminants in eggs of the yellow-legged gull (*Larus cachinnans michahellis*) in the North Eastern Mediterranean: is this gull a suitable biomonitor for the region? *Environ. Pollut.*, 126, 245, 2003.

72. Sakellarides Th., et al., Accumulation profiles of persistent organochlorines in liver and fat tissues of various waterbird species from Greece, *Chemosphere*, 63, 1392, 2006.

73. Klemens J.A., et al., A cross-taxa survey of organochlorine pesticide contamination in a Costa Rican wildland, *Environ. Pollut.*, 122, 245, 2003.

74. Erdogrul Ö., A. Covaci, P. Schepens, Levels of organochlorine pesticides, polychlorinated biphenyls and polybrominated diphenyl ethers in fish species from Kahramanmaras, Turkey, *Environ. Int.*, 31, 703, 2005.

75. Hoshi H.N., et al., Organochlorine pesticides and polychlorinated biphenyl congeners in wild terrestrial mammals and birds from Chubu region, Japan: interspecies comparison of the residue levels and compositions, *Chemosphere*, 36, 3211, 1998.

76. Naso B., et al., Organochlorine pesticides and polychlorinated biphenyls in European roe deer *Capreolus capreolus* resident in a protected area in Northern Italy, *Sci. Total Environ.*, 328, 83, 2004.

77. Sapozhnikova Y., et al., Evaluation of pesticides and metals in fish of the Dniester River, Moldova, *Chemosphere*, 60, 196, 2005.

78. COMPRENDO, Final Publishable Report, Executive Summary On The Project Results, Energy, Environment and Sustainable Development, March 2006.

79. Directive No. 76/464/EEC of the Council of the European Community of 4 May 1976, Pollution caused by certain dangerous substances discharged into the aquatic environment of the Community, *Off. J. Eur. Comm.*, 18/05/1976, p. 7.

80. EEC 1980, Council Directive 80/68/EEC of 17 December 1979 on the protection of groundwater against pollution caused by certain dangerous substances, *Off. J. Eur. Comm.*, L020, 26/01/1980.

81. Directive 2000/60/EC of the European Parliament and of the Council of 23 October 2000 establishing a framework for Community action in the field of water policy, *Off. J. Eur. Comm.*, L327, p. 1, 2000.

82. Lepper P., Identification of Quality Standards for Priority Substances in the Field of Water Policy. Towards the Derivation of Quality Standards for Priority Substances in the Context of the Water Framework Directive. Final Report of the Study Contract No. B4-3040/2000/30637/MAR/E1, Fraunhofer Institute, Germany, p. 124, 2002.

83. Allan I.J., et al., Strategic monitoring for the European Water Framework Directive, *Trends Anal. Chem.*, 25, 704, 2006.

84. Schnoor J.L., *Fate of Pesticides and Chemicals in the Environment*, John Wiley & Sons, New York, USA, 1992.

85. Wheeler W., *Pesticides in Agriculture and the Environment*, CRC Press, New York, USA, 2002.

86. Wania F., Multi-compartmental models of contaminant fate in the environment, *Biotherapy*, 11, 65, 1998.

87. Whitford F., *The Complete Book of Pesticide Management. Science, Regulation, Stewardship and Communication*, John Wiley & Sons, New York, USA, 2002.

88. Mackay D., *Multimedia Environmental Models. The Fugacity Approach*, Lewis Publ., Chelsea, MI, p. 257, 1991.

89. Van der Werf H.M.G., Assessing the impact of pesticides on the environment, *Agric. Ecosyst. Environ.*, 60, 81, 1996.

90. USEPA, U.S. Environmental Protection Agency, *Framework for Ecological Risk Assessment*, Risk Assessment Forum, Washington, D.C., 1992.

91. ECOFRAM, Ecological Committee on FIFRA Risk Assessment Methods, Aquatic and Terrestrial Final Draft Reports, USEPA, 1999, www.epa.gov/oppefed1/ecorisk/index.htm

92. USEPA, U.S. Environmental Protection Agency, *Guidelines for Ecological Risk Assessment*, Fed Reg 63:26846–26924, Washington, D.C., 1998.

93. Ritter L., et al., Sources, pathways and relative risks of contaminants in surface water and groundwater: a perspective prepared for the Walkerton inquiry, *J. Toxicol. Environ. Health*, 65, 1, 2002.

94. Hamer M., Ecological risk assessment for agricultural pesticides, *J. Environ. Monit.*, 2, 104N, 2000.

95. Linders J.B.H.J., R. Luttik, Uniform system for the evaluation of substances. V. ESPE, risk assessment for pesticides, *Chemosphere*, 31, 3237, 1995.

96. Sanchez-Bayo F., S. Baskaran, I.R. Kennedy, Ecological relative risk (EcoRR): another approach for risk assessment of pesticides in agriculture, *Agric. Ecosyst. Environ.*, 91, 37, 2002.

97. Reus J.A.W.A., G.A. Pak, *An Environmental Yardstick for Pesticides*, Med. Fac. Landbouww. Univ. Gent., Dept. Pub. No. 58, p. 249, 1993.

98. Gutsche V., The influence of pesticides and pest management strategies on wildlife. Integrated Crop Protection: Towards Sustainability, BCPC Symposium No. 63, 1995.

99. Swanson M., A. Socha, *Chemical Ranking and Scoring: Guidelines for Relative Assessment of Chemicals*, SETAC, Pensacola, FL, USA, 1997.

100. Snyder E.M., et al., SCRAM: a scoring and ranking system for persistent, bioaccumulative, and toxic substances for the North American Great Lakes. Part I. Structure of the scoring and ranking system, *Environ. Sci. Pollut. Res.*, 7, 1, 2000.

101. Kovach J., et al., *A Method to Measure the Environmental Impact of Pesticides*, New York's Food and Life Sciences Bulletin 139, Cornell University, New York, 1992.

102. Reus J., et al., Comparison and evaluation of eight pesticide environmental risk indicators developed in Europe and recommendations for future use, *Agric. Ecosyst. Environ.*, 90, 177, 2002.

103. Finizio A., S. Villa, Environmental risk assessment for pesticides. A tool for decision making, *Environ. Impact Assess. Rev.*, 22, 235, 2002.

104. Finizio A., M. Calliera, M. Vighi, Rating systems for pesticide risk classification on different ecosystems, *Ecotoxicol. Environ. Saf.*, 49, 262, 2001.

105. Van der Werf H.M.G., C. Zimmer, An indicator of pesticide environmental impact based on a fuzzy expert system, *Chemosphere*, 36, 2225, 1998.

106. Van Bol V., et al., Pesticide indicators, *Pestic. Outlook*, 14, 159, 2003.

107. Urban D.J., N.J. Cook, *Hazard Evaluation Division Standard Evaluation Procedure for Ecological Risk Assessment*, U.S. Environmental Protection Agency, EPA-540/9-85-001, Washington, D.C., 1986.

108. European Commission, 2002. Technical Guidance Document on Risk Assessment in Support of Council Directive 93/67/EEC for New Notified Substances and Commission Regulation 1488/94 on Risk Assessment for Existing Substances and Directive 98/8/EC of the European Parliament and the Council Concerning the Placing of Biocidal Products of the Market, EU, JRC, Brussels, Belgium.

109. USEPA, *Technical Progress Report: Implementation Plan for Probabilistic Ecological Assessments–Terrestrial Systems*, USEPA, 2000, www.epa.gov/scipoly/sap/

110. Kooijman S.A.L.M. A safety factor for LC50 values allowing for differences in sensitivity among species, *Wat. Res.*, 21, 269, 1987.

111. Van Straalen N.M., C.A.J. Denneman, Ecotoxicological evaluation of soil quality criteria, *Ecotoxicol. Environ. Saf.*, 18, 269, 1989.

112. Wagner C., H. Lokke, Estimation of ecotoxicological protection levels for NOEC toxicity data, *Water Res.*, 25, 1237, 1991.

113. Newman M.C., et al., Applying species sensitivity distributions in ecological risk assessment: assumptions of distribution type and sufficient number of species, *Environ. Toxicol. Chem.*, 19, 508, 2000.

114. Van der Hoeven N., Estimating the 5-percentile of the species sensitivity distributions without any assumptions about the distribution, *Ecotoxicology*, 10, 25, 2001.

115. Aldenberg T., J.S. Jaworska, Bayesian Statistical Analysis of Bimodality in Species Sensitivity Distributions. SETAC News, FL, USA, pp. 19–20, 1999.

116. GCPF (Global Crop Protection Federation), Framework for the Ecological Risk Assessment of Plant Protection Products, Technical Monograph No. 21, 1999.

117. Crane M., et al., Evaluation of probabilistic risk assessment of pesticides in the UK: chlorpyrifos use on top fruit, *Pest Manag. Sci.*, 59, 512, 2003.

118. Greco W., et al., Consensus on concepts and terminology for combined-action assessment: the Saariselka agreement, *Arch. Complex Environ. Stud.*, 4, 65, 1992.

119. Broderius S.J, M.D. Kahl, M.D. Hoglund, Use of joint toxic response to define the primary mode of toxic action for diverse industrial organic chemicals, *Environ. Toxicol. Chem.*, 9, 1591, 1995.

120. Solomon K., J. Giesy, P. Jones, Probabilistic risk assessment of agrochemicals in the environment, *Crop Prot.*, 19, 649, 2000.

121. Hart A., *Probabilistic Risk Assessment for Pesticides in Europe: Implementation and Research Needs*. Report of European Workshop on Probabilistic Risk Assessment for the Environmental Impacts of Plant Protection Products, The Netherlands, 2001.

122. Richardson G.M., Deterministic versus probabilistic risk assessment: strengths and weaknesses in a regulatory context, *Hum. Ecol. Risk Assess.*, 2, 44, 1996.

123. Postuma L., T. Traas, G.W. Suter, *Species Sensitivity Distributions in Risk Assessment*, Pensacola, SETAC Press, FL, 2001.

124. Boxall A.B.A., C.D. Brown, K.L. Barrett, Higher-tier laboratory methods for assessing the aquatic toxicity of pesticides, *Pest. Manag. Sci.*, 58, 637, 2002.

125. Aldenberg T., W. Slob, Confidence limits for hazardous concentrations based on logistically distributed NOEC toxicity data, *Ecotoxicol. Environ. Saf.*, 25, 48, 1993.

126. Giesy J.P., et al., Chlorpyrifos: ecological risk assessment in North American aquatic environments, *Rev. Environ. Contam. Toxicol.*, 160, 1, 1999.

127. Hall L.W.J., et al., A probabilistic ecological risk assessment of tributyltin in surface waters of the Chesapeake Bay, *Hum. Ecol. Risk Assess.*, 6, 141, 2000.

128. Solomon K.R., et al., Ecological risk assessment of atrazine in North American surface waters, *Environ. Toxicol. Chem.*, 15, 31, 1996.

129. USEPA, *Technical Progress Report of the Implementation Plan for Probabilistic Ecological Assessments–Aquatic Systems*, USEPA, 2000, www.epa.gov/scipoly

130. Stephan C.E., et al., *Guidelines for Deriving Numerical National Water Quality Criteria for the Protection of Aquatic Organisms and Their Uses*, EPA PB85-227049, U.S. Environmental Protection Agency, Washington, D.C., 1985.

131. Crommentuijn T., et al., Maximum permissible and negligible concentrations for some organic substances and pesticides, *J. Environ. Manag.*, 58, 297, 2000.

132. ANZECC (Australian and New Zealand Environment and Conservation Council), *Australian and New Zealand Guidelines for Fresh and Marine Water Quality, Volume 2, Aquatic Ecosystems—Rationale and Background Information*, Canberra, 2000.

133. SETAC, Pesticide Risk and Mitigation. Final Report of the Aquatic Risk Assessment and Mitigation Dialog Group, SETAC Foundation for Environmental Education, Pensacola, FL, 220 pp., 1994.

134. Mackenthun K.M., *Basic Concepts in Environmental Management*, Lewis Publications, Boca Raton, FL, USA, 1998.

135. Killeen S., Development and use of environmental quality standards (EQSs) for priority pesticides, *Pest. Sci.*, 49, 191, 1997.

136. CSTE (1994): EEC Water Quality Objectives for Chemicals Dangerous to Aquatic Environments (List 1). Scientific Committee on Toxicity and Ecotoxicity of Chemicals of the European Commission (CSTE(EEC)). *Rev. Environ. Contam. Toxicol.*, 137, Springer-Verlag, New York, pp. 3–110, 1994.

137. Rai S.N., et al., The use of probabilistic risk assessment in establishing drinking water quality objectives, *Hum. Ecol. Risk Assess.*, 8, 493, 2002.

138. Vighi M., et al., Water quality objectives for mixtures of toxic chemicals: problems and perspectives, *Ecotoxicol. Environ. Saf.*, 54, 139, 2003.

139. Van Straalen N.M., J. Schobben, T.P. Traas, The use of ecotoxicological risk assessment in deriving maximum acceptable half-lives of pesticides, *Pestic. Sci.*, 34, 227, 1992.

Index